Lecture Notes in Artificial Intelligence 9505

Subseries of Lecture Notes in Computer Science

LNAI Series Editors

Randy Goebel
  University of Alberta, Edmonton, Canada
Yuzuru Tanaka
  Hokkaido University, Sapporo, Japan
Wolfgang Wahlster
  DFKI and Saarland University, Saarbrücken, Germany

LNAI Founding Series Editor

Joerg Siekmann
  DFKI and Saarland University, Saarbrücken, Germany

More information about this series at http://www.springer.com/series/1244

Joe Suzuki · Maomi Ueno (Eds.)

Advanced Methodologies for Bayesian Networks

Second International Workshop, AMBN 2015
Yokohama, Japan, November 16–18, 2015
Proceedings

 Springer

Editors
Joe Suzuki
Osaka University
Osaka
Japan

Maomi Ueno
The University of Electro-Communications
Tokyo
Japan

ISSN 0302-9743 ISSN 1611-3349 (electronic)
Lecture Notes in Artificial Intelligence
ISBN 978-3-319-28378-4 ISBN 978-3-319-28379-1 (eBook)
DOI 10.1007/978-3-319-28379-1

Library of Congress Control Number: 2015958747

LNCS Sublibrary: SL7 – Artificial Intelligence

Printed on acid-free paper

This Springer imprint is published by SpringerNature
The registered company is Springer International Publishing AG Switzerland

Preface

Over the last few decades, graphical models such as Bayesian and Markov networks have become increasingly popular artificial intelligence (AI) approaches. In the International Workshop on Advanced Methodologies for Bayesian Networks (AMBN), we explore methodologies for enhancing the effectiveness of graphical models including modeling, reasoning, model selection, logic-probability relations, and causality. The exploration of methodologies is complemented by discussions of practical considerations for applying graphical models in real-world settings, covering concerns such as scalability, incremental learning, parallelization, and so on. The first AMBN was held in Tokyo (2010).

The second AMBN was held in Yokohama, Japan, during November 16–18, 2015; it was co-sponsored by the Japanese Society Artificial Intelligence (JSAI) and the National Institute of Advanced Industrial Science and Technology (AIST).

This AMBN was the first to celebrate the publication of accepted papers in the *Lecture Notes in Artificial Intelligence* series by Springer. Out of 29 submissions, 26 contributions were selected for presentation at the conference and 15 papers were selected for this volume. Each submission underwent a rigorous review by three members of the AMBN Program Committee, with each committee member reviewing at most two papers.

In addition to the 29 presentations, we were honored to have six invited speakers:

- Russell Almond (Florida State University, USA)
- Aapo Hyvarinen (University of Helsinki, Finland)
- Brandon Malone (Max Planck Institute, Germany)
- Changhe Yuan (Queens College/CUNY, USA)
- Cassio P. de Campos (Queens University, UK)
- Shohei Shimizu (Osaka University, Japan)

We are grateful for their highly inspiring presentations. The six abstracts and three full papers of the six invited talks are contained in this volume.

To conclude, we would like to thank the 23 Program Committee members and nine advisors for their efforts, and for their punctual and high-quality reviews. And, last but not least, we are most indebted to AIST for their financial support.

November 2015   Joe Suzuki
                                                          Maomi Ueno

Organization

AMBN 2015 was co-sponsored by the Japanese Society Artificial Intelligence (JSAI) and the National Institute of Advanced Industrial Science and Technology (AIST), Japan.

Program Chairs

Joe Suzuki	Osaka University, Japan
Maomi Ueno	University of Electro-Communications, Japan

Organizing Committee

Joe Suzuki	Osaka University, Japan
Maomi Ueno	University of Electro-Communications, Japan
Takashi Isozaki	SONY CSL, Japan
Yasuo Tabei	JST, Japan
Masakazu Ishihata	NTT, Japan

Advisors

Adnan Darwiche	UCLA, USA
David Heckerman	Microsoft, USA
Aapo Hyvarinen	University of Helsinki, Finland
Petri Myllmaki	University of Helsinki, Finland
Judea Pearl	UCLA, USA
Peter Spirtes	CMU, USA
Milan Studeny	Institute of Information Theory and Automation, Czech Republic
Takashi Washio	Osaka University, Japan

Program Committee

Russell Almond	Florida State University, USA
Alessandro Antonucci	IDSIA, Switzerland
Cassio P. de Campos	Queens University, UK
Hei Chan	The Institute of Statistical Mathematics, Japan
Arthur Choi	UCLA, USA
Robin Evans	University of Oxford, UK
Luca Faes	University of Trento, Italy
Antti Hyttinen	University of Helsinki, Finland
Seiya Imoto	University of Tokyo, Japan
Yoshinobu Kawahara	Osaka University, Japan
Manabu Kuroki	The Institute of Statistical Mathematics, Japan

Jose A. Lozano	University of the Basque Country UPV/EHU, Spain
Peter Lucas	Institute for Computing and Information Sciences, The Netherlands
Brandon Malone	Max Planck Institute, Germany
Alessio Moneta	Scuola Superiore Sant'Anna, Italy
Yoichi Motomura	National Institute of Advanced Industrial Science and Technology, Japan
Jose M. Pena	Linkoping University, Sweden
Hiroshi Sakamoto	Kyushu Institute of Technology, Japan
Shohei Shimizu	Osaka University, Japan
Yi Wang	Institute of High Performance Computing, Singapore
Changhe Yuan	Queens College/CUNY, USA
Jiji Zhang	Lingnan University, Hong Kong, SAR China
Kun Zhang	University of Southern California, USA

Sponsors

The Japanese Society Artificial Intelligence (JSAI)
The National Institute of Advanced Industrial Science and Technology (AIST)

Invited Paper Abstract

Advanced Search Algorithms for Learning Optimal Bayesian Network Structures

Changhe Yuan

Queens College/CUNY, USA

Abstract. Research on learning optimal Bayesian network structures from data, once thought impractical, have made strides in the last decade. One promising approach is based on admissible heuristic search. The approach formulates the learning problem as a shortest path problem, and uses search techniques such as A* or BFBnB to solve the problem in finding an optimal network structure. I will discuss how to create admissible heuristic functions for the algorithms. Moreover, I will discuss two techniques for extracting extra information from data in order to scale up the learning. The first technique can potentially lead to a decomposition of the learning problem to a set of smaller independent learning problems, and the second technique creates tighter heuristic functions that lead to much improved search efficiency.

Empirical Behavior of Bayesian Network Structure Learning Algorithms

Brandon Malone

Max Planck Institute for the Biology of Ageing

Abstract. Bayesian network structure learning (BNSL) is the problem of finding a BN structure which best explains a dataset. Score-based learning assigns a score to each network structure. The goal is to find the structure which optimizes the score. We review two recent studies of empirical behavior of BNSL algorithms.

The score typically reflects fit to a training dataset; however, models which fit training data well may generalize poorly. Thus, it is not clear that finding an optimal network is worthwhile. We review a comparison of exact and approximate search techniques. Sometimes, approximate algorithms suffice; for complex datasets, the optimal algorithms produce better networks.

BNSL is known to be NP-hard, so exact solvers prune the search space using heuristics. We next review problem-dependent characteristics which affect their efficacy. Empirical results show that, machine learning techniques based on these characteristics can often be used to accurately predict the algorithms' running times.

Keywords: Bayesian networks · Structure learning · Algorithm selection · Empirical hardness

Brandon Malone—This paper is based on Malone et al. (2014, 2015), with co-authors Matti Järvisalo, Petri Myllymäki, Kusta Kangas and Mikko Koiviso from HIIT and the Department of Computer Science at the University of Helsinki.

An Entropic Approach to Causal Discovery in Non-Gaussian and Non-linear Models

Aapo Hyvarinen

University of Helsinki, Finland

Abstract. Recent advances in machine learning have shown how it is possible to determine the causal direction, or direction of effect, between two continuous-valued random variables. We show how to use entropy to develop a simple and general framework for determining the causal direction. First, we consider the likelihood ratio under the linear non-Gaussian acyclic model (LiNGAM) and show how it gives rise to a non-Gaussianity measures based on differential entropy. Second, we develop a similar framework for the nonlinear additive noise model. We further discuss how to extend the framework to more than two variables, and how the framework is related to independent component analysis.

A Non-Gaussian Approach for Causal Discovery in the Presence of Hidden Common Causes

Shohei Shimizu

The Institute of Scientific and Industrial Research, Osaka University
8-1 Mihogaoka, Ibaraki, Osaka 5670047, Japan
https://sites.google.com/site/sshimizu06/

Abstract. We discuss the problem of estimating the causal direction between two observed variables in the presence of hidden common causes. Managing hidden common causes is essential when studying causal relations based on observational data. We previously proposed a Bayesian estimation method for estimating the causal direction using the non-Gaussianity of data. This method does not require us to explicitly model hidden common causes. The experiments on artificial data presented in this paper imply that Bayes factors could be useful for selecting a better causal direction when using a non-Gaussian method.

Keywords: Causal discovery · Hidden common causes · Structural equation models · Non-Gaussianity

Learning Bayesian Networks with Biomedical Applications

Cassio P. de Campos

Queen's University Belfast
Northern Ireland, UK
c.decampos@qub.ac.uk

Abstract. This talk presents a brief overview of methods for learning Bayesian networks. It discusses on recent methods and theoretical results to speed up computations and to improve accuracy, leading to an approach which can deal with many thousands of variables. Applications arising in biomedical problems are described, where it is argued that Bayesian networks can provide meaningful and interpretable results. In particular, we discuss on the use of Bayesian networks for data imputation, unsupervised clustering and classification using high-dimensional data sets of lymphoma patients.

Keywords: Bayesian networks · Structure learning · Data imputation · Clustering

Tips and Tricks for Building Bayesian Networks for Scoring Game-Based Assessments

Russell G. Almond

Educational Psychology and Learning Systems
College of Education
Florida State University
Tallahassee, FL 32306
ralmond@fsu.edu

Abstract. Game-based assessments produce multiple, dependent observations from student game play. Bayesian networks can model the dependence, but, typically, only a small amount of pilot data are available at the time the network is constructed. This paper examines the process of creating Bayesian network scoring models, focusing on several practical techniques that have been used in the construction of models for *Physics Playground*. In particular, the following techniques are helpful: (1) The use of evidence-centered assessment design to define latent competency variables and observable indicator variables. (2) The use of correlation matrixes to uncover and validate the conditional independence structure of the Bayes net. (3) The use of discrete IRT models to create large portion of the Bayesian networks from a single spreadsheet. (4) Adjusting the Bayes net parameters using both hand tuning and a generalized EM algorithm, creating networks which are a mixture of expert opinion and data. (5) Using expected classification accuracy matrixes to judge assessment validity and reliability. (6) Using evidence balance sheets to identify unusual subjects and observable indicators.

Keywords: Bayesian networks · Evidence-centered assessment design · Prior information · Weight of evidence · Classification consistency

Contents

Efficiently Learning Bayesian Network Structures Based on the B&B Strategy: A Theoretical Analysis

Joe Suzuki$^{(\boxtimes)}$

Osaka University, Toyonaka, Japan
suzuki@math.sci.osaka-u.ac.jp

Abstract. This paper addresses the problem of efficiently finding an optimal Bayesian network structure w.r.t. maximizing the posterior probability and minimizing the description length. In particular, we focus on the branch and bound strategy to save computational effort. To obtain an efficient search, a larger lower bound of the score is required (when we seek its minimum). We generalize an existing lower bound (Campose and Ji, 2011) for the Bayesian Dirichlet BDeu (Bayesian Dirichlet equivalent uniform) to one for the BD (Bayesian Dirichlet) and mathematically prove that the number of variables in each parent set cannot be bounded for maximizing the posterior probability.

Keywords: Parent set · B&B (branch and bound) · Lower bound · Score · MDL (minimum description length) · Maximizing posterior probability · Learning Bayesian network structures

1 Introduction

In many tasks associated with statistical analysis and machine learning, we often need to find a relation between variables given a dataset in which n-tuples of examples for N variables are stored in an $n \times N$ data frame. If we wish to obtain a relation that maximizes the posterior probability given the dataset, the task itself may be straightforward, but the computation is intractable if the number of variables N is large.

In this paper, assuming that the probabilistic relation is expressed by a Bayesian network [12], we find an efficient search for the optimal solution. Based on the data, we connect edges that express probabilistic relations between vertices that represent variables until the structure is completed.

By the best Bayesian network structure, we refer to two things: one that maximizes the posterior probability (Cooper and Herskovits 1991 [5]) and one that minimizes the description length (Suzuki 1993 [14]). The latter is used to describe the given data in terms of a rule and its exceptions in many ways and to choose the rule that renders the total length the shortest (minimum description length (MDL) principle [8]). Although constraint-based algorithms [11] that statistically test conditional independence for each combination of variables are

© Springer International Publishing Switzerland 2015
J. Suzuki and M. Ueno (Eds.): AMBN 2015, LNAI 9505, pp. 1–14, 2015.
DOI: 10.1007/978-3-319-28379-1_1

available, in this paper, we focus on the two criteria based on the associated scores.

Currently, the algorithm developed by Silander and Myllymaki [9] is considered to be the fastest, and it can deal with both the posterior probability and the description length. However, this method utilizes dynamic programming (DP) and requires a lot of memory; therefore, Yuan and Malone [19], for example, proposed consuming a partial amount of resources using the idea of the A^* algorithm.

In contrast, in the early era of learning Bayesian network structures, Suzuki [13] proposed applying the branch and bound (B & B) technique to save computational effort. We can do so if the current solution reaches the lower bound of the solution set when we aim for minimization. The idea was adopted by many authors such as Tian 2000 [17] and recently by Campose and Ji 2011 [6]. The branch and bound technique can be applied to both DP and A^* algorithms.

In particular, Campos and Ji [6] noted that the optimal parent set contains at most $O(\log n)$ variables for a minimum description length. However, no upper bound of the number of variables in each parent set has been obtained for maximizing the posterior probability. Although they derived a lower bound for the posterior probability, the bound was only obtained for BDeu [2] (not the whole BD). Moreover, we are not sure whether the bound is tight enough, even if we know that the value obtained is a lower bound.

Our contributions are as follows:

1. an extension of the lower bound [6] from BDeu to BD (Theorem 1)
2. the number of variables in each parent set cannot be upper bounded for maximizing the posterior probability (Theorem 2).

This paper is organized as follows: Sect. 2 provides a background for efficient searches of optimal Bayesian network structures. Because the algorithms for the problem are quite complicated, we make special efforts to precisely understand the previous results. Section 3 states and proves two theorems. The results occupy only a few pages because we have found a shorter proof, which is preferable in the sense of the MDL principle. Section 4 concludes the paper and summarizes the results.

2 Background

2.1 Bayesian Network

Let $X^{(1)}, \cdots, X^{(N)}$ ($N \geq 1$) be random variables that take on a finite number of values. We define a Bayesian network (BN) using a directed acyclic graph (DAG) that expresses the factorization of the distribution $P(X^{(1)}, \cdots, X^{(N)})$. For example, suppose $N = 3$. If we express the three variables by X, Y, Z, then the three factorizations

$$P(X)P(Y|X)P(Z|Y) \ , \ P(Y)P(X|Y)P(Z|Y) \ , \ P(Z)P(Y|Z)P(X|Y)$$

and the factorization $P(X)P(Y|XZ)P(Z)$ are expressed as

$$\frac{P(XY)P(YZ)}{P(Y)} \text{ and } \frac{P(X)P(Z)P(XYZ)}{P(ZX)} ,$$

respectively. We can check that only two of the 27 directed graphs with three vertices contain a cycle, where the 25 DAGs are categorized into eleven classes (see Fig. 1):

$$P(X)P(Y)P(Z)$$

$$P(X)P(YZ), P(Y)P(ZX), P(Z)P(XY)$$

$$\frac{P(ZX)P(XY)}{P(X)} , \frac{P(XY)P(YZ)}{P(Y)} , \frac{P(ZX)P(XY)}{P(Z)}$$

$$\frac{P(Y)P(Z)P(XYZ)}{P(YZ)} , \frac{P(Z)P(X)P(XYZ)}{P(ZX)} , \frac{P(X)P(Y)P(XYZ)}{P(XY)} ,$$

$$\text{and } P(XYZ)$$

Hereafter, we denote the eleven equations as (1)–(11).

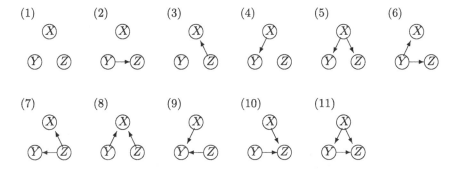

Fig. 1. The eleven Bayesian networks

2.2 Learning Bayesian Network Structures

Suppose that we wish to test whether random variables X and Y are independent[1] from n pairs of examples $x^n = (x_1, \cdots, x_n)$ and $y^n = (y_1, \cdots, y_n)$ emitted from X and Y.

Let α and β be the cardinalities of the sets in which X and Y take values, respectively. Suppose that we define the prior probability $0 < p < 1$ of $X \perp\!\!\!\perp Y$ and some alternatives $Q^n(X), Q^n(Y), Q^n(X,Y)$ of probabilities of $x^n, y^n, (x^n, y^n)$ and that we decide that $X \perp\!\!\!\perp Y$ if and only if

$$pQ^n(X)Q^n(Y) \geq (1-p)Q^n(X,Y). \tag{12}$$

---

[1] We denote $X \perp\!\!\!\perp Y|Z$ if X and Y are independent given Z.

If we estimate the conditional probability of x_i by $\dfrac{c_{i-1}(x_i) + a}{i - 1 + a\alpha}$ when $X = x_i$ occurs $c_{i-1}(x_i)$ times in $(x_1, \cdots, x_{i-1})$, $i = 1, \cdots, n$, the probability of $(x_1, \cdots, x_n)$ can be expressed as [7]

$$Q^n(X) = \prod_{i=1}^{n} \frac{c_{i-1}(x_i) + a}{i - 1 + \alpha a},$$

where $a > 0$ is a constant. Alternatively, the formula can be obtained by weighting the probability θ_x of $X = x$ by

$$w(\theta) = K \prod_x \theta_x^{a-1}$$

where K is a normalization constant:

$$Q^n(X) = \int \prod_x \theta^{c_n(x)} w(\theta) d\theta = K \int \prod_x \theta^{c_n(x) + a - 1} d\theta = \frac{\Gamma(\alpha a) \prod_x \Gamma(c_n(x) + a)}{\Gamma(a)^\alpha \Gamma(n + \alpha a)},$$

(13)

where x ranges over all of the values that X takes. If we require the constant a in (13) to depend on x, such as $a(x)$, the formula (13) can be expressed as

$$Q^n(X) = \frac{\Gamma(\sum_x a(x)) \prod_x \Gamma(c_n(x) + a(x))}{\prod_x \Gamma(a(x)) \Gamma(n + \sum_x a(x))}$$

(14)

(BD, Bayesian Dirichlet [5]).

Similar constructions are possible for $Q^n(Y)$ and $Q^n(X, Y)$. For example, we can obtain $Q^n(X, Y)$ by replacing the occurrence $c_n(x)$ of $X = x$ with their simultaneous occurrences $c_n(x, y)$ of $(X, Y) = (x, y)$ and replacing α with $\alpha\beta$. Then, the decision (12) is made based on the values $Q^n(X)Q^n(Y)$ and $Q^n(X, Y)$ multiplied by the prior probabilities p and $1 - p$ of $X \perp\!\!\!\perp Y$ and $X \not\perp\!\!\!\perp Y$, respectively, and we choose the quantity with the larger posterior probability. Suzuki 2012 [15] showed

$$(12) \Longleftrightarrow X \perp\!\!\!\perp Y$$

for large n with probability one as $n \to \infty$.

Now, given n-tuples of examples

$$X^{(1)} = x_{1,1}, \cdots, X^{(N)} = x_{1,N}$$

$$\cdots, \cdots, \cdots$$

$$X^{(1)} = x_{n,1}, \cdots, X^{(N)} = x_{n,N}$$

of variables $X^{(1)} \cdots, X^{(N)}$, we learn the BN structure. For example, if $N = 3$, the problem is to choose one of the eleven DAGs in Fig. 1. We assume the following:

1. no missing values in the n-tuples of examples, and
2. the prior probabilities are given.

If $N = 3$, for (1)–(11), we compare the values

$$Q^n(X)Q^n(Y)Q^n(Z)$$

$$Q^n(X)Q^n(Y,Z), Q^n(Y)Q^n(Z,X), Q^n(Z)Q^n(X,Y)$$

$$\frac{Q^n(Z,X)Q^n(X,Y)}{Q^n(X)}, \quad \frac{Q^n(X,Y)Q^n(Y,Z)}{Q^n(Y)}, \quad \frac{Q^n(Z,X)Q^n(X,Y)}{Q^n(Z)}$$

$$\frac{Q^n(Y)Q^n(Z)Q^n(X,Y,Z)}{Q^n(Y,Z)}, \quad \frac{Q^n(Z)Q^n(X)Q^n(X,Y,Z)}{Q^n(Z,X)}, \quad \frac{Q^n(X)Q^n(Y)Q^n(X,Y,Z)}{Q^n(X,Y)},$$

$$Q^n(X,Y,Z)$$

multiplied by the prior probabilities, and we obtain a BN structure with the maximum posterior probability (Cooper and Herskovits, 1992 [5]).

We term values such as $Q^n(X), Q^n(Y,Z)$ *local scores* and denote by $Q^n(W)$ the local score of the variables in $W \subseteq V$. For example, if $W = \{X, Y, Z\}$, $Q(W)$ expresses $Q(X, Y, Z)$.

In contrast, we term $\dfrac{Q^n(Z,X)Q^n(X,Y)}{Q^n(X)}, \dfrac{Q^n(X,Y)Q^n(Y,Z)}{Q^n(Y)}$ multiplied by the prior probabilities as *global scores*. The posterior probabilities are proportional to the global scores, and maximizing either one is equivalent to the other. For example, if $N = 3$, there are eight local and eleven global scores.

However, for general N, the computation has been already proven to be NP-hard [4].

In 2006, assuming that the prior probabilities are equal, Silander and Myllymaki [9][2] proposed a way of efficiently finding a BN structure with the maximum posterior probability.

If we define the parent set $\pi_W(X)$ of $X \in W$ w.r.t. $W \subseteq V$ to be the U that maximizes $Q^n(X|U) := \dfrac{Q^n(\{X\} \cup U)}{Q^n(U)}$ for $U \subseteq W \backslash \{X\}$, then, we recursively obtain the parent set $Q^n(X|\pi_W(X))$ by

$$Q^n(X|\pi_W(X)) = \max\{Q^n(X|W) , \max_{Y \in W\backslash\{X\}} Q^n(X|\pi_{W\backslash\{Y\}}(X))\} , \qquad (15)$$

where W is assumed to be sorted in lexicographic order. Furthermore, we obtain the global score $R^n(V)$ of a structure with the maximum posterior probability recursively by $R^n(\{X\}) = Q^n(\{X\})$, $X \in V$, and

$$R^n(W) := \max_{X \in W}[Q^n(X|\pi_W(X)) \cdot R^n(W\backslash\{X\})] \qquad (16)$$

for $W \subseteq V$. The BN structure with vertices

$$X_N := \text{argmax}_{X \in V} R^n(V) , \quad X_{N-1} := \text{argmax}_{X \in V\backslash\{X_N\}} R^n(V\backslash\{X_N\}) , \quad \cdots$$

---

[2] The idea of using dynamic programing was invented by A. P. Singh & A. W. Moore (2005).

directed from the vertices included in

$$\pi_N := \pi_{V_N}(X_N) \, , \ \pi_{N-1} := \pi_{V_{N-1}}(X_{N-1}) \, , \cdots , \ \pi_1 := \pi_{V_1}(X_1) = \{\}$$

maximizes the posterior probability, where

$$V_N := V \, , \ V_{N-1} := V \backslash \{X_N\} \, , \cdots , \ V_1 = \{X_1\}.$$

In other words, we have

$$R^n(V) = Q^n(X_N|\pi_N) \cdot Q^n(X_{N-1}|\pi_{N-1}) \cdots Q^n(X_2|\pi_2)Q^n(X_1),$$

and the probability distribution is factorized into

$$P(X_N|\pi_N)P(X_{N-1}|\pi_{N-1}) \cdots P(X_2|\pi_2)P(X_1). \tag{17}$$

For example, if $W = \{Y, Z\}$ and $V = \{X, Y, Z\}$, then

$$
\begin{aligned}
&Q^n(X|\pi_{\{Y,Z\}}(X)) \\
&= \max\{Q^n(X|\{Y,Z\}), Q^n(X|\pi_{\{Y\}}(X)), Q^n(X|\pi_{\{Z\}}(X))\} \\
&= \max\{Q^n(X|\{Y,Z\}), \max\{Q^n(X|\{Y\}), Q^n(X)\}, \max\{Q^n(X|\{Z\}), Q^n(X)\}\} \\
&= \max\{Q^n(X|\{Y,Z\}), Q^n(X|\{Y\}), Q^n(X|\{Z\}), Q^n(X)\}
\end{aligned}
$$

in (15) and

$$
\begin{aligned}
&R^n(\{X, Y, Z\}) \\
&= \max\{Q^n(X|\pi_{\{Y,Z\}}(X))R^n(\{Y, Z\}), Q^n(Y|\pi_{\{Z,X\}}(Y))R^n(\{Z, X\}), \\
&\quad Q^n(Z|\pi_{\{X,Y\}}(Z))R^n(\{X, Y\})\} \\
&= \max\{Q^n(X|\{Y,Z\})\max\{Q^n(Y|\pi_{\{Z\}}(X))R^n(\{Z\}), Q^n(Z|\pi_{\{Y\}}(X))R^n(\{Y\})\}, \\
&\quad Q^n(Y|\pi_{\{Z,X\}}(Y))\max\{Q^n(Z|\pi_{\{X\}}(Z))R^n(\{X\}), Q^n(X|\pi_{\{Z\}}(X))R^n(\{Z\})\}, \\
&\quad Q^n(Z|\pi_{\{X,Y\}}(Z))\max\{Q^n(X|\pi_{\{Y\}}(X))R^n(\{Y\}), Q^n(Y|\pi_{\{X\}}(Y))R^n(\{X\})\}\} \\
&= \max\{Q^n(X|\pi_{\{Y,Z\}}(X))Q^n(Y|\pi_{\{Z\}}(Y))Q^n(\{Z\}), Q^n(X|\pi_{\{Y,Z\}}(X))Q^n(Z|\pi_{\{Y\}}(Z))Q^n(\{Y\}), \\
&\quad Q^n(Y|\pi_{\{Z,X\}}(Y))Q^n(Z|\pi_{\{X\}}(Z))Q^n(\{X\}), Q^n(Y|\pi_{\{Z,X\}}(Y))Q^n(X|\pi_{\{Z\}}(X))Q^n(\{Z\}), \\
&\quad Q^n(Z|\pi_{\{X,Y\}}(Z))Q^n(X|\pi_{\{Y\}}(X))Q^n(\{Y\}), Q^n(Z|\pi_{\{X,Y\}}(Z))Q^n(Y|\pi_{\{X\}}(Y))Q^n(\{X\})\}
\end{aligned}
$$

in (16) can be computed. For general N, $O(N^2 2^N)$ time and $O(N2^N)$ memory in (15) and $O(N2^N)$ time and $O(2^N)$ memory in (16) are required, as stated by Silander and Myllymaki [9].

Learning Bayesian network structures based on the MDL principle

Thus far, we have considered maximizing the posterior probability of the Bayesian network structure given the data.

However, for learning Bayesian network structures, there is another option: minimizing the description length of the given examples. The idea is to describe the given data in terms of a rule and its exceptions in many ways and to choose the rule that makes the total length the shortest (minimum description length (MDL) principle [8]). The idea of applying the MDL principle to learning Bayesian networks was first developed by Suzuki 1993 [14][3].

Although they are closely related, there are clear merits in applying the MDL principle over maximizing the posterior probability, which will be explained after applying the MDL principle to the current problem.

Let α, β, γ be the cardinalities of the sets in which variables X, Y, Z take values. If we calculate $-\log Q^n(X)$ with $a = 0.5$ in (13) using Stirling's formula, then we find [8] that the difference between $-\log Q^n(X)$ and

$$L^n(X) = H^n(X) + \frac{\alpha - 1}{2} \log n, \tag{18}$$

is at most $O(1)$, where the first term

$$H^n(X) := \sum_x c_n(x) \log \frac{n}{c_n(x)},$$

is called the *empirical entropy* of $x^n = (x_1, \cdots, x_n)$ w.r.t. X, and $\alpha - 1$ is the *number of parameters*.

When we find the parent set $\pi_{\{Y,Z\}}(X)$ of X w.r.t. $\{Y, Z\}$, we compare the values of description lengths $L^n(X)$, $L^n(X|\{Y\})$, $L^n(X|\{Z\})$, and $L(X|\{Y, Z\})$ given the candidates $\{\}, \{Y\}, \{Z\}$, and $\{Y, Z\}$, respectively. For example, given $(x_1, \cdots, x_n)$ and $(y_1, \cdots, y_n)$, $L^n(X|\{Y\})$ can be computed as follows.

For each $Y = y$, if we extract x_i such that $y_i = y$ to obtain $(\tilde{x}_1, \cdots, \tilde{x}_{c_n(y)})$ from $(x_1, \cdots, x_n)$, where $c_n(y)$ is the number of occurrences of y in $(y_1, \cdots, y_n)$, then we can obtain a description length similar to (18) with $n = c_n(y)$ for the sequence $(\tilde{x}_1, \cdots, \tilde{x}_{c_n(y)})$ instead of $(x_1, \cdots, x_n)$:

$$L^n(X|y) = H^n(X|y) + \frac{\alpha - 1}{2} \log c_n(y),$$

where $H^n(X|y)$ is the empirical entropy of $(\tilde{x}_1, \cdots, \tilde{x}_{c_n(y)})$. Thus, in total, the description length will be

$$\sum_y L^n(X|y) = \sum_y \{H^n(X|y) + \frac{\alpha - 1}{2} \log c_n(y)\},$$

where the approximation is still reasonable at this point. However, for convenience of computation, we usually set the upper bounds using $c_n(y) \leq n$ for each $Y = y$:

$$L^n(X|\{Y\}) = H^n(X|Y) + \frac{(\alpha - 1)\beta}{2} \log n,$$

where $H^n(X|Y) = \sum_y H^n(X|y)$. It is considered that the last approximation causes a problem when $c_n(y)$ is small for some y. In a similar manner, we obtain the values of $L^n(X)$, $L^n(X|\{Y\})$, $L^n(X|\{Z\})$, and $L(X|\{Y, Z\})$ for which the number of parameters are $\alpha - 1$, $(\alpha - 1)\beta$, $(\alpha - 1)\gamma$, and $(\alpha - 1)\beta\gamma$, respectively.

---

[3] At the same conference (Uncertainty in Artificial Intelligence 1993), Wai and Bucchus [20] presented another approach for MDL-based Bayesian network learning.

Despite the approximation loss in the MDL formulae, there are merits to be gained over maximizing the exact posterior probability. One merit is that we can adjust the balance between the empirical entropy and the number of parameters in terms of a function $d(n)$ of n. For example, if the AIC [1] is preferred over MDL in a situation, we can replace MDL with $d(n) = 0.5 \log n$ by setting $d(n) = 1$ for AIC.

2.3   Efficiency of Learning Bayesian Network Structures w.r.t. MDL

The other merit, on which we focus in this paper, is that we can save the computational effort of finding the best structure w.r.t. MDL by using the branch and bound technique.

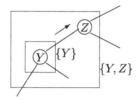

Fig. 2. Choose either $\{Y\}$, $\{Y, Z\}$ or their supersets for the parent set of X.

In 1996, Suzuki [13] showed that when we find the parent set of variable X, for any $Z \notin W \subseteq V$, if the value of

$$H^n(X|W) + \frac{k(X|W)}{2} \log n$$

has been obtained and is smaller than $\dfrac{k(X|W \cup \{Z\})}{2} \log n$, before computing

$$H^n(X|W \cup \{Z\}) + \frac{k(X|W \cup \{Z\})}{2} \log n$$

we find that neither $W \cup \{Z\}$ nor its supersets are optimal parent sets. Suzuki proposed using a cutting rule for searching for the best Bayesian network structure w.r.t. MDL.

For example, suppose that the parent set of X is $\{Y\}$ and that we need to determine whether Z should be included in the current parent set (Fig. 2). Then, the rule is that if

$$H^n(X|Y) \leq \frac{(\alpha - 1)\beta(\gamma - 1)}{2} \log n, \tag{19}$$

we should exclude both $\{Y, Z\}$ and its supersets from the parent set of X. This is because $H^n(X|Y, Z) \geq 0$ and

$$H^n(X|Y) + \frac{(\alpha - 1)\beta}{2} \log n \leq H^n(X|Y, Z) + \frac{(\alpha - 1)\beta\gamma}{2} \log n,$$

and the number $(\alpha-1)\beta\gamma$ of parameters further increases if an additional variable is joined to $\{Y, Z\}$.

In 2011, Cassio de Campose and Qiang Ji [6] noted the following:

$$\beta \geq n \Longrightarrow (19). \qquad (20)$$

This is because from $H^n(X|Y) \leq n \log \alpha$ and $\gamma \geq 2$, we have

$$\beta \geq n \geq \frac{2n \log \alpha}{(\alpha-1) \log n} \geq \frac{2H^n(X|Y)}{(\alpha-1)(\gamma-1) \log n}.$$

Hereafter, we use the term *state* for the combination of the variables in a parent set. For example, if $\{Y, Z\}$ is the parent set of X, there are $\beta\gamma$ states for X:

$$(1,1), \cdots, (1,\gamma), \cdots, (\beta, 1), \cdots, (\beta, \gamma) .$$

When the parent set consists of more than one variable other than one such as $\{Y\}$, the value β is replaced by the number of states of the parent set π for X.

In contrast, if a parent set π contains $|\pi|$ variables, then we have $\beta \geq 2^{|\pi|}$. In combination with (20), we have

$$|\pi| \geq \lceil \log_2 n \rceil \Longrightarrow 2^{|\pi|} \geq n \Longrightarrow \beta \geq n \Longrightarrow (19)$$

Proposition 1 (Campse and Ji, 2011 [6]). The number of variables in the best parent set w.r.t. MDL is at most $\lceil \log_2 n \rceil$.

Proposition 1 is attractive because only $O(N^2 n)$ time and $O(Nn)$ memory are required in (15), which is the computationally most intensive process in Silander and Myllymaki [9].

For (16), we currently cannot determine the merit of Proposition 1 in Silander and Myllymaki [9], and it is possible to obtain an efficient A^* search strategy from the algorithm approach by Yuan and Malone [19] using Proposition 1.

2.4 Efficiency of Learning Bayesian Network Structures w.r.t. Maximizing the Posterior Probability

A similar bound has been found for maximizing the posterior probability.

Suppose that we have already obtained the value of $-\log Q^n(X|\pi(X))$ and hope to avoid the computation of $-\log Q^n(X|\pi(X) \cup \{Z\})$ if

$$-\log Q^n(X|\pi(X)) \leq -\log Q^n(X|\pi(X) \cup \{Z\})$$

for each of $Z \notin \pi \cup \{X\}$. The computation can be avoided if $-\log Q^n(X|\pi(X))$ is smaller than a lower bound of $-\log Q^n(X|\pi \cup \{Z\})$. The lower bound can be obtained easily, and the computation (overhead) should be as small as possible and at least less than that involved when computing $-\log Q^n(X|\pi(X) \cup \{Z\})$ itself.

Campose and Ji [6] considered a specific case (BDeu [2,18]) of the BD (Bayesian Dirichlet) model [5] in which the constant a in $Q^n(X|\pi(X))$ does not depend on $X = x$ and $\pi(X) = s$, where s is a state of the parent set $\pi(X)$ for X:

$$Q^n(X|\pi(X)) = \prod_s \frac{\Gamma(\alpha a) \prod_x \Gamma(c_n(x,s) + a)}{\Gamma(a)^\alpha \Gamma(c_n(s) + \alpha a)}, \qquad (21)$$

where $c_n(s) := \sum_x c_n(x,s)$ is the number of occurrences of state $\pi(X) = s$ for X in the n-tuples of examples.

They derived the following lower bound:

Proposition 2 (Campse and Ji [6], J. Cussen [3]).

$$-\log Q^n(X|\pi(X)) \geq S_n \log \alpha,$$

where S_n is the number of actually occurred states s in the n-tuples of examples $(c_n(s) \geq 1)$.

However, the current result does not answer the following questions:

1. what is the exact form of Proposition 2 for the general BD rather than BDeu?
2. how tight is the bound obtained in Proposition 2?
3. how many variables should be prepared in the parent set for the purposes of conditions such as Proposition 1?

3   Contributions

In this section, we focus on the score $-\log Q^n(X|\pi(X))$. We first generalize Proposition 2 for a general BD and then provide a negative result (main result) for the second and third problems stated at the end of the previous section.

3.1   A General Lower Bound for the Score

To obtain the main results, we derive some simple mathematical statements. They can be obtained easily by using elementary calculus and the property $z\Gamma(z) = \Gamma(z+1)$ of the Gamma function $\Gamma(x) = \int_0^\infty t^{z-1}e^{-t}dt$.

Proposition 3. We define x' such that $a(x') = \max_x a(x)$. We have $Q^n(X) = 1$ for $n = 0$ and

$$Q^n(X) \leq \frac{a(x')}{\sum_x a(x)}$$

for $n \geq 1$.

Proof. We claim the following for $n \geq 0$"

$$Q^n(X) \leq Q_*^n(X) := \frac{\Gamma(n + a(x'))}{\Gamma(a(x'))} \cdot \frac{\Gamma(\sum_x a(x))}{\Gamma(n + \sum_x a(x))} \cdot \qquad (22)$$

In fact, for $n = 0$, both sides are one. If the claim is true for n, then we have

$$Q^{n+1}(X) = Q^n(X) \cdot \frac{c_n(x_{n+1}) + a(x_{n+1})}{n + \sum_x a(x)}$$

$$\leq Q^n_*(X) \cdot \frac{c_n(x_{n+1}) + a(x_{n+1})}{n + \sum_x a(x)}$$

$$\leq \frac{\Gamma(n + a(x'))}{\Gamma(a(x'))} \cdot \frac{\Gamma(\sum_x a(x))}{\Gamma(n + \sum_x a(x))} \cdot \frac{n + a(x')}{n + \sum_x a(x)}$$

$$= \frac{\Gamma(n + 1 + a(x'))}{\Gamma(a(x'))} \cdot \frac{\Gamma(\sum_x a(x))}{\Gamma(n + 1 + \sum_x a(x))}$$

$$= Q^{n+1}_*(X) ,$$

which means (22), where the first inequality follows from the assumption of induction at n, and the second follows from $c_n(x_{n+1}) \leq n$ and $a(x_{n+1}) \leq a(x')$ for any x_{n+1}. Finally, from $a(x') \leq \sum_x a(x)$, we have for $n \geq 1$,

$$Q^n_*(X) = \frac{n + a(x')}{n + \sum_x a(x)} \cdots \frac{1 + a(x')}{1 + \sum_x a(x)} \cdot \frac{a(x')}{\sum_x a(x)} \leq \frac{a(x')}{\sum_x a(x)} . \qquad (23)$$

Equations (22) and (23) imply the proposition.

Next, we consider the conditional probability of X given $\pi(X)$ expressed by

$$Q^n(X|\pi(X)) = \prod_s \frac{\Gamma(\sum_x a(x, s)) \prod_x \Gamma(c_n(x, s) + a(x, s))}{\prod_x \Gamma(a(x, s)) \Gamma(c_n(s) + \sum_x a(x, s))} , \qquad (24)$$

when given a pair of examples $(x_1, \cdots, x_n)$ and $(s_1, \cdots, s_n)$, where $c_n(s)$ and $c_n(x, s)$ are the numbers of occurrences of $\pi(X) = s$ and $(X, \pi(X)) = (x, s)$, respectively, in the two sequences.

Then, we obtain the first result:

Theorem 1. We consider x_s such that $a(x_s, s) = \max_x a(x, s)$ for each s. We have

$$- \log Q^n(X|\pi(X)) \geq - \sum_{s : c_n(s) \geq 1} \log \frac{a(x_s, s)}{\sum_x a(x, s)}$$

for $n \geq 0$.

Proof. If we define

$$Q^n(X|s) = \frac{\Gamma(\sum_x a(x, s)) \prod_x \Gamma(c_n(x, s) + a(x, s))}{\prod_x \Gamma(a(x, s)) \Gamma(c_n(s) + \sum_x a(x, s))} ,$$

then we have from Proposition 3 $Q^n(X|s) = 1$ for $n = 0$ and

$$Q^n(X|s) \leq \frac{a(x', s)}{\sum_x a(x, s)}$$

for $n \geq 1$. By multiplying all s such that $c_n(s) \geq 1$ and taking $- \log$, we obtain the theorem.

Note that if the value of $a(x, s)$ does not depend on $X = x$ and $\pi(X) = s$ in Theorem 1, then Proposition 2 is obtained.

3.2 Main Result

Let $\alpha^{(1)}, \cdots, \alpha^{(N)}$ be the cardinalities of the sets to which $X^{(1)}, \cdots, X^{(N)}$ belong. Then, for the n-tuples of the examples consisting of N variables, the $(\alpha^{(1)} \cdots \alpha^{(N)})^n$ datasets can be obtained.

Our claim is that for at least one of the datasets and at least one variable $X = X^{(j)}$, the score $-\log Q^n(X|\pi(X))$ will not be lower than the lower bounds stated in Proposition 2 and Theorem 1. Hence, we cannot bound the number of variables in the parent sets.

Proposition 4. There exists at least one $(x_1, \cdots, x_n)$ such that

$$Q^n(X) \leq \alpha^{-n}.$$

Proof. From (14), the sum of $Q^n(X)$ over all the α^n sequences is one. Suppose we have $Q^n(X) > \alpha^{-n}$ where all the α^n are sequences. This contradicts $\sum Q^n(X) = 1$, which completes the proof.

Theorem 2. At least one dataset and one variable X exist such that the score $-\log Q^n(X|\pi(X))$ is not lower than the lower bound of $-\log Q^n(X|\pi(X) \cup \{Y\})$ with $Y \notin \pi(X) \cup \{X\}$ given by Theorem 1.

Proof. From Proposition 4, there exists at least one pair of $(x_1, \cdots, x_n)$ for each s such that

$$Q^n(X|s) \leq \alpha^{-c_n(s)}.$$

By multiplying all y, we have

$$Q^n(X|\pi(X)) = \prod_s Q^n(X|s) \leq \prod_s \alpha^{-c_n(s)} = \alpha^{-\sum_s c_n(s)} = \alpha^{-n}.$$

Thus, there exists at least one pair of $(x_1, \cdots, x_n)$ and $(s_1, \cdots, s_n)$ such that

$$-\log Q^n(X|\pi(X)) \geq n \log \alpha.$$

Let S_n be the number of states S associated with parent set $\pi(X) \cup \{Y\}$, $Y \notin \pi(X) \cup \{X\}$, such that $c_n(s) \geq 1$. Then, from $n \geq S_n$, we have

$$n \log \alpha \geq S_n \log \alpha \geq - \sum_{s:c_n(s) \geq 1} \log \frac{a(x_s, s)}{\sum_x a(x, s)},$$

where s ranges over all the states associated with parent set $\pi(X) \cup \{Y\}$. This means that at least for one pair $(x_1, \cdots, x_n)$ and $(s_1, \cdots, s_n)$, $-\log Q^n(X|\pi(X))$ does not reach the lower bound of $-\log Q^n(X|\pi(X) \cup \{Y\})$ for any $\pi(X)$ and $Y \notin \pi(X) \cup \{X\}$.

Hence, we conclude that the lower bounds given by Theorem 1 as well as Proposition 2 are too loose to specify at most how many variables are to be prepared in each parent set before the dataset is available.

4   Concluding Remarks

We have obtained the two results given by Theorems 1 and 2.

The generalization (Theorem 1) is useful because thus far, the lower bound of the score for maximizing the posterior probability was given only for BDeu [6]. We have now provided a general lower bound for arbitrary BD.

The main result negatively answers the question of whether a counterpart of Proposition 1 is available, which, however, does not necessarily mean the lower bounds (Proposition 2 and Theorem 1) are not useful. In fact, from (23), we can increase the lower bound by

$$\sum_s \sum_{i=1}^{c_n(s)} \log\{\frac{i + \sum_x a(x,s)}{i + a(x_s, s)}\} \tag{25}$$

if we check the counts $c_n(s)$ each time, though the overhead may increase. Theorem 2 does not claim that no data exist that do not reach the original bound plus (25). In this sense, we need to consider an improved lower bound in the future.

For MDL, the approximation error is large for states s with small counts $c_n(s)$, and the formula is simple enough to evaluate the lower bound. In fact, in MDL, the number of states accounts for even the states s such that $c_n(s) = 0$. Conversely, for maximizing the exact posterior probability, it is more difficult to obtain an appropriate lower bound of the score.

We will continue to develop an efficient search for finding the best Bayesian network structures for both criteria.

References

1. Akaike, H.: Information theory and an extension of the maximum likelihood principle. In: 2nd International Symposium on Information Theory, Budapest, Hungary (1973)
2. Buntine, W.: Theory refinement on Bayesian networks. In: Uncertainty in Artificial Intelligence, Los Angels, CA pp. 52–60 (1991)
3. Cussens, J., Bartlett, M.: GOBNILP 1.6.2 User/Developer Manual1, University of York (2015)
4. Chickering, D.M., Meek, C., Heckerman, D.: Large-sample learning of Bayesian networks is NP-hard. In: Uncertainty in Artificial Intelligence, Acapulco, Mexico, pp. 124–133 (2003)
5. Cooper, G.F., Herskovits, E.: A Bayesian method for the induction of probabilistic networks from data. Mach. Learn. **9**(4), 309–347 (1992)
6. de Campos, C.P., Ji, Q.: Efficient structure learning of bayesian networks using constraints. J. Mach. Learn. Res. **12**, 663–689 (2011)
7. Krichevsky, R.E., Trofimov, V.K.: The performance of universal encoding. IEEE Trans. Inf. Theory **IT–27**(2), 199–207 (1981)
8. Rissanen, J.: Modeling by shortest data description. Automatica **14**, 465–471 (1978)

9. Silander, T., Myllymaki, P.: A simple approach for finding the globally optimal bayesian network structure. In: Uncertainty in Artificial Intelligence (2006)
10. Singh, A.P., Moore, A.W.: Finding optimal Bayesian networks by dynamic programming. Research Showcase@CMU (2005)
11. Spirtes, P., Glymour, C., Scheines, R.: Causation, Prediction and Search. Springer, Berlin (1993)
12. Pearl, J.: Probabilistic Reasoning in Intelligent Systems: Networks of Plausible Inference (Representation and Reasoning), 2nd edn. Morgan Kaufmann Pub., San Mateo (1988)
13. Suzuki, J.: Learning bayesian belief networks based on the minimum description length principle: an efficient algorithm using the B & B technique. In: International Conference on Machine Learning, pp. 462–470 (1996)
14. Suzuki, J.: A construction of bayesian networks from databases on an MDL principle. In: The Ninth Conference on Uncertainty in Artificial Intelligence, Washington D.C., pp. 266–273 (1993)
15. Suzuki, J.: The bayesian Chow-Liu algorithm. In: The Proceedings of the Sixth European Workshop on Probabilistic Graphical Models, Granada, Spain (2012)
16. Suzuki, J.: Consistency of learning bayesian network structures with continuous variables: an information theoretic approach. Entropy 2015 **17**(8), 5752–5770 (2015)
17. Tian, J.: A branch-and-bound algorithm for MDL learning Bayesian networks. In: Uncertainty in Artificial Intelligence, Palo Alto, pp. 580–588 (2000)
18. Ueno, M.: Robust learning Bayesian networks for prior belief. In: Uncertainty in Artificial Intelligence, Corvallis, Oregon, pp. 698–707 (2011)
19. Yuan, C., Malone, B.: Learning optimal bayesian networks: a shortest path perspective. J. Mach. Learn. Res. **48**, 23–65 (2013)
20. Lam, W., Bacchus, F.: Using causal information and local measures to learn bayesian networks. In: The Ninth Conference on Uncertainty in Artificial Intelligence, Washington D.C., pp. 243–250 (1993)

Constraint-Based Learning Bayesian Networks Using Bayes Factor

Kazuki Natori$^{(\boxtimes)}$, Masaki Uto, Yu Nishiyama,
Shuichi Kawano, and Maomi Ueno

Graduate School of Information Systems,
The University of Electro-Communications,
1-5-1, Chofugaoka, Chofu-shi, Tokyo 182-8585, Japan
{natori,uto,yu.nishiyama,skawano,ueno}@ai.is.uec.ac.jp

Abstract. A score-based learning Bayesian networks, which seeks the best structure with a score function, incurs heavy computational costs. However, a constraint-based (CB) approach relaxes this problem and extends the available learning network size. A severe problem of the CB approach is its lower accuracy of learning than that of a score-based approach. Recently, several CI tests with consistency have been proposed. The main proposal of this study is to apply the CI tests to CB learning Bayesian networks. This method allows learning larger Bayesian networks than the score based approach does. Based on Bayesian theory, this paper addresses a CI test with consistency using Bayes factor. The result shows that Bayes factor with Jeffreys' prior provides theoretically and empirically best performance.

Keywords: Bayesian networks · Conditional independence test · Jeffreys' prior · Learning Bayesian networks

1 Introduction

A Bayesian network is a probabilistic graphical model that represents relations of random variables using a directed acyclic graph (DAG) and a conditional probability table (Heckerman 1995; Pearl 1988). When a joint probability distribution has the DAG probabilistic structure, it can be decomposed exactly into a product of the conditional probabilities of variables given their parent variables. Therefore, a Bayesian network is guaranteed to provide a good approximation of the joint probability distribution. When we use a Bayesian network, it is necessary to estimate the structure of a Bayesian network from data because it is generally unknown. Estimating the structure is called "learning Bayesian network".

Two approaches can be used for learning Bayesian networks. First are score-based (SB) approaches (Chickering 2002; Cooper and Herskovits 1992; Heckerman 1995; Heckerman *et al.* 1995). The SB approach seeks the best structure with a score function that has consistency with the true DAG structure.

© Springer International Publishing Switzerland 2015
J. Suzuki and M. Ueno (Eds.): AMBN 2015, LNAI 9505, pp. 15–31, 2015.
DOI: 10.1007/978-3-319-28379-1_2

Therefore, this approach is called score-based learning. A popular Bayesian network learning score is the marginal likelihood (ML) score (using a Dirichlet prior over model parameters), which finds the maximum a posteriori (MAP) structure, as described by Buntine (1991) and Heckerman *et al.* (1995). In addition, the Dirichlet prior is known as a distribution which is only likelihood equivalent when certain conditions hold (Heckerman *et al.* 1995); this score is known as "Bayesian Dirichlet equivalence (BDe)" (Heckerman *et al.* 1995). Given no prior knowledge, the Bayesian Dirichlet equivalence uniform (BDeu), as proposed earlier by Buntine (1991), is often used. Actually, BDeu requires an "equivalent sample size (ESS)", which is the value of a user-specified free parameter. Moreover, it has been demonstrated in recent studies that the ESS plays an important role in the resulting network structure estimate.

Several learning algorithms in this approach have been developed based on dynamic programming (Cowell 2009; Koivisto and Sood 2004; Silander and Myllymaki 2006), A* search (Yuan *et al.* 2011), branch and bound (Malone *et al.* 2011), and integer programming (Cussens 2011; Jaakkola *et al.* 2010). However, the Bayesian network score-based learning is adversely affected by exponential time and NP hard problems (Chickering 1996). Consequently, the SB approach makes it difficult to apply a large network.

Second is a constraint-based (CB) approach. Fundamentally, the solution of the CB approach sequentially checks conditional independence relations among all variables by statistical testing (CI), and directs edges of the structure from observed data. Actually, the CB approach can relax computational cost problems and can extend the available learning network size for learning. Recently, Yahezkel et al. (2009) proposed the recursive autonomy identification (RAI) algorithm. The RAI algorithm decomposes into autonomous sub-structures after the basic solution of CB approaches. This sequence is performed recursively for each sub-structure. The advantage of the RAI algorithm is to be able to minimize the number of parent nodes when using CI tests in the CB approach. The RAI algorithm is, therefore, the highest accuracy in CB approaches. However, the CB approach depends on the threshold of CI test. It has no consistency with the true DAG structure. Traditional CI tests use G^2 or χ^2 test, and mutual information (MI). Recently, several CI tests with a score function have been proposed for learning Bayesian networks. For example, de Campos (2006) proposed a new score function based on MI for CI tests (de Campos 2006). MI shows consistency for the conditional independence relations between two nodes, but it has not proved the strong consistency (van der Vaart 2000).

On the other hand, a Bayes factor is known to have a strong consistency (van der Vaart 2000). The Bayes factor indicates the ratio of the marginal likelihoods for two hypotheses. The marginal likelihood finds the maximum a posteriori (MAP) structure, as described by Buntine (1991) and Heckerman *et al.* (1995). Steck and Jaakkola (2002) proposed a CI test using a Bayes factor that set of BDeu as the marginal likelihood. The CI test does not address the orientation of edges between two variables. To detect the orientation correctly, BDeu adjusts the number of parameters to be constant. However, this adjustment entails bias of the prior distribution (Ueno 2011).

In addition, Suzuki (2012) proposed a CI test that has strong consistent estimator of mutual information. As the result of the research, the proposed method corresponds to asymptotically a Bayes factor. But the method is only applied in the Chou–Liu algorithm and is not used in the learning Bayesian networks. Suzuki (2015) also proposed a CI test but he did not write how to use the test for learning Bayesian networks.

This study proposes constraint-based learning Bayesian networks using Bayes factor. A Bayes factor consists of the marginal likelihood for conditional joint probability distributions between two variables in Bayesian networks. This paper also shows that the Bayes factor using Jeffreys' prior is theoretically optimal for CI tests of Bayesian network. Clarke and Barron (1994) derived that the minimum risk value of the hyperparameter of Dirichlet prior is $1/2$, which is Jeffreys' prior because it minimizes the entropy risk of prior. For a score-based learning Bayesian network, the Jeffreys' prior works worse than BDe(u) does because it does not satisfy the likelihood equivalence property. However, this study shows theoretically that Jeffreys' prior is the optimal for the proposed Bayes factor. In addition, some numerical experiments underscore the effectiveness of the proposed method. This study gives score-based learning for a large Bayesian network including more than 60 variables.

This paper is organized as follows. First, we introduce the learning Bayesian networks in Sect. 2. Section 3 shows traditional CI tests. Section 4 presents the CI test using the Bayes factor with consistency. Section 5 presents the theoretical analyses about the proposed method that is introduced into Sect. 4. Section 6 introduces the recursive autonomy identification algorithm, which is the state-of-the-art algorithm in the CB approach. Section 7 shows experimental evaluations using the RAI algorithm. In these experiments, we review the learning accuracy of the RAI algorithm according to comparison of each CI tests. Section 8 concludes the paper and suggests avenues of future work.

2   Learning Bayesian Networks

Let $\{x_1, x_2, \cdots, x_N\}$ be a set of N discrete variables; each can take values in the set of states $\{1, \cdots, r_i\}$. Actually, $x_i = k$ means that x_i is state k. According to the Bayesian network structure $g \in G$, the joint probability distribution is given as

$$p(x_1, x_2, \cdots, x_N \mid g) = \prod_{i=1}^{N} p(x_i \mid \Pi_i, g), \tag{1}$$

where G is the possible set of Bayesian network structures, and Π_i is the parent variable set of x_i.

Next, we introduce the problem of learning a Bayesian network. Let θ_{ijk} be a conditional probability parameter of $x_i = k$ when the j-th instance of the parents of x_i is observed (we write $\Pi_i = j$). Buntine (1991) assumed the Dirichlet prior and used an expected a posteriori (EAP) estimator as the parameter estimator $\hat{\Theta} = (\hat{\theta}_{ijk})$ $(i = 1, \cdots, N, j = 1, \cdots, q_i, k = 1, \cdots, r_i - 1)$:

$$\hat{\theta}_{ijk} = \frac{\alpha_{ijk} + n_{ijk}}{\alpha_{ij} + n_{ij}}, \quad (k = 1, \cdots, r_i - 1). \tag{2}$$

Therein, n_{ijk} represents the number of samples of $x_i = k$ when $\Pi_i = j$, $n_{ij} = \sum_{k=1}^{r_i} n_{ijk}$, α_{ijk} denotes the hyperparameters of the Dirichlet prior distributions (α_{ijk} is a pseudo-sample corresponding to n_{ijk}), $\alpha_{ij} = \sum_{k=1}^{r_i} \alpha_{ijk}$, and $\hat{\theta}_{ijr_i} = 1 - \sum_{k=1}^{r_i-1} \hat{\theta}_{ijk}$.

The marginal likelihood is obtained as

$$p(\mathbf{X} \mid g, \alpha) = \prod_{i=1}^{N} \prod_{j=1}^{q_i} \frac{\Gamma(\alpha_{ij})}{\Gamma(\alpha_{ij} + n_{ij})} \prod_{k=1}^{r_i} \frac{\Gamma(\alpha_{ijk} + n_{ijk})}{\Gamma(\alpha_{ijk})}. \tag{3}$$

Here, q_i signifies the number of instances of Π_i, where $q_i = \prod_{x_l \in \Pi_i} r_l$ and $\mathbf{X}$ is a dataset. The problem of learning a Bayesian network is to find the MAP structure that maximizes the score (3).

Particularly, Heckerman *et al.* (1995) presented a sufficient condition for satisfying the likelihood equivalence assumption in the form of the following constraint related to hyperparameters of (3):

$$\alpha_{ijk} = \alpha p(x_i = k, \Pi_i = j \mid g^h). \tag{4}$$

Here, α is the user-determined equivalent sample size (ESS); g^h is the hypothetical Bayesian network structure that reflects a user's prior knowledge. This metric was designated as the Bayesian Dirichlet equivalence (BDe) score metric.

As Buntine (1991) described, $\alpha_{ijk} = \alpha/(r_i q_i)$ is regarded as a special case of the BDe metric. Heckerman *et al.* (1995) called this special case "BDeu". Actually, $\alpha_{ijk} = \alpha/(r_i q_i)$ does not mean "uniform prior," but "is the same value of all hyperparameters for a variable".

These methods are called a "score based approach." Score-based learning Bayesian networks are hindered by heavy computational costs. However, a conditional independence (CI) based approach is known to relax this problem and to extend the available learning network size.

3   CI Tests

Common means of CI testing are by thresholding conditional mutual information (CMI) or a statistic that measures statistical independence between variables (in Pearson's chi-square or likelihood ratio G-test).

Mutual Information. Mutual Information (MI) between variables X and Y measures the amount of information shared between these variables, which is provided as

$$\mathrm{MI}(X;Y) = \sum_{x \in X, y \in Y} P(x,y) \log\{P(x,y)/(P(x)P(y))\}. \tag{5}$$

It also measures the degree to which uncertainty about Y decreases when X is observed (and vice versa) (Cover and Thomas 1991). Actually, MI is the Kullback–Leibler (KL) divergence between $P(x, y)$ and $P(x)P(y)$ (Cover and Thomas 1991), measuring how much the joint differs from the marginals' product, or how much the variables can be regarded as not independent.

The CMI between X and Y, given a conditioning set $\mathbf{Z}$, is given as

$$\mathrm{CMI}(X; Y \mid \mathbf{Z}) = \sum_{x \in X, y \in Y, z \in \mathbf{Z}} P(x, y, z) \log\{P(x, y \mid z) / (P(x \mid z) P(y \mid z))\}. \tag{6}$$

By definition, $\mathrm{MI}(X; Y)$ and $\mathrm{CMI}(X; Y \mid \mathbf{Z})$ are non-negative. $\mathrm{MI}(X; Y) = 0$ ($\mathrm{CMI}(X; Y \mid \mathbf{Z}) = 0$) if and only if X and Y are independent (given $\mathbf{Z}$). The true MI is unknown. The estimated $\widehat{\mathrm{MI}}$ is larger than MI (Treves and Panzeri 1995), and therefore for independent variables larger than 0. Practically, $\widehat{\mathrm{MI}}$ is compared to a small threshold, ε, to distinguish pairs of dependent and pairs of independent variables (Aliferis *et al.* 2010; Besson 2010; Cheng *et al.* 1999; 2002). If $\widehat{\mathrm{MI}}(X; Y) < \varepsilon$, X and Y are regarded as independent and the edge connecting them is removed. The test for CI using CMI is similar.

Pearson's chi-square and G^2 test Statistical tests compare the null hypothesis that two variables are independent of the alternative hypothesis. If the null is rejected (cannot be rejected), then the edge is learned (removed). A statistic that is asymptotically chi-square distributed is calculated and compared to a critical value. If it is greater (smaller) than the critical value, then the null is rejected (cannot be rejected) (Agresti 2002; Spirtes *et al.* 2000). In Pearson's chi-square test, the statistic X_{st}^2 is

$$X_{st}^2 = \sum_{x \in X, y \in Y} (O_{xy} - E_{xy})^2 / E_{xy} \sim \chi_{\mathrm{d.f}=(|X|-1)(|Y|-1)}^2, \tag{7}$$

where $O_{xy}(E_{xy})$ is the number of records (expected to be if the null was correct) for which $X = x$, $Y = y$, and $|X|$ and $|Y|$ are the corresponding cardinalities. If the null is correct, $P(x, y) = P(x) \cdot P(y), \forall x \in X, y \in Y$. We expect that $E_{xy}/N = (E_x/N) \cdot (E_y/N), \forall x \in X, y \in Y$ and $E_{xy} = E_x \cdot E_y/N$ for E_x and E_y, which are the numbers of records in which $X = x$ and $Y = y$, respectively, and where N is the total number of records. If X_{st}^2 is greater than a critical value for a significance value α, $X_{st}^2 > X_{\mathrm{d.f}=(|X|-1)(|Y|-1),\alpha}$, then we reject the null hypothesis.

Instead, based on maximum likelihood, if the statistic

$$G_{st}^2 = 2 \sum_{x \in X, y \in Y} O_{xy} \log(O_{xy}/E_{xy}) \sim \chi_{\mathrm{d.f}=(|X|-1)(|Y|-1)}^2 \tag{8}$$

is larger than the previous critical value $G_{st}^2 > X_{\mathrm{d.f}=(|X|-1)(|Y|-1),\alpha}^2$, then we reject the null hypothesis.

However, the learning accuracy of the CB approach is less than that of score-based learning because these CI tests have no strong consistency (van der Vaart 2000).

4 Bayes Factor for CI Test

Traditional CI tests have used statistical tests without consistency. Therefore, the traditional CI tests are not guaranteed to obtain the correct structure even when the data size becomes large. In this paper, we propose a CI test with consistency using the Bayes factor to improve the traditional CI test.

The Bayes factor is the ratio of the marginal likelihood (ML) (Kass and Raftery 1995), which finds the maximum a posteriori (MAP) of the statistical model. Therefore, the Bayes factor has asymptotic consistency. For example, the Bayes factor is given as $p(\mathbf{X} \mid g_1)/p(\mathbf{X} \mid g_2)$, where g_1 and g_2 are the hypothetical structures from observed data $\mathbf{X}$. If the value is larger than 1.0, then g_1 is favored more than g_2, else g_2 is favored more than g_1.

Steck and Jaakkola (2002) proposed a CI test using the Bayes factor. In this method, $\mathbf{X}$ presents observed data for only two variables X_1 and X_2 given conditional variables as

$$\log \frac{p(\mathbf{X} \mid g_1)}{p(\mathbf{X} \mid g_2)}. \tag{9}$$

In the CI test, g_1 shows a dependent model in Fig. 1; g_2 shows an independent model in Fig. 2, where C is the conditional variables. When the log-Bayes factor takes a negative value, then the edge between x_1 and x_2 is deleted.

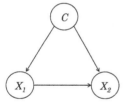

Fig. 1. g_1; dependent model.

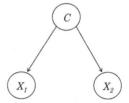

Fig. 2. g_2; independent model.

Steck and Jaakkola (2002) applied BDeu as the marginal likelihoods of the Bayes factor. However, Ueno (2010, 2011) pointed out that BDeu's prior is not non-informative. Especially, BDeu is not guaranteed to optimize CI tests because it was developed for score-based learning Bayesian network. The CI test does not address the orientation of edge between two variables. To detect the orientation correctly, BDeu adjusts the number of parameters to be constant. However, this adjustment causes the bias of the prior distribution (Ueno 2011).

To solve this problem, our approach uses a joint probability distribution of X_1 and X_2 because it is unnecessary to consider the orientation of edge between X_1 and X_2. Let $\theta_{jk_1k_2}$ represent $p(x_1 = k_1, x_2 = k_2 \mid \Pi_{(x_1,x_2)} = j, g_1)$, where $\Pi_{(x_1,x_2)}$ represents a set of common parents variables of x_1 and x_2. Here, $n_{jk_1k_2}$ denotes the number of samples of $x_1 = k_1$ and $x_2 = k_2$ when $\Pi_{(x_1,x_2)} = j$, $n_{k_1k_2} = \sum_{k_1=1}^{r_1} \sum_{k_2=1}^{r_2} n_{jk_1k_2}$. It is noteworthy that $\theta_{jr_1r_2} = 1 - \sum_{k_1=1}^{r_1-1} \sum_{k_2=1}^{r_2-1} \theta_{jk_1k_2}$. Assuming a uniform prior $\alpha_{jk_1k_2} = \alpha$, the marginal likelihood is obtained as

$$p(\mathbf{X}|g_1) = \frac{\Gamma(r_1 r_2 \alpha)}{\Gamma(\alpha)} \prod_{j=1}^{q_i} \prod_{k_1=1}^{r_1} \prod_{k_2=1}^{r_2} \frac{\Gamma(\alpha + n_{jk_1k_2})}{\Gamma(r_1 r_2 \alpha + n_{k_1k_2})}, \tag{10}$$

$$p(\mathbf{X}|g_2) = \prod_{i=1,2} \frac{\Gamma(r_i \alpha)}{\Gamma(\alpha)} \prod_{j=1}^{q_i} \prod_{k_i=1}^{r_i} \frac{\Gamma(\alpha + n_{jk_i})}{\Gamma(r_i \alpha + n_{k_i})}. \tag{11}$$

The remaining problem is determination of the value of hyper-parameter α. Clarke and Barron (1994) described that the optimal minimum risk value of the hyperparameter of the Dirichlet prior is $1/2$, which is Jeffreys' prior because it minimizes the entropy risk of prior. Ueno (2010, 2011) claimed that Jeffreys' prior is not efficient for score-based learning Bayesian network. However, this study specifically examines CI tests. The Jeffreys' prior is theoretically optimum for this problem.

Suzuki (2012) proposed a Bayes estimator of the mutual information for extending the Chow–Liu algorithm. The estimator is almost identical to the proposed Bayes factor in this paper. However, their purposes differ because Suzuki (2012) learned probabilistic tree structures to maximize the Bayes estimator.

Suzuki (2015) also proposed a CI test but he did not write how to use the test for learning Bayesian networks. The main proposal of this study is to apply the Bayes factor CI test to CB learning Bayesian networks.

5   Theoretical Analyses

In this section, we present results from some theoretical analyses of CI tests using the proposed method. From (3), the sum of hyperparameters α of BDeu is constant for the number of parents because $\alpha_{ijk} = \alpha/(r_i q_i)$, but that of the proposed method increases as the number of parents increases. For example, one might consider two binary variables with the empty set of C, as shown in Figs. 1 and 2. Then the proposed score for g_1 is calculable by

$$p(\mathbf{X} \mid g_1) = \frac{\Gamma(4\alpha)}{\Gamma(\alpha)} \prod_{k_1=1}^{2} \prod_{k_2=1}^{2} \frac{\Gamma(\alpha + n_{k_1k_2})}{\Gamma(4\alpha + n_{k_1k_2})}.$$

The proposed score for g_2 is obtained as

$$p(\mathbf{X} \mid g_2) = \frac{\Gamma(2\alpha)}{\Gamma(\alpha)} \prod_{k_1=1}^{2} \prod_{k_2=1}^{2} \frac{\Gamma(\alpha + n_{k_1k_2})}{\Gamma(2\alpha + n_{k_1k_2})}.$$

The proposed score for g_1 is equivalent to the BDeu score where ESS = 4α, but the proposed score for g_2 is equivalent to the BDeu score where ESS = 2α. Consequently, from the view of BDeu, the proposed score changes the ESS value according to the number of parameters. From this, the reader might suspect that the proposed method is affected by estimation bias.

To clarify the mechanisms of marginal likelihood of Bayesian network, Ueno (2010) analyzed the log-marginal likelihood asymptotically and derived the following theorem.

Theorem 1. *(Ueno 2010) When $\alpha + n$ is sufficiently large, log-marginal likelihood converges to*

$$\log p(\mathbf{X} \mid g, \alpha) = \log p(\widehat{\Theta} \mid \mathbf{X}, g, \alpha) - \frac{1}{2} \sum_{i=1}^{N} \sum_{j=1}^{q_i} \sum_{k=1}^{r_i} \frac{r_i - 1}{r_i} \log \left(1 + \frac{n_{ijk}}{\alpha_{ijk}}\right) + const.,$$

(12)

where

$$\log p(\widehat{\Theta} \mid \mathbf{X}, g, \alpha) = \sum_{i=1}^{N} \sum_{j=1}^{q_i} \sum_{k=1}^{r_i} (\alpha_{ijk} + n_{ijk}) \log \frac{(\alpha_{ijk} + n_{ijk})}{(\alpha_{ij} + n_{ij})},$$

and const. is the term that is independent of the number of parameters.

From (12), the log-marginal likelihood can be decomposed into two factors: (1) a log-posterior term $\log p(\widehat{\Theta} \mid \mathbf{X}, g, \alpha)$ and (2) a penalty term $\frac{1}{2} \sum_{i=1}^{N} \sum_{j=1}^{q_i} \sum_{k=1}^{r_i} \frac{r_i - 1}{r_i} \cdot \log \left(1 + \frac{n_{ijk}}{\alpha_{ijk}}\right)$. $\sum_{i=1}^{N} \sum_{j=1}^{q_i} \sum_{k=1}^{r_i} \frac{r_i - 1}{r_i}$ is the number of parameters.

This well known model selection formula is generally interpreted (1) as reflecting the fit to the data and (2) as signifying the penalty that blocks extra arcs from being added. This result suggests that a tradeoff exists between the role of α_{ijk} in the log-posterior (which helps to block extra arcs) and its role in the penalty term (which helps to add extra arcs).

From (12), the value of hyperparameter α_{ijk} should not be changed because the change of α_{ijk} strongly affects the penalty term of the score. The difference between BDeu and the proposed marginal likelihood is that the value of α_{ijk} in BDeu decreases as the number of parameters increases because $\alpha_{ijk} = \alpha/(r_i q_i)$ in BDeu, but that of the proposed method is constant for the different number of parameters. However, we use $\alpha_{ijk} = \alpha/(r_i q_i)$ only for correct orientation identification. Therefore, generally, the decrease of α_{ijk} leading to the increase the number of parameters in BDeu cannot be justified. Consequently, BDeu might show somewhat unstable performance in the CI test.

6   Recursive Autonomy Identification Algorithm

The remaining problem is which CB algorithm we employ to implement the Bayes factor CI test. In this study, we use the recursive autonomy identification (RAI) algorithm which is the state-of-art algorithm for the CB approach. In this section, we present the definition and procedure of the RAI algorithm.

Yehezkel and Lerner (2009) proposed the RAI algorithm to reduce unnecessary CI tests. They show that X and Y which are the variables of structure are independent conditioned on a set of conditional variables S using $X \perp Y \mid S$, and make use of d-separation (Pearl 1988). Also, they define d-separation resolution as the purpose to evaluate d-separation for different the number of conditional variables, and an autonomous substructure.

D-Separation Resolution. The resolution of a d-separation relation between a pair of non-adjacent nodes in a graph is the size of the smallest condition set that d-separates the two nodes.

Exogenous Causes. A node Y in $g(V, E)$ is an exogenous cause to $g'(V', E')$, where $V' \subset V$ and $E' \subset E$, if $Y \notin V'$ and $X \in V', Y \in Pa(X, g)$ or $Y \notin Adj(X, g)$ (Pearl 2000).

Autonomous Sub-structure. In DAG $g(V, E)$, a sub-structure $g^A(V^A, E^A)$ such that $V^A \subset V$ and $E^A \subset E$ is said to be autonomous in g given a set $V_{ex} \subset V$ of exogenous causes to g^A if $\forall X \in V^A$, $Pa(X, g) \subset \{V^A \cup V_{ex}\}$. If V_{ex} is empty, we say the sub-structure is (completely) autonomous.

They define sub-structure autonomy in the sense that the sub-structure holds the Markov property for its nodes. Given a structure g, any two non-adjacent nodes in an autonomous sub-structure g^A in g are d-separated given nodes either included in the sub-structure g^A or exogenous causes to g^A.

In this method, starting from a complete undirected graph and proceeding from low to high graph d-separation resolution, the RAI algorithm uncovers the correct pattern of a structure by performing the following sequence of operations.

First, all relations between nodes in the structure are checked using the CI test. Second, the edges are directed by orientation rules. Third, structure decomposes autonomous sub-structures. For each sub-structure, the RAI algorithm is applied recursively, while increasing the order of the CI tests. The important idea is that the entire structure decomposes autonomous sub-structures. By performing that procedure, decrease the high order of the CI tests. In the experimentally obtained results, the RAI algorithm was shown to be significant in comparison with other algorithms of the CB approach.

By the procedure, the RAI algorithm is able to realize the computational cost smaller than any other algorithm in the CB approach.

7   Numerical Experiments

This section presents some numerical experiments used to evaluate the effectiveness of our proposed method. For this purpose, we compare the learning accuracy of the proposed method with the other methods.

7.1 Experimental Design

We conducted some simulation experiments to evaluate the effectiveness of the proposed method. In the experiments, we compare the performances of Bayes factor with $\alpha_{ijk} = 1/2$, those with $\alpha_{ijk} = 1$, those with BDeu ($\alpha = 1$) (Steck and Jaakkola 2002), those of de Campos's method (2006), and those of the mutual information with the threshold of 0.003 which is derived as best value by Yehezkel and Lerner (2009). These methods are presented in Table 1.

In Sect. 7.2, we evaluate the performances of CI tests using three small network structures with binary variables. First structure shows a strongly skewed conditional probability distribution. Second has a skewed conditional probability distribution. Third has a uniform conditional probability distribution.

In Sects. 7.3 and 7.4, we present learning results obtained using large networks. We use the Alarm network in Sect. 7.3 and the win95pts network in Sect. 7.4. These benchmark networks were used from the *bnlearn repository* (Scutari 2010).

Table 1. Comparison of methods.

#	Methods
1	$\alpha_{ijk} = \frac{1}{2}$
2	$\alpha_{ijk} = 1$
3	BDeu ($\alpha = 1$)(Steck and Jaakkola 2002)
4	MI & χ^2 (de Campos 2006)
5	MI (Yehezkel and Lerner 2009)

7.2 Experimentation with Small Network

First, we evaluated the learning accuracy using a five-variable structure. Figure 3 has a strongly skewed conditional probability distribution. Figure 4 has a skewed conditional probability distribution. Figure 5 has a uniform conditional probability distribution.

The procedures of this experiment are described below.

1. We generated 100, 200, 300, 400, 500, 600, 700, 800, 900, and 1,000 samples from the three structures.
2. Using CI tests in Table 1, Bayesian network structures were estimated from 100, 200, 300, 400, 500, 600, 700, 800, 900, and 1,000 samples.
3. We repeated procedure 2 for 10 iterations for each number of samples.

We presented the average of the total learning errors for each CI test. The learning error shows the difference between the learned structure and the true structure, which is called the structure Hamming distance (SHD).

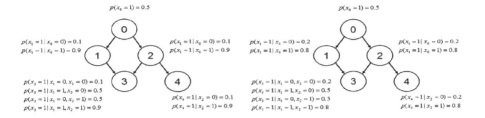

Fig. 3. Strongly skewed distribution.   **Fig. 4.** Skewed distribution.

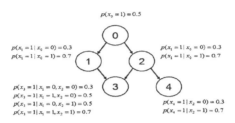

Fig. 5. Uniform distribution.

Tsamardinos *et al.* (2009) proposed the evaluation of the accuracy of the learning structure using the SHD, which is the most efficient metric between the learned and the true structure.

The results are depicted in Fig. 6. The results show that our proposed method (#1) produces the best performance. For a strongly skewed distribution (Fig. 3), our proposed method decreases the learning error faster than $\alpha_{ijk} = 1$ as the sample size becomes large. For a skewed distribution (Fig. 4), our proposed method decreases the learning error faster than $\alpha_{ijk} = 1$ as the sample size becomes large. For a uniform distribution (Fig. 5), all CI tests tend to be adversely affected, showing somewhat unstable behaviors. However, only the method with $\alpha_{ijk} = 1/2$ converges to zero error for a uniform distribution.

From Fig. 6, for a small network, performances with de Campos's method and MI are more adversely affected than those with the other methods because they have no strong consistency.

7.3 Experimentally Obtained Result with the Alarm Network

To evaluate a large network, we first used the Alarm network because it is widely known as a benchmark structure for the evaluation of learning Bayesian networks. The Alarm network includes 37 variables and 46 edges. The maximum in-degree is four. In this experiment, we determined the number of states of all variables as two.

To evaluate the CI test accuracy, we used learning errors of three types (Spirtes *et al.* 2000; Tsamardinos *et al.* 2006). An extra edge (EE) is a learned

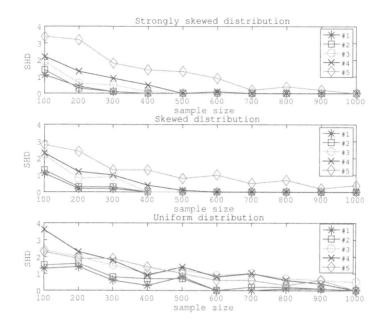

Fig. 6. Results of the learning small network.

edge, although it does not exist in the true graph. A missing edge (ME) is a missed edge from learning, although it exists in the true graph. Additionally, we used SHD.

For evaluation of learning of the Alarm network, we generated $N = 10,000$, 20,000, 50,000, 100,000, and 200,000 samples. Then we let the RAI algorithm with each CI test learn the structure using these samples. We repeated this procedure 10 times. We plot the MEs, EEs, and SHDs of the methods for each sample size to evaluate the learning accuracy in Figs. 7, 8, and 9. Additionally, we show the average of run-time in comparison with the method presented in Table 2.

Table 2. Comparison of the average run-time for each CI method in the Alarm network.

N	Average run-time results (s)				
	#1	#2	#3	#4	#5
10,000	80.9469	80.9974	80.2859	0.7680	0.5558
20,000	167.6280	168.2110	169.8730	1.1758	0.7945
50,000	423.5380	423.7020	424.3510	2.3933	1.6321
100,000	1.8034E+03	1.8283E+03	1.7869E+03	5.8928	4.3668
200,000	4.3404E+03	4.3984E+03	4.3753E+03	9.5591	7.1311

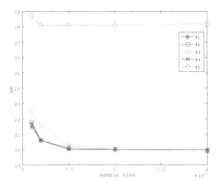

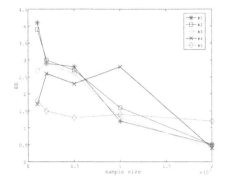

Fig. 7. Average numbers of MEs   **Fig. 8.** Average numbers of EEs

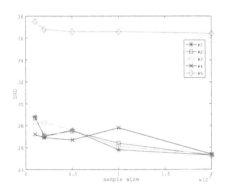

Fig. 9. Average numbers of SHDs

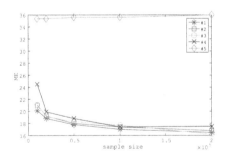

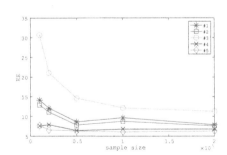

Fig. 10. Average numbers of MEs.   **Fig. 11.** Average numbers of EEs.

In Table 2, the proposed methods are shown to consume more run-time than the traditional MI methods do. In addition, the run-time of the proposed methods increases linearly as the sample size increases.

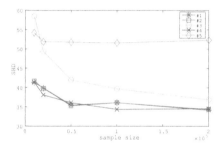

Fig. 12. Average numbers of SHDs.

From Figs. 7 and 9, Bayes factor with $\alpha_{ijk} = 1/2$ outperforms other methods in many cases. Our proposed method tends to be adversely affected more by extra edges for small sample sizes. As the sample size becomes larger than 100,000, the EEs of the proposed method show the best results.

7.4 Experimentally Obtained Results with the Win95pts Network

In the SB approach, Cussens (2011) proposed a learning algorithm using the integer programming and achieved the learning structure with 60 variables. To prove that our proposed method can learn a structure with more than 60 variables, we used the win95pts network. The network includes 76 variables and 112 edges. In addition, the maximum number of degrees is seven.

In this experiment, we also evaluated our proposed method using the same method as that used for learning the Alarm network. We compared the performances of the CI tests for $N = 10,000, 20,000, 50,000, 100,000$, and 200,000 samples. The procedure was repeated 10 times.

In Figs. 10, 11, and 12, we depict the experimentally obtained results from using MEs, EEs, and SHDs. Additionally, we show the average of run-time in comparison with the method presented in Table 3.

Table 3. Comparison of the average run-time for each CI method in the win95pts network.

N	Average run-time results (s)				
	#1	#2	#3	#4	#5
10,000	1.0222e+03	1.0642e+03	986.1200	7.7304	5.2507
20,000	2.1132e+03	1.9826e+03	2.0241e+03	13.5052	8.3541
50,000	4.9998e+03	5.1772e+03	4.7857e+03	21.9171	13.4316
100,000	1.5838e+04	1.5379e+04	1.4425e+04	39.9133	23.6153
200,000	3.3139e+04	3.2942e+04	3.2829e+04	66.8592	36.7520

From Fig. 10, our proposed method (#1) is shown to be the best. From Fig. 11, our proposed method (#1) tends to be adversely affected by extra edges. However, de Campos's method produces fewer extra edges. From Fig. 12, for a small sample size, the Bayes factor with $\alpha_{ijk} = 1$ exhibits superior performance. However, regarding the performance of the proposed method, the Bayes factor with $\alpha_{ijk} = 1$, and de Campos's method show almost identical performance when the sample size becomes large. Actually, de Campos's method without strong consistency provides the best performance because the sample size in this experiment is insufficiently large for this network.

From Table 3, the proposed methods consume more run-time than the traditional MI methods do. In addition, the run-time of the proposed methods increases linearly as the sample size increases. The run-time of the traditional MI methods increases rapidly as the network size increases. Consequently, the proposed method is expected to be applicable to extremely large networks.

8 Conclusion

As described herein, we proposed a new CI test using the Bayes factor with $\alpha_{ijk} = 1/2$ for learning Bayesian networks. Additionally, we provided some theoretical analyses of the proposed method. The results show that the prior distribution of BDeu for score-based learning is not non-informative, and it might cause biased and unstable estimations. The proposed CI test based on Jeffreys' prior minimizes the entropy risk of the prior and optimum the learning results. Using some experiments, we demonstrated that our proposed method improves learning accuracy compared with the other CI tests. Although the CI tests using Bayes Factor based on BDeu (Steck and Jaakkola 2002) have already been proposed, our proposed CI test worked better than the other CI tests did. However, for a large network, we were unable to find a significant difference from the other methods. For a large network, the proposed method requires a large sample size because it has asymptotic consistency.

On a different note, this work indicates that it begins taking a modest step towards improving the theory of the CB approach. A future work is to investigate the performance of the proposed method for larger networks and huge samples.

References

Agresti, A.: Categorical Data Analysis, Wiley Series in Probability and Statistics (2002)

Aliferis, C.F., Statnikov, A., Tsamardinos, I., Mani, S., Koutsoukos, X.D.: Local causal and Markov blanket induction for causal discovery and feature selection for classification. Part I: algorithms and empirical evaluation. J. Mach. Learn. Res. **11**, 171–234 (2010)

van der Vaart, A.W.: Asymptotic Statistics, Cambridge Series in Statistical and Probabilistic Mathematics (2000)

Besson, P.: Bayesian networks and information theory for audio-visual perception modeling. Biol. Cybern. **103**, 2013–2226 (2010)

Buntine, W.L.: Theory refinement on Bayesian networks. In: Proceedings of the 7th International Conference on Uncertainty in Artificial Intelligence, pp. 52–60 (1991)

de Campos, L.M.: A scoring function for learning Bayesian networks based on mutual information and conditional independence tests. J. Mach. Learn. Res. **7**, 2149–2187 (2006)

Cheng, J., Greiner, R.: Comparing Bayesian network classifiers. In: Proceedings of the 15th International Conference on Uncertainty in Artificial Intelligence, pp. 101–107 (1999)

Cheng, J., Greiner, R., Kelly, J., Bell, D., Liu, W.: Learning Bayesian networks from data: an information-theory based approach. Artif. Intell. **137**, 43–90 (2002)

Chickering, D.M.: Learning Bayesian networks is NP-complete. In: Learning from Data: Artificial Intelligence and Statistics V, pp. 121–130 (1996)

Chickering, D.M.: Optimal structure identification with greedy search. J. Mach. Learn. Res. **3**, 507–554 (2002)

Clarke, B.S., Barron, A.R.: Jeffreys' prior is asymptotically least favorable under entropy risk. J. Stat. Plann. Infer. **41**, 37–60 (1994)

Cover, T.M., Thomas, J.A.: Elements of Information Theory (2nd ed.), Wiley Series in Telecommunications and Signal Processing (1991)

Cooper, G.F., Herskovits, E.A.: Bayesian method for the induction of probabilistic networks from data. Mach. Learn. **9**, 309–347 (1992)

Cowell, R.G.: Efficient maximum likelihood pedigree reconstruction. Theor. Popul. Biol. **76**(4), 285–291 (2009)

Cussens, J.: Bayesian network learning with cutting planes. In: Cozman, F.G., Pfeffer, A. (eds.), Proceedings of the 27th International Conference on Uncertainty in Artificial Intelligence, pp. 153–160. AUAI Press (2011)

Heckerman, D.: A tutorial on learning with Bayesian networks. In: Technical Report: TR-95-06, Microsoft Research (1995)

Heckerman, D., Geiger, D., Chickering, D.M.: Learning Bayesian networks: The combination of knowledge and statistical data. Mach. Learn. **20**, 197–243 (1995)

Jaakkola, T., Sontag, D., Globerson, A., Meila, M.: Learning Bayesian network structure using LP relaxations. In: 13th International Conference on Artificial Intelligence and Statistics, vol. 9, pp. 358–365 (2010)

Kass, R.E., Raftery, A.E.: Bayes factors. J. Am. Stat. Assoc. **90**, 773–795 (1995)

Koivisto, M., Sood, K.: Exact Bayesian structure discovery in Bayesian networks. J. Mach. Learn. Res. **5**, 549–573 (2004)

Malone, B., Yuan, C., Hansen, E., Bridges, S.: Improving the scalability of optimal Bayesian network learning with external-memory frontier breadth-first branch and bound search. In: Cozman, F.G., Pfeffer, A. (eds.), Proceedings of the 27th International Conference on Uncertainty in Artificial Intelligence, pp. 479–488. AUAI Press (2011)

Pearl, J.: Probabilistic Reasoning in Intelligent Systems: Networks of Plausible Inference. Morgan-Kaufmann, San Francisco (1988)

Pearl, J.: Causality: Models, Reasoning, and Inference. Cambridge University Press, New York (2000)

Scutari, M.: Learning Bayesian networks with the bnlearn R package. J. Stat. Softw. **35**(3), 1–22 (2010)

Spirtes, P., Glymour, C., Scheines, R.: Causation, Prediction and Search, 2nd edn. MIT Press, Cambridge (2000)

Silander, T., Myllymaki, P.: A simple approach for finding the globally optimal Bayesian network structure. In: Proceedings of the 22nd International Conference on Uncertainty in Artificial Intelligence, pp. 445–452 (2006)

Steck, H., Jaakkola, T.S.: On the Dirichlet prior and Bayesian regularization. In: Advances in Neural Information Processing Systems, pp. 697–704. MIT Press, Vancouve (2002)

Suzuki, J.: The Bayesian Chow-Liu algorithm. In: Sixth European Workshop on Probabilistic Graphical Models, pp. 315–322 (2012)

Suzuki, J.: Consistency of learning Bayesian network structures with continuous variables: an information theoretic approach. Entropy **17**(8), 5752–5770 (2015)

Treves, A., Panzeri, S.: The upward bias in measures of information derived from limited data samples. Neural Comput. **7**, 399–407 (1995)

Tsumardinos, I., Brown, L.E., Aliferis, C.F.: The max-min hill-climbing Bayesian network structure learning algorithm. Mach. Learn. **65**(1), 31–78 (2006)

Ueno, M.: Learning networks determined by the ratio of prior and data. In: Grunwald, P., Spirtes, P. (eds.), Proceedings of the 26th Conference on Uncertainty in Artificial Intelligence, pp. 598–605. AUAI Press (2010)

Ueno, M.: Robust learning Bayesian networks for prior belief. In: Cozman, F.G., Pfeffer, A. (eds.), Proceedings of the 27th International Conference on Uncertainty in Artificial Intelligence, pp. 698–707. AUAI Press (2011)

Yehezkel, R., Lerner, B.: Bayesian network structure learning by recursive autonomy identification. J. Mach. Learn. Res. **10**, 1527–1570 (2009)

Yuan, C., Malone, B., Wu, X.: Learning optimal Bayesian networks using A* search. In: Proceedings of the 22nd International Joint Conference on Artificial Intelligence (2011)

Learning Bayesian Network Parameters from Small Data Set: A Spatially Maximum a Posteriori Method

Zhi-gao Guo, Xiao-guang Gao$^{(\boxtimes)}$, Ruo-hai Di, and Yu Yang

Department of System Engineering,
Northwestern Polytechnical University, Xian, China
cxg2012@nwpu.edu.cn

Abstract. To learn accurate BN parameters from small data set, combined with data, domain knowledge is often incorporated into the learning process as parameter constraints. Currently, most of the existing parameter learning methods take parameter learning problem as an exact optimization problem and regard the optimal solutions as the final parameters. However, due to the scarcity of data, objective functions constructed from the data, like likelihood function and entropy function, are not accurate. Therefore, parameters derived from the objective functions do not approach the true parameters well while some suboptimal parameters fit the true parameters better. Thus, searching more reasonable suboptimal parameters is a possible approach to learn better BN parameters. In this paper, we propose to visualize suboptimal parameters with parallel coordinate system and propose a Spatially Maximum a Posteriori (SMAP) method. Experimental results reveal that the proposed method outperforms most of the existing parameter learning methods.

Keywords: Bayesian Networks · Parameter learning · Small data set · Convex optimization · Linear programming

1 Introduction

Bayesian Network (BN) is a type of directed acyclic graph (DAG) with parameters, which is the combination of probability theory and graphical model theory [1]. It was systematically introduced in 1988 [1]. After about 30 years development, it has become a powerful tool for uncertainty analysis and is applied on wide issues like gene analysis [2], fault diagnosis [3], robot control [4], target tracking [5], signal processing [6], ecosystem modeling [7], etc. Generally, to construct a BN, relevant data is required and the number is decided by complexity of the problem to be solved. When sufficient data is available, constructing a good BN model from training data can be accomplished by traditional methods, like Maximum Likelihood [8] for parameter learning. Unfortunately, for domains like earthquake prediction and new-emerging disease diagnosis, collecting sufficient data is tough. In that situation, domain knowledge is often merged into modelling process of the network as supplement information.

© Springer International Publishing Switzerland 2015
J. Suzuki and M. Ueno (Eds.): AMBN 2015, LNAI 9505, pp. 32–45, 2015.
DOI: 10.1007/978-3-319-28379-1_3

In this paper, we focus on BN parameter learning and assume structure of the network is fixed. For parameter learning, domain experts feel more comfortable to provide qualitative domain knowledge [9], which can be transformed into parameter constraints, in the form like $p_1 > 0.8$, $p_1 \approx p_2$, $p_1 > p_2$, $(p_1 + p_2) > (p_3 + p_4)$, etc. These constraints look simple, but are very effective for improving BN model accuracy, especially when the data set is small. Generally, research about parameter learning from small data set experienced two stages: learning with single type of parameter constraints and learning with multiple types of parameter constraints. In the first stage: Wittig [10] proposed a Constrained Adaptive Probabilistic Networks (CAPN) method, which is suitable for qualitative influence constraints. First, parameter constraints are transformed from the qualitative expert knowledge. Then, an optimization model consisting of maximum entropy function and parameter constraints is built. Finally, the optimization model is solved by Adaptive Probabilistic Networks. Altendorf [11] proposed a Gradient-descent Estimation (GDE) method, which also applies to qualitative influence constraints. The difference is that the parameter constraints are integrated into the maximum entropy function as a penalty function, which evolves into a new objective function. The new objective function is then solved using gradient-descent algorithm. Feelders [12] proposed an Isotonic Regression Estimation (IRE) method, which also fits qualitative influence constraints. First, initial rough parameters are learnt by ML method. Then, parameter orders are constructed from qualitative influence constraints. Finally, the initial parameters are regulated by Isotonic Regression method and ultimately obey all the parameter orders. Isozaki [13] proposed a Minimum Free Energy (MFE) method, which is effective on the basic normalization constraints. First, a minimum free energy function, which consists of Kullback-Leibler divergence and entropy function, is taken as the objective optimization function. Then the minimum free energy function and normalization constraints are combined by Lagrange multipliers. Finally, the gradient method is employed to optimize the model.

As single type of parameter constraints can only restrict the optimal parameters into a broad possible parameter space, constraining force of those parameter constraints is weak. As a result, parameters computed based on the data and single type of constraints are far from accurate. On the other hand, methods suiting one type of constraints generally do not work on other types of constraints. Therefore, methods applicable to multiple types of parameter constraints are preferred and studied. In this stage: Niculescu [14] and Campos [15] proposed a Constrained Maximum Likelihood (CML) method. First, multiple types of parameter constraints are transformed from expert knowledge. Then, a convex optimization problem consisting of the likelihood function and parameter constraints is built. Finally, the optimization model is solved by convex optimization method. Campos [16] proposed a Constrained Maximum Entropy (CME) method. First, an Imprecise Dirichlet Model incorporating both data and prior information is constructed. Then a convex optimization model consisting of an entropy function and convex parameter constraints is presented. Finally, the model is also solved by the convex optimization method. Rui [17] proposed a Qualitative Maximum a Posterior (QMAP) method. First, certain amount of possible parameters are sampled from possible parameter space constructed by parameter constraints using rejection-acceptance sampling method. Then, hyper-parameters of Dirichlet distribution prior are specified by a virtual sampling number. Finally, optimal parameters are computed as the maximum a posteriori estimation of real data and pseudo data. Guo [18] proposed a Dually Constrained Estimation (DCE) method. First, parameters in the network are separated into two classes: parameters referring to different child states but the same parent configuration state and parameters referring to different parent configuration states but the same child state. Then, a beta distribution approximation based Bayesian estimation method is proposed to learn parameters of the first class. Finally, isotonic regression estimation method is employed to compute the second class parameters. Zhou [19] proposed a Multinomial Parameter Learning with Constraints (MPL-C) method. First, frequency number of different child node states but fixed parent node configuration state are counted. Then, an auxiliary BN model incorporating both data and parameter constraints is constructed. Finally, the optimal parameters are computed as the mean values of the probability distribution, which is inferred by a dynamic discretization junction tree method.

Generally, existing methods take the global optimal solution of the constrained optimization problem as the final parameters. However, when the available data is limited, objective function constructed from the data, like likelihood function, will overfit the data. As a result, parameters calculated by the existing methods often fail to approach the true parameters well while some suboptimal parameters approach the true parameters better. For the above reasons, in this paper, we analyze BN parameter learning problem from a spatial viewpoint and propose a Spatially Maximum a Posteriori (SMAP) method.

This paper is organized as follows: In Sect. 2, we describe the studied problem. In Sect. 3, we introduce the principle of the proposed method. In Sect. 4, the proposed Spatially Maximum a Posteriori (SMAP) method is introduced in detail. In Sect. 5, we perform some experiments and analyze the experiment results. In Sect. 6, we give some conclusions and point out some interesting future research directions.

2 Preliminaries

2.1 Bayesian Network

Bayesian network is a probabilistic graphic model, whose foundation are graph theory and probability theory. A Bayesian network consists of structure and parameters. Figure 1 is a typical and well-known Bayesian network – Asia BN. In that network, nodes like VA, S and TB, represent disease symptoms or diagnoses. Arrows from one node to another represent the influence imposed from the top node onto the bottom node. Conditional probability, like $P(D|LCTB, B)$, represents the strength of joint influence imposed by the symptom nodes $LCTB$

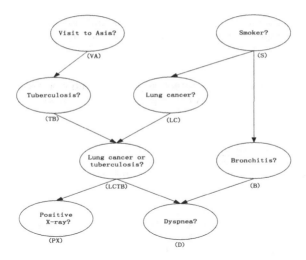

Fig. 1. The Asia Bayesian Network

and B onto the diagnosis node D. In this paper, we focus on learning of parameters in a network, especially discrete Bayesian network, whose structure is fixed in advance.

2.2   Parameters Learning in Bayesian Network

Learning parameters in a BN is to estimate parameters from a given sample data set. In this paper, samples with missing values are not considered. For a network with n node variables, parameter estimation can be expressed as a maximization problem of the log-likelihood function $\ell(p|D)$, where

$$\ell(p|D) = \sum_{i=1}^{n} \sum_{x_i, x_{\pi(i)}} n(x_i, x_{\pi(i)}) * \log p(x_i \mid x_{\pi(i)}) \tag{1}$$

According to the decomposability of BN, parameter estimation of a network can be decomposed into the product of independent estimation of each variable node X_i and the Maximum Likelihood (ML) estimation of the parameter $p(x_i|x_{\pi(i)})$ is

$$p(x_i \mid x_{\pi(i)}) = \frac{n(x_i, x_{\pi(i)})}{n(x_{\pi(i)})} \tag{2}$$

2.3   The Sample Complexity of Parameters Learning in Fixed-Structure Bayesian Networks

When the given data is sufficient, ML method is an ideal approach for accurate parameter learning. However, when the data set is small, learning results of

ML method are not satisfactory. Therefore, definition on sample complexity of BN parameter learning helps to decide whether ML is superior enough to learn expected accuracy parameters. If not, more information such as parameter constraints are required. About this topic, Dasgupta [20] defined a calculation of sample complexity bounds for parameters learning under a fixed BN structure. For a network with n boolean nodes, if no node has more than k parents, with confidence $1 - \delta$, sample complexity is upper-bounded by:

$$\frac{288 * n^2 * 2^k}{\epsilon^2} \ln^2(1 + \frac{3n}{\epsilon}) \ln \frac{1 + 3n/\epsilon}{\epsilon\delta} \qquad (3)$$

Constant ϵ is the error rate and is often set as $\epsilon = \alpha n$, for a small constant α.

2.4   Common Parameter Constraints

Generally, eight types of parameter constraints can be provided by domain experts as qualitative domain knowledge. The constraints are defined as below:

(1) Axiomatic Constraint: It describes relation between parameters referring to a fixed parent configuration state. It is a very basic constraint, which means, domain experts are not required to provide them.

$$\sum_{k=1}^{r_i} \theta_{ijk} = 1, 0 \leq \theta_{ijk} \leq 1, \forall i, j, k \qquad (4)$$

(2) Range Constraint: It defines the upper and lower values of a parameter, which is very common in reality. Also, domain experts feel more comfortable to provide such constraints.

$$0 \leq \alpha_{ijk} \leq \theta_{ijk} \leq \beta_{ijk} \leq 1 \qquad (5)$$

(3) Intra-distribution Constraint: It describes the comparative relation between two parameters referring to the same parent configuration state j but different child node states k and k'.

$$\theta_{ijk} \leq \theta_{ijk'}, \forall k \neq k' \qquad (6)$$

(4) Cross-distribution Constraint: It describes the comparative relation between two parameters referring to the same child node state k but different parent node configuration states j and j'.

$$\theta_{ijk} \leq \theta_{ij'k}, \forall j \neq j' \qquad (7)$$

(5) Inter-distribution Constraint: It describes the comparative relation between two absolutely different parameters.

$$\theta_{ijk} \leq \theta_{i'j'k'}, \forall i \neq i', j \neq j', k \neq k' \qquad (8)$$

(6) Approximate-Equality Constraint: It describes the close relation between any two parameters.

$$\theta_{ijk} \approx \theta_{i'j'k'}, \forall i \neq i', j \neq j', k \neq k' \tag{9}$$

Since the form of the above constraints is intractable for calculation, it can be transformed into the form as below:

$$\mid \theta_{ijk} - \theta_{i'j'k'} \mid \leq \epsilon, \forall i \neq i', j \neq j', k \neq k' \tag{10}$$

(7) Additive Synergy Constraint: It describes the comparative relation between sums of each two parameters under different distributions.

$$\theta_{ij_1k} + \theta_{ij_2k} \leq \theta_{ij_3k} + \theta_{ij_4k}, \forall i, k \tag{11}$$

(8) Product Synergy Constraint: It describes the comparative relation between products of each two parameters under different distributions.

$$\theta_{ij_1k} * \theta_{ij_2k} \leq \theta_{ij_3k} * \theta_{ij_4k}, \forall i, k \tag{12}$$

To be noticed, parameter constraints of type (1–7) are all convex, while product synergy constraints are non-convex.

3   Principle of the Proposed Method

To illustrate the principle of the proposed method, we take a local network (Fig. 2) of Asian network as the sample BN and explain the parameter learning process. First, a random set of parameters are generated and assumed as the true parameters. Then, small amount of synthetic data is sampled from the network.

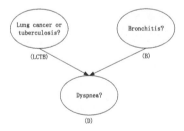

Fig. 2. An independent example Bayesian Network

Usually, parameters in BN are "high-dimensional" (at least four dimensions with one Boolean child node and one Boolean parent node). So, it is tough to visualize parameters by X-Y-Z axis system. For that reason, we propose to visualize parameters by parallel coordinate system. For simplicity, we assign an index for each parameter of the sample network in Table 1 and the indexes instead of parameters are plotted on the x axis of the parallel coordinate graph.

Table 1. Indexes for parameters

Index	Parameter	Index	Parameter
1	$P(D = 0 \mid LCTB = 0, B = 0)$	5	$P(D = 1 \mid LCTB = 0, B = 0)$
2	$P(D = 0 \mid LCTB = 0, B = 1)$	6	$P(D = 1 \mid LCTB = 0, B = 1)$
3	$P(D = 0 \mid LCTB = 1, B = 0)$	7	$P(D = 1 \mid LCTB = 1, B = 0)$
4	$P(D = 0 \mid LCTB = 1, B = 1)$	8	$P(D = 1 \mid LCTB = 1, B = 1)$

Table 2. Parameter constraints

Index	Constraint	Index	Constraint	Index	Constraint
1	$P_1 \geq P_5$	4	$P_8 \geq P_4$	7	$P_8 \geq P_7$
2	$P_6 \geq P_2$	5	$P_8 \geq P_5$	8	$P_6 \geq P_5$
3	$P_7 \geq P_3$	6	$P_8 \geq P_6$	9	$P_7 \geq P_5$

Parameter constraints transformed from the medical expert knowledge are listed in Table 2.

Then, we compute the maximum and minimum values of each parameter by linear programming with constraints and we plot them as green lines (one line for one set of parameters) in Fig. 3. Finally, we plot the true parameters as the red line in Fig. 3, which satisfy all the parameter constraints.

To better approach the true parameters, we try to reduce the possible parameter space, where the true parameters lie in. In Fig. 4, we assume that the reduced possible parameter space is the area between the black line and the blue line, which respectively represent center parameters and border parameters.

If the center and border parameters can be correctly determined (their computation will be explained in later section), then, we can find a set of parameters, which better approach the true parameters.

Before the calculation of more optimal parameters, let us consider the principle indicated in Fig. 5: (1) If the point we search for lies between point A and point B, then point A is closer to the searched point than point M and point

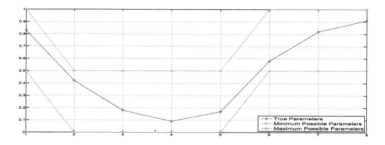

Fig. 3. Visualization of the True Parameters

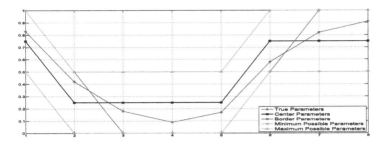

Fig. 4. Visualization of the Border and Center Parameters

D. (2) If the point we search for lies between point B and point C, then point M is closer to the searched point than point A and point D. (3) If the point we search for lies between point C and point D, then point D is closer to the searched point than point A and point M. Uniformly, the searched point more likely lies between point B and point C (probability 50 %) than that between point A and point B (probability 25 %) and that between point C and point D (probability 25 %).

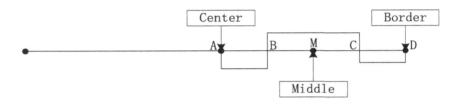

Fig. 5. Principle of the proposed method

Based on the above indicated principle, in Fig. 6, under uniform distribution, expected parameters (in pink) have more possibility (50 %) to better approach the true parameters (in red) than the border parameters (in blue) (25 %) and center parameters (in black) (25 %).

To compute expected parameters in Fig. 6, border parameters and center parameters should be calculated first. In Sect. 4, we will introduce the computation of those parameters.

4   Spatially Maximum a Posteriori Method

For BN parameter estimation, it has a trait, which is: (1) when the given data set contains much parameter information, parameters calculated by ML method may satisfy all the parameter constraints. In that case, parameters θ_{ijk}^{ML} can be taken as the final optimal parameters. (2) when the given data set contains not much information, parameters calculated by ML method fail to satisfy all the parameter constraints. In that case, better parameters are preferred.

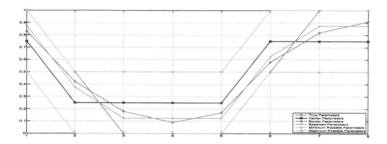

Fig. 6. Visualization of the true parameters and the optimal parameters (Color figure online)

For Spatially Maximum a Posteriori (SMAP) method, to learn better parameters, border parameters and center parameters should be computed first.

4.1   Border Parameter Calculation

When parameters θ_{ijk}^{ML} do not satisfy all the parameter constraints, parameters derived from constrained maximum likelihood problem lie on border of the possible parameter space. Therefore, the border parameter can be acquired [21] as solutions of the convex optimization model defined by Eqs. (13) and (14).

$$Maximize \quad logP(D \mid \theta, G) \qquad (13)$$

$$Subject\ to \quad \Omega(\theta_{ijk}) \leq 0 \qquad (14)$$

$\Omega(\theta_{ijk})$ denotes all parameter constraints.

4.2   Center Parameter Calculation

To calculate the center parameter, maximum value θ_{ijk}^{max} and minimum value θ_{ijk}^{min} of parameter θ_{ijk} should be calculated beforehand. With models in Eqs. (15) and (16), the maximum value θ_{ijk}^{max} can be computed by linear programming.

$$Maximize \quad \theta_{ijk} \qquad (15)$$

$$Subject\ to \quad \Omega(\theta_{ijk}) \leq 0 \qquad (16)$$

Likewise, the minimum value θ_{ijk}^{min} can be figured out from the following model by linear programming.

$$Maximize \quad -\theta_{ijk} \qquad (17)$$

$$Subject\ to \quad \Omega(\theta_{ijk}) \leq 0 \qquad (18)$$

Based on parameters θ_{ijk}^{max} and θ_{ijk}^{min}, the center parameter θ_{ijk}^{C} can be computed by Eq. 19.

$$\theta_{ijk}^{C} = \frac{1}{2} * (\theta_{ijk}^{max} + \theta_{ijk}^{min})$$ (19)

4.3   Spatially Maximum a Posteriori Parameter Calculation

Based on the above analysis and calculation, SMAP method can be summarized as bellow:

Step 1: Calculate parameter θ_{ijk}^{ML} from training data set by ML method (Eq. 2).

Step 2: If the parameter θ_{ijk}^{ML} satisfies all the parameter constraints, then it is taken as the final optimal parameter

$$\theta_{ijk}^{SMAP} = \theta_{ijk}^{ML}$$ (20)

If not, go to step 3.

Step 3: Calculate border parameter θ_{ijk}^{B} by Eqs. (13) and (14) and center parameter θ_{ijk}^{C} by Eqs. (15)–(19) and go to step 4.

Step 4: Compute SMAP parameter by following equation:

$$\theta_{ijk}^{SMAP} = \frac{1}{2} * (\theta_{ijk}^{B} + \theta_{ijk}^{C})$$ (21)

To be noticed, since ML method converges, calculation in Step 1 guarantees the convergence of SMAP method.

5   Experiments

We perform experiments on the network in Fig. 2 with seven parameter learning schemes: Maximum Likelihood (ML), Constrained Maximum Likelihood (CML), Maximum Entropy (ME), Constrained Maximum Entropy (CME), Maximum a Posteriori (MAP), Qualitatively Maximum a Posteriori (QMAP) and Spatially Maximum a Posteriori (SMAP). All the experiments are implemented under Matlab software environment.

5.1   Parameter Learning with Different Sample Sizes

Under different sample size, we evaluate learning results of different algorithms by Kullback-Leibler divergence [22] of the learnt parameters to the true parameters. Since the true parameters of the network are unknown, we assign random probability values as the true parameters. For such assignment, it has an advantage, that is, network embedded with different parameter assignment stands for different reality network. Thus, such assignments can test the generality of different algorithms. For each parameter assignment, we sample synthetic data sets of different sample size, i.e. 10, 20, 30, 40, 50, 60, 70, 80, 90, and 100. The generated data is used to learn the original assigned parameters. In each learning task, MAP method is executed with Dirichlet parameter $\alpha_{ijk} = 1$. We repeated each learning process with different parameter assignment. Final results for each learning scheme with various training samples are shown in Fig. 7.

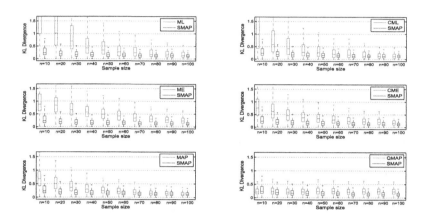

Fig. 7. Parameter learning results under different sample sizes

Seen from the experimental results in Fig. 7, we can conclude that SMAP method outperforms all the existing parameter learning methods under almost any sample size except extremely small one, like 10. The explanation is, with extremely small size data set, border parameters can not be correctly determined by Eqs. (13) and (14). That makes parameters calculated by SMAP method deviate from the true parameters. However, with slightly more data, calculation results of Eqs. (13) and (14) become more trustable and further improve the performance of SMAP method.

5.2   Parameter Learning to Achieve Certain KL Divergences

We calculate sample sizes needed by different algorithms to achieve certain KL divergences, like 0.1, 0.2, 0.3, 0.4, and 0.5. Final results are shown in Fig. 8.

Seen from the experimental results in Fig. 8, we can find that, to achieve any KL divergence, SMAP method requires fewer samples than any other algorithms. Besides, algorithm showing good performance in the first experiment may performs terrible and require much more samples to achieve certain KL divergence. For example, QMAP can learn notably low KL divergence parameters with small data set. However, when higher accuracy parameters are preferred (like parameters of KL divergence 0.1, 0.2), extremely large number of samples (more than 100) are needed by QMAP algorithm.

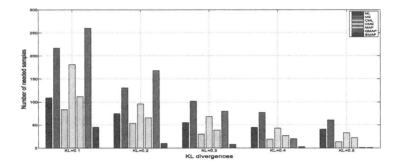

Fig. 8. Sample sizes needed to achieve different KL divergences (Color figure online)

5.3 Time Consumption Analysis

To show the time consumption of each algorithm, we continue the experiments and calculate the average running time of algorithm. Final results are shown in Fig. 9.

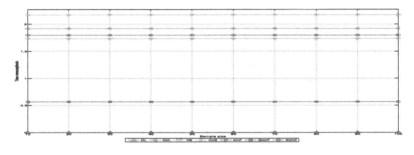

Fig. 9. Time consumption of different algorithms

Seen from the experiment results in Fig. 9, we can find that, MLE algorithm and MAP algorithm have much lower time consumption than other algorithms. ME, CML and CME consume more time, because those algorithms involve optimization of entropy and likelihood function, which is very time-consuming. Averagely, SMAP is more time-consuming than QMAP algorithm because it involves calculation of both center and border parameters, while QMAP algorithm includes only center parameters calculation. However, compared with QMAP algorithm, SMAP algorithm is less time-consuming on center parameters calculation because linear programming technology in SMAP algorithm consumes much less time than rejection-acceptance sampling method in QMAP algorithm.

6   Conclusions

From spatial analysis viewpoint, a new BN parameter learning method – SMAP is proposed in this paper. The proposed method has following features:

Advantages:

(1) The method is proposed based on high-dimensional analysis, which makes it possible to observe the learning processes and results of different parameter learning algorithms. Furthermore, distance between the learnt parameters and true parameters is more intuitive under high-dimensional visualization than KL divergence.

(2) Linear programming technology is embedded into SMAP algorithm to calculate the center parameters, which is less time-consuming than rejection-acceptance sampling method in QMAP algorithm.

Disadvantages:

(1) Border parameters derived from Eqs. (13) and (14) is occasionally incorrect when sample size is extremely small. So, better border parameter calculation methods will be one of the future research directions, which can enormously improve the learning performance of SMAP algorithm.

(2) Convex optimization of likelihood function is time-consuming, which makes SMAP algorithm slow. So, better likelihood function optimization methods will be another future research direction, which would tremendously reduce the time consumption of SMAP algorithm.

(3) As border parameters are calculated by convex optimization method, SMAP algorithm does not suit non-convex parameter constraints, like product synergy constraints in Sect. 2. So, extension of SMAP algorithm making it applicable to no-convex constraints will also be a future research direction, which may also improve the learning performance of SMAP algorithm.

Acknowledgements. This work is supported in part by National Nature Science Foundation of China (grant 61573285) and the Doctoral Fund of Ministry of Education of China (grant 20116102110026).

References

1. Pearl, J.: Probabilistic Reasoning in Intelligent Systems. Morgan Kaufmann, Burlington (1988)
2. Tamda, Y., Imoto, S., Araki, H.: Estimating genome-wide gene networks using nonparametric Bayesian network models on massively parallel computers. IEEE Trans. Comput. Biol. Bioinform. **3**, 683–697 (2011)
3. Ibrahim, W., Beiu, V.: Using Bayesian networks to accurately calculate the reliability of complementary metal oxide semiconductor gates. IEEE Trans. Reliab. **60**, 538–549 (2011)
4. Infantes, G., Ghallab, M., Ingrand, F.: Learning the behavior model of a robot. Auton. Robot. **30**, 157–177 (2011)

5. Steven, M., Ann, N., Kevin, K.: Anomaly detection in vessel tracks using Bayesian networks. Int. J. Approximate Reasoning **55**, 84–98 (2014)
6. Neil, W., Mahmood, R., Kevin, K.: Detection and classification of non-stationary transient signals using sparse approximations and Bayesian networks. IEEE/ACM Trans. Audio Speech Lang. Process. **22**, 1750–1764 (2014)
7. Dries, L., Steven, B., Rob, D., Guy, E., Joris, A.: A review of Bayesian belief networks in ecosystem service modeling. Environ. Model. Softw. **46**, 1–11 (2013)
8. Redner, R., Walker, H.: Mixture densities, maximum likelihood and the EM algorithm. SIAM Rev. **26**, 195–239 (1984)
9. Helsper, E., Gaag, L., Groenendal, F.: Designing a procedure for the acquisition of probability constraints for Bayesian networks. In: Proceedings of the Fourteenth Conference on Engineering Knowledge in the Age of the Semantic Web, pp. 280–292 (2004)
10. Wittig, F., Jameson, A.: Exploiting qualitative knowledge in the learning of conditional probabilities of Bayesian networks. In: Proceedings of the Sixteenth International Conference on Uncertainty in Artificial Intelligence, pp. 644–652 (2000)
11. Altendorf, E., Restificar, A., Dietterich, T.: Learning from sparse data by exploiting monotonicity constraints. In: Proceedings of the Twenty First International Conference on Uncertainty in Artificial Intelligence, pp. 18–26 (2005)
12. Feelders, A., Gaag, L.: Learning Bayesian networks parameters under order constraints. Int. J. Approximate Reasoning **42**, 37–53 (2006)
13. Isozaki, T., Kato, N., Ueno, M.: Data temperature in minimum free energies for parameter learning of Bayesian networks. Int. J. Artif. Intell. Tools **18**, 653–671 (2009)
14. Niculescu, R., Mitchell, T., Rao, B.: Bayesian network learning with parameter constraints. J. Mach. Learn. Res. **7**, 1357–1383 (2006)
15. Campos, C., Yan, T., Qiang, J.: Constrained maximum likelihood learning of Bayesian networks for facial action recognition. In: Proceedings of the Tenth European Conference on Computer Vision, pp. 168–181 (2008)
16. Campos, C., Qiang, J.: Improving Bayesian network parameter learning using constraints. In: Proceedings of the Nineteenth International Conference on Pattern Recognition, pp. 1–4 (2008)
17. Rui, C., Shoemaker, R., Wei, W.: A novel knowledge-driven systems biology approach for phenotype prediction upon genetic intervention. IEEE Trans. Comput. Biol. Bioinform. **1**, 1170–1181 (2011)
18. Zhi-gao, G., Xiao-guang, G., Ruo-hai, D.: Learning Bayesian network parameters under dual constraints from small data set. Acta Automatica Sinica **40**, 1509–1516 (2014)
19. Yun, Z., Fenton, N., Neil, M.: Bayesian network approach to multinomial parameter learning using data and expert judgments. Int. J. Approximate Reasoning **55**, 1252–1268 (2014)
20. Dasgupta, S.: The sample complexity of learning fixed-structure Bayesian networks. Mach. Learn. **29**, 165–180 (1997)
21. Boyd, S.: CVX: Matlab software for disciplined convex programming (2015). http://cvxr.com/cvx
22. Kullback, S., Leibler, R.: On information and sufficiency. Ann. Math. Stat. **22**, 79–86 (1951)

Hashing-Based Hybrid Duplicate Detection for Bayesian Network Structure Learning

Niklas Jahnsson[1], Brandon Malone[2(✉)], and Petri Myllymäki[1,3]

[1] University of Helsinki, Helsinki, Finland
[2] Max Planck Institute for the Biology of Ageing, Cologne, Germany
brandon.malone@age.mpg.de
[3] Helsinki Institute for Information Technology, Esbo, Finland

Abstract. In this work, we address the well-known score-based Bayesian network structure learning problem. Breadth-first branch and bound (BFBnB) has been shown to be an effective approach for solving this problem. Delayed duplicate detection (DDD) is an important component of the BFBnB algorithm. Previously, an external sorting-based technique, with complexity $O(m \log m)$, where m is the number of nodes stored in memory, was used for DDD. In this work, we propose a hashing-based technique, with complexity $O(m)$, for DDD. In practice, by removing the $O(\log m)$ overhead of sorting, over an order of magnitude more memory is available for the search. Empirically, we show the extra memory improves locality and decreases the amount of expensive external memory operations. We also give a bin packing algorithm for minimizing the number of external memory files.

Keywords: Bayesian networks · Structure learning · State space search · Delayed duplicate detection

1 Introduction

Bayesian networks (BNs) are a widely-used formalism for capturing uncertainty among variables in a domain of interest. When the relationship among the variables is not known *a priori*, we must learn those relationships from data. In this work, we present a novel approach to significantly reduce both the time and memory complexity for an existing structure learning algorithm.

In the *score-based* framework, the BN structure learning problem (BNSL) is cast as an optimization problem in which the goal is to find a BN structure which optimizes a scoring function. The scoring function is typically a penalized log-likelihood function which trades off the fit of a BN to the data with its complexity. Even though BNSL is known to be NP-hard (Chickering 1996), many algorithms have been proposed which solve the problem exactly (Ott et al. 2004; Koivisto and Sood 2004; Silander and Myllymäki 2006; de Campos and Ji 2011; Yuan and Malone 2013; Bartlett and Cussens 2015; van Beek and Hoffmann 2015).

State space search algorithms have been shown to be among the state-of-the-art approaches to solving BNSL (Malone et al. 2014). In particular, breadth-first branch and bound (BFBnB) takes advantage of regularities in the search

© Springer International Publishing Switzerland 2015
J. Suzuki and M. Ueno (Eds.): AMBN 2015, LNAI 9505, pp. 46–60, 2015.
DOI: 10.1007/978-3-319-28379-1_4

space to efficiently find the optimal BN. Nevertheless, the size of the space is still exponential in the size of the network, and, in the worst case, exploring it requires exponential time and memory. Previous work (Malone et al. 2011) has shown that, in practice, the exponential memory requirement is especially challenging. The algorithm can always be given more time; however, if it exceeds the available memory resources, nothing can be done to solve the instance.

Previously (Malone et al. 2011), a sorting-based hybrid duplicate detection (sHDD) strategy was used to allow BFBnB to efficiently use external memory, such as hard disk. This approach uses a hash table of size m to efficiently detect duplicates in the search space. When the size of the hash table grows too large, it is sorted and written to disk; duplicates are detected later using an external memory merge-sort operation. Of course, sorting the hash table requires $O(m \log m)$ memory[1]. Thus, sHDD actually requires $M = O(m \log m)$ memory due to sorting. In some sense, then, a factor of $O(\log m)$ memory is wasted.

In the heuristic search community, hashing-based delayed duplicate detection (Korf 2008) has emerged as a linear-complexity alternative to sorting-based approaches. In this approach, one hash function is used to write nodes to disk, while a second hash function is used to identify duplicates in memory. The first hash function must ensure the number of unique nodes written to any file does not exceed memory resources. Importantly, this hashing-based approach replaces the $O(m \log m)$ sorting with linear-complexity operations.

In this work, we develop a hashing-based hybrid duplicate detection (HHDD) strategy for use in BFBnB for solving BNSL. In particular, we construct an appropriate hash function for writing nodes to files which respects memory limitations. Additionally, we propose a bin packing algorithm for minimizing the number of files. Experimentally, we show that, for a fixed amount of memory, HHDD has better locality than sHDD because HHDD does not use memory for sorting. We also show that much of the locality inherent in sHDD can be preserved by expanding disk files in a particular order.

The rest of the paper is structured as follows. In Sect. 2, we give background on BNSL and the state space formulation. Section 3 presents our main contributions: the necessary hashing functions and bin packing algorithms for HHDD. We experimentally evaluate sHDD and HHDD in Sect. 4; discussion in Sect. 5 concludes the paper.

2   Background

2.1   Bayesian Networks

A Bayesian network (Pearl 1988) is a compact representation of a joint probability distribution over the random variables $\mathbf{V} = \{X_1, \ldots, X_n\}$. It consists

---

[1] In this work, we use "memory" to refer to fast-access storage, such as RAM; by "external memory," we mean storage with slower access, such as hard disks and network storage. All of the theoretical complexity analysis, such as $O(\cdot)$, refers to fast-access storage.

of a directed acyclic graph (DAG) in which each vertex corresponds to one of the random variables; the edges in the graph indicate conditional independencies among the variables. Additionally, each variable X_i has an associated probability distribution, conditioned on its parents in the DAG, PA_i. The joint probability distribution given by the network is

$$P(\mathbf{V}) = \prod_i^n P(X_i | PA_i). \tag{1}$$

Given a dataset $\mathcal{D} = \{D_1, \ldots D_N\}$, where each D_i is a complete instantiation of $\mathbf{V}$, the goal of structure learning is to find a Bayesian network $\mathcal{N}$ which best fits $\mathcal{D}$. The fit of $\mathcal{N}$ to $\mathcal{D}$ is quantified by a scoring function s. Many scoring functions have been proposed in the literature, including Bayesian scores (Cooper and Herskovits 1992; Heckerman et al. 1995), MDL-based scores (Suzuki 1999; Silander et al. 2008), and independence-based scores (de Campos and Huete 2000), among others. The scoring functions can typically be interpretted as penalized log-likelihood functions. All commonly used scoring functions are *decomposable* (Heckerman et al. 1995); that is, they decompose into a sum of *local scores* for each variable, its parents, and the data,

$$s(\mathcal{N}; \mathcal{D}) = \sum_i^n s_i(PA_i; \mathcal{D}), \tag{2}$$

where $s_i(PA_i)$ gives the score of X_i using PA_i as its parents and is non-negative. We omit $\mathcal{D}$ when it is clear from context.

A variety of pruning rules (Suzuki 1999; Tian 2000; Teyssier and Koller 2005; de Campos and Ji 2011) can be used to demonstrate that some parent sets are never optimal for some variables. Additionally, in practice, large parent sets are often pruned *a priori* (Malone et al. 2015). We refer to parent sets remaining after such pruning as *candidate parent sets* and denote all candidate parent sets of X_i as $\mathcal{P}_i$.

The *Bayesian network structure learning* problem (BNSL) is then defined as follows[2].

The BNSL Problem

Input: A set $\mathbf{V} = \{X_1, \ldots, X_n\}$ of variables and a local score $s_i(PA_i)$ for each $PA_i \in \mathcal{P}_i$ for each X_i.

Task: Find a DAG N^* such that

$$N^* \in \arg\min_N \sum_{i=1}^n s_i(PA_i),$$

where PA_i is the parent set of X_i in N and $PA_i \in \mathcal{P}_i$.

---

[2] The problem can also be defined as a maximization using non-positive local scores.

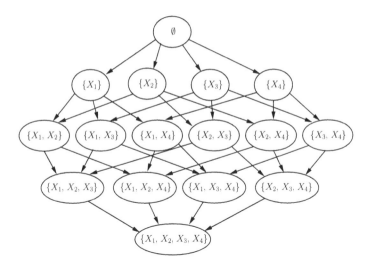

Fig. 1. The order graph for four variables

2.2   State Space Search for BNSL

State space search algorithms are one state-of-the-art technique for solving BNSL (Yuan and Malone 2013; Malone et al. 2014). Figure 1 shows the *implicit* space for four variables. Each node in the space corresponds to an optimal network over a subset of the variables; we refer to nodes in the graph and subsets of the variables interchangeably. The top-most node, containing no variables, is the *start* node, and the bottom-most node with all variables is the *goal* node.

An arc from $\mathbf{U}$ to $\mathbf{U} \cup \{X_i\}$ in the graph indicates generating a *successor* node by adding the variable X_i as a leaf of the optimal subnetwork for $\mathbf{U}$. The cost of the arc is the score of the optimal parent set for X_i out of $\mathbf{U}$,

$$BestScore(X_i, \mathbf{U}) = \min_{PA_i \subset \mathbf{U}} s_i(PA_i). \qquad (3)$$

Each node $\mathbf{U}$ stores the cost from *start* to $\mathbf{U}$, $g(\mathbf{U})$, which is the sum of the arc costs on the shortest path from *start* to $\mathbf{U}$.

Expanding a node amounts to adding each remaining variable as a leaf of its current subnetwork. Thus, a path in the implicit graph from *start* to *goal* corresponds to a total ordering over the variables; consequently, this graph is called the *order graph*. The cost of the path exactly gives the cost of the corresponding network. Thus, BNSL can be solved by finding the shortest path from *start* to *goal*.

Several algorithms, such as A* (Yuan and Malone 2013) and depth-first search (Malone and Yuan 2014), have been used to solve this problem. Breadth-first branch and bound (Malone et al. 2011; Fan et al. 2014) (BFBnB) has been shown to be an effective search strategy for this space. In BFBnB, nodes are expanded in a layer-wise fashion, where a layer corresponds to all subnetworks of the same size. The optimal network can be reconstructed using standard back-pointer techniques (Russell and Norvig 2003) or more memory-efficient recursive strategies (Zhou and Hansen 2003).

Pruning. Much like depth-first branch and bound, BFBnB can benefit from pruning (Zhou and Hansen 2006). Before beginning the search, an upper bound b on the cost of the optimal solution is found with an approximation algorithm, such as greedy hill climbing. Then, for each node $\mathbf{U}$, an admissible heuristic function h is used to estimate the distance from $\mathbf{U}$ to the goal, $h(\mathbf{U})$. If the sum $g(\mathbf{U}) + h(\mathbf{U})$ is worse than b, then $\mathbf{U}$ can be pruned. For BNSL, pattern databases (Yuan and Malone 2012; Fan et al. 2014) are effective admissible heuristics.

Immediate Duplicate Detection. One of the main operators of BFBnB is *duplicate detection* (DD). In the context of BNSL, duplicates are subnetworks over the same set of variables but with different orderings. Due to Eq. 3, DD entails selecting the subnetwork with the minimum score.

A typical approach to DD is to use an in-memory hash table to detect nodes for the same subnetwork. The hash table retains the best copy of a node found so far. Since the hash table is used to detect duplicates as soon as they are generated, we refer to this strategy as *immediate duplicate detection* (IDD). For BNSL, the in-memory hash table stores, in the worst case, all of the nodes in one layer. The largest layer of the order graph contains $\mathrm{O}\left(C(n, \frac{n}{2})\right)$, where $C(\cdot, \cdot)$ is the binomial coefficient. Thus, the size of the largest layer is still exponential in the number of variables.

Delayed Duplicate Detection. Due to the exponential worst-case memory requirement, previous work (Malone et al. 2011) augmented IDD with *delayed duplicate detection* (DDD) (Korf 2004). The essence of DDD is to use external memory, such as hard disk, to store nodes. Then, efficient disk access techniques are used to remove duplicates from disk.

2.3   Hybrid Duplicate Detection

The previous BFBnB algorithm for BNSL (Malone et al. 2011) combined IDD with DDD. We refer to this combined strategy as *hybrid duplicate detection* (HDD). In particular, a sorting-based HDD (sHDD) strategy was used. In sHDD, IDD is performed until the hash table reaches a given memory limit m; then, the nodes in the hash table are sorted and written to an external memory file. After expanding all of the nodes in one layer, DDD is performed via an external-memory merge sort operation on the disk files (Korf 2004). As a result of the sorting, nodes for each layer are always expanded in sorted order. The space complexity of sHDD is dominated by sorting, which requires $\mathrm{O}\left(m \log m\right)$ space.

An empirically important aspect of HDD is *locality*. In this context, locality refers to duplicates detected with IDD. Specifically, we define the locality of search strategy S as follows.

$$\ell_M^S = \frac{n_u}{n_w},$$

(4)

where S is the HDD strategy, M is the memory bound, n_u is the number of unique nodes in the search space and n_w is the number of nodes written to disk. So, algorithms with a high locality remove most duplicates with IDD. If IDD removes all duplicates, then $\ell_M^S = 1$. In our analysis, we define the locality of HDD for a single layer of the order graph analogously.

Increasing locality reduces the time spent on expensive external memory read and write operations. Larger in-memory hash tables (larger m) allow for more effective IDD and improve locality. (We empirically verify this in Sect. 4.)

As previously mentioned, sorting the hash table before writing it to disk requires $\mathrm{O}\,(m \log m)$ space.[3] Thus, in order to ensure sorting does not exceed the memory bound M, the size of the hash table is restricted to $\frac{M}{\log m}$. So, the sorting operation impairs locality because the hash table cannot use all of the available memory.

3   Hashing-Based HDD for BNSL

In this work, we propose to use hashing-based DDD (Korf 2008) for the DDD operation in HDD. In hashing-based DDD, rather than writing nodes to multiple sorted files, a hash function is used to distribute nodes to files. The hash function must ensure all duplicates are written to the same file. Additionally, it must ensure at most M unique nodes are written to any single file. After expanding all of the nodes in a layer, each external file is sequentially read back into an in-memory hash table to identify and remove duplicates. The space complexity of hashing-based DDD is $\mathrm{O}\,(m)$. Since hashing-based DDD avoids the $\mathrm{O}\,(\log m)$ overhead of sorting, it allows the use of the entire M space for the hash table.

In this section, we describe the hash functions necessary for hashing-based DDD for BNSL. Additionally, we give a bin packing algorithm for minimizing the number of external memory files.

3.1   Dividing Nodes into Families

The basic idea for dividing order graph nodes into files is based on the use of subcombinations, somewhat similar to the approach used by Tamada *et al.*, for parallelization (Tamada et al. 2011). We refer to $\mathbf{S} \subset \mathbf{U}$ as a *subcombination* of $\mathbf{U}$. Furthermore, when $\mathbf{S}$ contains (exactly) the lexicographically first k elements of $\mathbf{U}$, we refer to $\mathbf{S}$ as a *first-k subcombination* of $\mathbf{U}$. Additionally, when $\mathbf{S}$ is a first-k subcombination of $\mathbf{U}$, we refer to $\mathbf{U}$ as an *extension* of $\mathbf{S}$. We denote all size-l extensions of $\mathbf{S}$ as $\mathcal{F}_l^{\mathbf{S}}$ (the "family" of $\mathbf{S}$). In general, $\mathcal{F}_l^{\mathbf{S}}$ contains $\mathrm{O}\,(C(n-k, l-k))$ elements, where $C(\cdot, \cdot)$ is the binomial coefficient.

---

[3] Efficient sorting implementations, such as the g++ version of std::sort, often do not exhaust the additional $\mathrm{O}\,(\log m)$ space; however, it is difficult to *a priori* estimate the required overhead, so $\mathrm{O}\,(m \log m)$ must be used to ensure stable algorithm behavior.

We assign each node to the lexicographically first family to which it belongs. For example, $\mathbf{U} = \{X_1, X_2, X_3\}$ belongs to $\mathcal{F}_3^{\{X_1, X_2\}}$ rather than $\mathcal{F}_3^{\{X_1, X_3\}}$.

3.2 Optimizing Family Size

In order to minimize hashing-related overhead, we aim to minimize the number of families. As mentioned in Sect. 3.1, at layer l with families based on first-k subcombinations, the largest family contains $C(n - k, l - k)$ nodes.

As a resource restriction, we assume the in-memory hash table can store at most m nodes[4]. Thus, we select k such that the largest family contains no more than m nodes. At each layer of the order graph, we select k by solving

$$\underset{k}{\operatorname{argmin}}\, C(n - k, l - k) \leq m, \tag{5}$$

where n, l and m are fixed and $0 < k < l$. Empirically, we found that a linear scan of values of k outperformed more sophisticated optimization strategies, such as Newton's method, for minimizing Eq. 5.

3.3 Distributing Families to Files

A simple approach, which we empirically evaluate in Sect. 4, assigns each family to a separate external memory file. Thus, the hash function extracts the first-k subcombination of a node and directly maps that to a file on disk. During the search, the files are expanded in lexicographic order of the corresponding first-k subcombinations; within the files, though, the nodes are unordered. We refer to this strategy as HHDD.

3.4 Packing Families into Files

After choosing k by solving Eq. 5, the *maximum* number of nodes in any family is $C(n-k, l-k)$. However, as mentioned in Sect. 3.1, nodes are distributed to the lexicographically first family to which they can belong. For example, $\mathcal{F}_3^{\{X_1, X_2\}}$ "steals" $\{X_1, X_2, X_3\}$ from $\mathcal{F}_3^{\{X_1, X_3\}}$. Therefore, $\mathcal{F}_3^{\{X_1, X_3\}}$ will contain one node less than given by the bound.

In fact, because nodes are assigned to the lexicographically first family possible, $\mathcal{F}_l^{\mathbf{S}}$ is assigned only nodes which follow $\mathbf{S}$ lexicographically. This is, if we take X_i as the lexicographically last element of $\mathbf{S}$, then all extensions assigned to $\mathcal{F}_l^{\mathbf{S}}$ use only variables X_j such that $j > i$, other than those in $\mathbf{S}$. So, due to stealing, $\mathcal{F}_l^{\mathbf{S}}$, with X_i as the lexicographically last element, is assigned $C'(n - i, l - k)$ nodes, where $C'(n - i, l - k)$ is defined to be 0 when $0 < n - i < l - k$ and the binomial coefficient otherwise. We denote the number of nodes assigned to $\mathcal{F}_l^{\mathbf{S}}$ as $|\mathcal{F}_l^{\mathbf{S}}|$.

---

[4] For this analysis, we do not consider the load factor of the hash table.

Algorithm 1. Bin packing algorithm for packing families into files

---

procedure PACK(memory limit M, families $\mathcal{F}$)
    $files \leftarrow \emptyset$ ▷ map from families to disk files
    $file \leftarrow \emptyset$ ▷ families are greedily added to the current file
    for $\mathcal{F}_l^S$ in $\mathcal{F}$ **do**   ▷ iterate in reverse lexicographic order
        if $|\mathcal{F}_l^S| + \sum_{S' \in file} |\mathcal{F}_l^{S'}| \leq M$ **then**
            $file \leftarrow file \cup \mathcal{F}_l^S$
        else
            $files \leftarrow files \cup file$
            $file \leftarrow \mathcal{F}_l^S$
        end if
    end for
    return $files$
end procedure

Furthermore, since the lexicographically last element of the subcombination for the family is X_i, the other $k-1$ elements must come from the $i-1$ elements which precede X_i. Thus, $C(i-1, k-1)$ families end with X_i.

For a variety of reasons, such as operating system constraints on open file handles and latencies associated with accessing many hard disk files, we aim to minimize the number of files used for storing the families. Thus, we assign many families to a single file. The constraint on assigning families is that the sum of the sizes of families assigned to a single file cannot exceed M.

Based on the previous analysis, the number of subcombinations and the number of items belonging to each subcombination is known. Thus, we can solve a bin packing problem to assign families to files in order to minimize the number of required files. Bin packing is known to be NP-hard (Garey and Johnson 1979); however, a variety of efficient approximation algorithms are available for this problem (Johnson 1973; Korf 2002). We use Algorithm 1 for assigning families to files.

Thus, for this strategy, the hash function for distributing a node to disk first extracts the first-k subcombination to find the family of the node. Then, the results of the bin packing algorithm give the file to which that family is written. The bin packing algorithm does not attempt to assign lexicographically similar families to the same files, so the order of node expansions is close to random. We refer to this strategy as HHDD-PACK.

4   Experiments

Locality is an important factor in the performance of HDD. Thus, we designed a set of experiments to better understand how different HDD strategies affect locality. In particular, we tested the following three hypotheses.

1. The proposed bin packing algorithm is comparable to state-of-the-art techniques.

2. Given a fixed hash table size m, SHDD has a higher locality than HHDD and HHDD-PACK.
3. Given a fixed amount of memory M, HHDD and HHDD-PACK have a higher locality than SHDD.
4. For a fixed packing strategy, locality and m are positively correlated.

4.1 Datasets and Environment

All of the experiments were run on dual quad-core Intel Xeon E5540 processors with 32 GB of RAM. Hard disk space was limited to 16 GB.

As described in Sect. 2.2, pruning is very important for the performance of BFBnB; however, pruning is dependent both on the quality of the bound and the quality of the heuristic. These are both data-dependent and difficult to analytically characterize. Consequently, in order to remove this confounding factor from our analysis, we do not using pruning.

Our analysis uses a 29-variable dataset from a previous study (Malone and Yuan 2013). We do not use pruning, so the dataset generation parameters investigated in the previous study do not affect locality. Since the locality behavior of the HDD strategies is completely deterministic, we report results on just one dataset.

4.2 Bin Packing Performance

We first evaluated the performance of the HHDD-PACK bin packing algorithm in Sect. 3.4 by considering the number of files used by different strategies. For comparison, we include the basic approach of assigning each family to a single file, HHDD. Additionally, we include the commonly-used first-fit decreasing approximation algorithm(FFD) (Johnson 1973), which is known to be an $\frac{11}{9}$-approximation, and an optimal packing strategy (OPT)[5]. Figure 2 shows that the packing algorithm reduces the number of files by over an order of magnitude compared to HHDD across all bounds for m. Indeed, the HHDD-PACK packing strategy only results in one extra file for two layers compared to OPT. Thus, we conclude that our packing strategy is appropriate for reducing the number of files.

4.3 Locality, Fixed Hash Table Size

We next compared the locality of SHDD, HHDD and HHDD-PACK when all algorithms are allowed to use the same maximum hash table size $m = 25e6$. Thus, due to the overhead of sorting, this experiment actually allows SHDD to use more memory than the other approaches. In Fig. 3, we show how locality varies across layers in the order graph from 29 variables.

---

[5] The strategy is optimal in that it minimizes the number of files. We solve the optimization problem using an integer linear programming formulation.

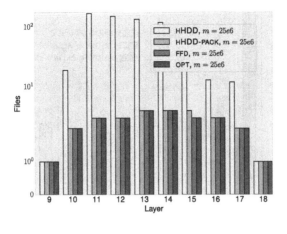

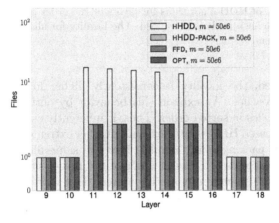

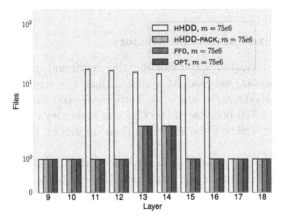

Fig. 2. The number of files resulting from different bin packing strategies. There is only 1 file for all strategies in the unshown layers.

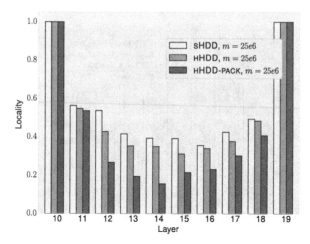

Fig. 3. The locality of HDD algorithms for layers of the order graph for 29 variables when the size of the hash table is $m = 25e6$. The locality for all HDD algorithms is 1 for all unshown layers.

As we expected, the locality is consistently higher for sHDD compared to the other approaches. We explain this behavior by noting that, by design, sHDD expands nodes in sorted order. Thus, it inherently exhibits more locality than the hashing-based HDD strategies. On the other extreme, HHDD-PACK sacrifices locality by packing many families into the same file. As mentioned in Sect. 3.3, HHDD expands the files in order of the first-k subcombinations, but the nodes within the same family are unordered. So it represents a compromise between the completely-sorted expansion strategy of sHDD and the nearly-random strategy of HHDD-PACK. The figure shows that its locality is closer to that of sHDD than HHDD-PACK.

4.4   Locality, Fixed Maximum Memory

We then compared locality when holding the maximum memory requirement constant. In particular, we used $M = 25e6$; thus, the size of the hash tables for HHDD and HHDD-PACK are $m = 25e6$, while the size of the hash table for sHDD is $m = 1.7e6$ because of the $O(\log m)$ memory overhead for sorting. Unsurprisingly, Fig. 4 shows the same relationship among HHDD and HHDD-PACK as in Fig. 3.

Of course, due to the smaller hash table size, sHDD exhibits worse locality for most layers than the other two strategies. Unexpectedly, though, sHDD has *better* locality than HHDD-PACK for layers 13 and 14. We again attribute this behavior to the locality inherent in expanding nodes in sorted order.

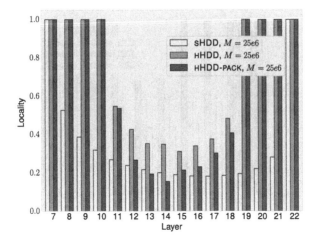

Fig. 4. The locality of HDD algorithms for layers of the order graph for 29 variables when the maximum amount of memory is constant at $M = 25e6$. For sHDD, $m = 1.7e6$; for hHDD, $m = 25e6$. The locality for all HDD algorithms is 1 for all unshown layers.

4.5   Locality and Memory Correlation

We finally evaluated the relationship between the hash table size m and locality for all three HDD strategies. Figure 5 reveals several differences among the relationships for the three strategies. Compared to the other HDD strategies, HHDD-PACK undergoes the most drastic locality changes as the available memory increases. In particular, it exhibits an almost linear relationship between m and locality. We explain this behavior with the somewhat random order of node expansions due to packing; in contrast, both sHDD and HHDD inherently expand nodes in sorted order, so even a modest-sized hash table removes many of the duplicates. Therefore, the locality inherent in these strategies reduces the need for larger hash tables. Interestingly, for the largest layers (14 and 15), the locality of sHDD improves markedly from $m = 25e6$ to $m = 50e6$, but the improvement is less drastic when the size is again increased to $m = 75e6$. On the other hand, for HHDD, there is little improvement from $m = 25e6$ to $m = 50e6$, but the locality significantly improves when the size of the hash table is increased to $m = 75e6$.

5   Discussion

In this paper, we have presented a novel approach for hybrid duplicate detection using hashing-based delayed duplicate detection for solving Bayesian network structure learning with breadth-first branch-and-bound search. The main contribution of this work is a hash function which is used to distribute the nodes in the search space to files on disk; importantly, the hash function ensures that no single file contains more unique nodes than will fit in memory. Compared

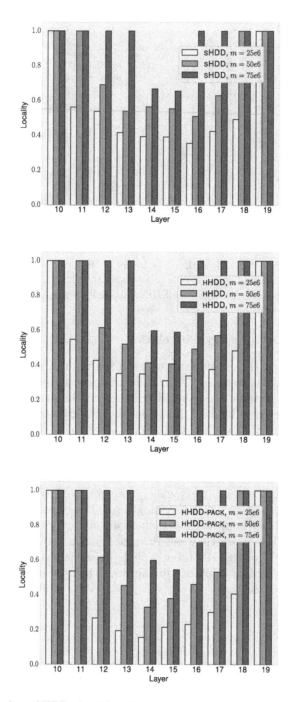

Fig. 5. The locality of HDD algorithms for layers of the order graph for 29 variables as the hash table size increases. The locality for all HDD algorithms is 1 for all unshown layers.

to previous sorting-based techniques (sHDD), HHDD reduces the the memory complexity by a factor of $O(\log m)$. In practice, m is on the order of tens to hundreds of millions, so the reduction is quite substantial. Experimentally, we verified that HHDD can significantly improve locality and reduce the number of expensive external memory operations used for delayed duplicate detection.

Furthermore, we developed a bin packing algorithm for minimizing the number of external memory files. Empirically, we showed that the bin packing approach often optimally minimizes the number of files, but sacrifices locality for doing so. Minimizing the number of files and maximizing locality amounts to a dual-objective discrete optimization problem. Nevertheless, locality-aware packing strategies could improve this behavior. The hash function and bin packing algorithm could also be used to distribute nodes for parallel processing.

References

Bartlett, M., Cussens, J.: Integer linear programming for the Bayesian network structure learning problem. Artif. Intell. (2015)

Chickering, D.M.: Learning Bayesian networks is NP-complete. In: Fisher, D., Lenz, H.-J. (eds.) Learning from Data: Artificial Intelligence and Statistics V, pp. 121–130. Springer, New York (1996)

Cooper, G.F., Herskovits, E.: A Bayesian method for the induction of probabilistic networks from data. Mach. Learn. **9**, 309–347 (1992)

de Campos, C.P., Ji, Q.: Efficient learning of Bayesian networks using constraints. J. Mach. Learn. Res. **12**, 663–689 (2011)

de Campos, L.M., Huete, J.F.: A new approach for learning belief networks using independence criteria. Int. J. Approximate Reasoning **24**(1), 11–37 (2000)

Fan, X., Yuan, C., Malone, B.: Tightening bounds for Bayesian network structure learning. In: Proceedings of the 28th AAAI Conference on Artificial Intelligence (2014)

Garey, M.R., Johnson, D.S.: Computers and Intractability: A Guide to the Theory of NP-Completeness. W. H. Freeman & Co., New York (1979)

Heckerman, D., Geiger, D., Chickering, D.M.: Learning Bayesian networks: the combination of knowledge and statistical data. Mach. Learn. **20**, 197–243 (1995)

Johnson, D.: Near-Optimal Bin Packing Algorithms. Ph.D. thesis, Massachusetts Institute of Technology (1973)

Koivisto, M., Sood, K.: Exact Bayesian structure discovery in Bayesian networks. J. Mach. Learn. Res. **5**, 549–573 (2004)

Korf, R.E.: A new algorithm for optimal bin packing. In: Proceedings of the 18th AAAI Conference on Artificial Intelligence (2002)

Korf, R.E. Best-first frontier search with delayed duplicate detection. In: Proceedings of the 19th AAAI Conference on Artificial Intelligence (2004)

Korf, R.E.: Linear-time disk-based implicit graph search. J. ACM **35**(6) (2008)

Malone, B., Järvisalo, M., Myllymäki, P.: Impact of learning strategies on the qual packing Bayesian networks: an empirical evaluation. In: Proceedings of the 31st Conference on Uncertainty in Artificial Intelligence (2015)

Malone, B., Kangas, K., Järvisalo, M., Koivisto, M., Myllymäki, P.: Predicting the hardness of learning Bayesian networks. In: Proceedings of the 28th AAAI Conference on Artificial Intelligence (2014)

Malone, B., Yuan, C.: Evaluating anytime algorithms for learning optimal Bayesian networks. In: Proceedings of the 29th Conference on Uncertainty in Artificial Intelligence (2013)

Malone, B., Yuan, C.: A depth-first branch and bound algorithm for learning optimal Bayesian networks. In: Croitoru, M., Rudolph, S., Woltran, S., Gonzales, C. (eds.) GKR 2013. LNCS, vol. 8323, pp. 111–122. Springer, Heidelberg (2014)

Malone, B., Yuan, C., Hansen, E.: Memory-efficient dynamic programming for learning optimal Bayesian networks. In: Proceedings of the 25th AAAI Conference on Artifical Intelligence (2011)

Ott, S., Imoto, S., Miyano, S.: Finding optimal models for small gene networks. In: Proceedings of the Pacific Symposium on Biocomputing (2004)

Pearl, J.: Probabilistic Reasoning in Intelligent Systems: Networks of Plausible Inference. Morgan Kaufmann Publishers Inc., San Mateo (1988)

Russell, S.J., Norvig, P.: Artificial Intelligence: A Modern Approach. Pearson Education, Upper Saddle River (2003)

Silander, T., Myllymäki, P.: A simple approach for finding the globally optimal Bayesian network structure. In: Proceedings of the 22nd Conference on Uncertainty in Artificial Intelligence (2006)

Silander, T., Roos, T., Kontkanen, P., Myllymäki, P.: Factorized normalized maximum likelihood criterion for learning Bayesian network structures. In: Proceedings of the 4th European Workshop on Probabilistic Graphical Models (2008)

Suzuki, J.: Learning Bayesian belief networks based on the MDL principle: an efficient algorithm using the branch and bound technique. IEICE Trans. Inf. Syst. **E82–D**(2), 356–367 (1999)

Tamada, Y., Imoto, S., Miyano, S.: Parallel algorithm for learning optimal Bayesian network structure. J. Mach. Learn. Res. **12**, 2437–2459 (2011)

Teyssier, M., Koller, D.: Ordering-based search: a simple and effective algorithm for learning Bayesian networks. In: Proceedings of the 21st Conference on Uncertainty in Artificial Intelligence (2005)

Tian, J.: A branch-and-bound algorithm for MDL learning Bayesian networks. In: Proceedings of the 16th Conference on Uncertainty in Artificial Intelligence (2000)

van Beek, P., Hoffmann, H.-F.: Machine learning of Bayesian networks using constraint programming. In: Pesant, G. (ed.) CP 2015. LNCS, vol. 9255, pp. 429–445. Springer, Heidelberg (2015)

Yuan, C., Malone, B.: An improved admissible heuristic for finding optimal Bayesian networks. In: Proceedings of the 28th Conference on Uncertainty in Artificial Intelligence (2012)

Yuan, C., Malone, B.: Learning optimal Bayesian networks: a shortest path perspective. J. Artif. Intell. Res. **48**, 23–65 (2013)

Zhou, R., Hansen, E. A.: Sparse-memory graph search. In: Proceedings of the 18th International Joint Conference on Artificial Intelligence (2003)

Zhou, R., Hansen, E.A.: Breadth-first heuristic search. Artif. Intell. **170**, 385–408 (2006)

A Bayesian Network Approach for Predicting Purchase Behavior via Direct Observation of In-store Behavior

Yi Zuo[1]($\boxtimes$), Katsutoshi Yada[2], and Eisuke Kita[3]

[1] Institute of Innovation for Future Society, Nagoya University, Nagoya, Japan
zuo@coi.nagoya-u.ac.jp
[2] Faculty of Commerce, Kansai University, Osaka, Japan
[3] Graduate School of Information Sciences, Nagoya University, Nagoya, Japan

Abstract. In strategic management of retail industry, the advanced investigation by using radio frequency identification (RFID) technology to capture customers' in-store behavior has been dramatically attracted scholars and practitioners in past ten years. As a small RFID tag attached to the shopping carts can be recognized as surrogates instead of enumerators to trail the customers, it can provide us an objective and direct perspective to observe and measure the in-store behavior of customers. In this article, we present a study on this new type of in-store behavior data named RFID data, which can improve the understanding of purchase behavior of customers with emphasis on meaningful knowledge via analysis of RFID data. In contrast to prior studies in this research field, this paper has paid special attention to shopping time that customers spent in supermarket (so-called stay time), and presents methodological analysis into two folds. First, we develop a bayesian network (BN) model to combine both of purchase behavior and in-store behavior as features. As BN is a probabilistic graphical model, it can provide an quantitative analysis process of purchase behavior decision over stay time and also allow us to interpret the decision process of purchasing in a much more intuitive measurement. The results show BN has a better accuracy than other typical prediction models (linear discriminant analysis, logistic regression and support vector machine). Second, due to BN can estimate and predict in a nonlinear correlation between purchase intention and stay time, we examine a tedium effect on purchase behavior. During the customers wander in shopping, purchase intention represents a non-monotonic phenomena accounting for the long stay time. Finally, we also investigate the sensitivity and specificity of purchase behavior predicted by our proposal in adjustment of decision threshold and implement several business decision-making implications in actual business situations.

Keywords: RFID data · In-store behavior · Tedium effect · Bayesian network · Shopping session

© Springer International Publishing Switzerland 2015
J. Suzuki and M. Ueno (Eds.): AMBN 2015, LNAI 9505, pp. 61–75, 2015.
DOI: 10.1007/978-3-319-28379-1_5

1   Introduction

In management of retail industry, consumer behavior has been increasingly recognized as a key strategic resource. Most of business models and loyalty programs in this field are being applied to lead the customer to being loyalty and profitability [1,5,12]. In most of previous studies [4,6,7,16], consumer behavior is generally considered as purchase behavior of customer which is based on the historical point of sale (POS) data. Via analyzing the POS data can help retailers to enhance the activity and loyalty of their customers so as to increase their sales and profits. However, it is impossible to shed any light on the decision-making process of purchase only depending on POS data.

Recently, a wireless no-contact technology named radio frequency identification (RFID) has brought a new perspective on this situation. By using the RFID tag attached to the customer, their in-store behavior can be tracked accurately [18]. When the customers walk along the shelves in supermarket, the RFID tag emits signals every second giving the location information as coordinates (x, y). Then these signals are received and sent to the back-end server via an RFID receptor at the bottom (on the top) of shelves, and transformed into RFID data automatically in the back-end server. The utility of RFID data has been investigated by Larson et al. [10], also in other previous studies [8,13,19,20]. In these studies, through analyzing the in-store behavior of customers individually or in groups, several canonical shopping path and visiting patterns are discovered, and behavioral hypotheses are tested.

In contrast to prior innovators in this research field, this paper has paid special attention to time spent on shopping in a target area rather than the whole supermarket, which can allow us to interpret the decision process of purchasing one product or a series of products in a much more intuitive and precise measurement. Also, in these studies, shopping time is used only as a clustering indicator, and its effect on the decision-making process of purchase behavior and time-based predictions of purchase behavior are not considered. Therefore, this article develops an integrated model to combine POS data and RFID data. Using the historical POS data, we generate an attitudinal feature - purchase background, which can represent past purchase incidences of individual customer. Using the RFID data, we generate a behavioral feature - stay time, which can represent customers' purchase intention over time. This paper employ a stochastic graphical approach - bayesian network (BN) [2,14,21] to combine these two types of consumers' features as explanatory variables and inspect their affect on the purchase decision. As BN can represent a set of random variables and their conditional dependencies via a directed acyclic graph, The use of the BN enables the purchase decision to be estimated with probability in quantitative process and also a nonlinear approach. These advantages of BN can support us to investigate the tedium effect [11] between the purchase decision and the time spent on the decision to be estimated with probability in a quantitative process and to revise the traditional results from a nonlinear perspective. Additionally, due to BN can treat the variables into discrete values (states) by using the clustering algorithm, we examine an investigation across multiple shopping sessions during the customers wander in shopping.

In the experiments, the optimal cluster number of stay time and purchase background is examined for maximizing the performance accuracy, and the results also show BN has a better accuracy than other typical prediction models (linear discriminant analysis [4, 16], logistic regression [6, 7] and support vector machine [3, 9, 15]). Finally, we also investigate the sensitivity and specificity of purchase behavior predicted by our proposal in adjustment of decision threshold and implement several business decision-making implications in actual business situations.

The remaining part of the article is as follows. In Sect. 2, a overview of RFID system on consumer in-store behavior and the preliminary stage of POS data and RFID data are described briefly. Bayesian network application and the present algorithm is explained in Sect. 3. Numerical results and discussion are shown in Sect. 4. The results are summarized again in Sect. 5.

2 System Overview of RIFD Data

2.1 Collection of RFID Data

In this section, we demonstrate how this system can be used to capture actual movement data of customers. This system is implemented at a mid-sized supermarket in the Chubu region of Japan. During the period of experiment, the shopping carts used by customers are equipped with RFID tags (Fig. 1(a)), which allow the carts as surrogates to track customer movements within the store precisely. The RFID system consists of 5 steps as shown in Fig. 1.

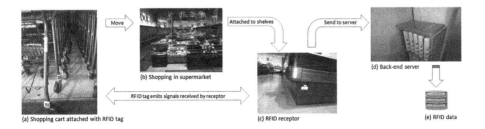

Fig. 1. Overview of RFID system for data collecting procedure in supermarket

(a) RFID tags are attached the shopping carts with an unique ID, individually.
(b) When the customers walk through shelves with this cart, RFID tag emits signals per second which can express the position information.
(c) These signals are received and sent to the back-end server via a RFID receptor at the bottom (on the top) of shelves.

(d) In the back-end server, a tracking system is employed to identify the signals and save them as the raw data.

(e) By using another preprocessing system in the back-end server, the raw data is transformed into RFID data in XML form.

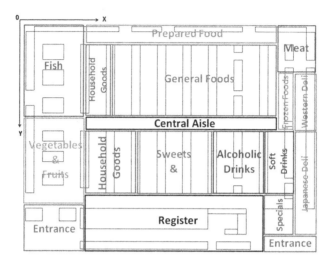

Fig. 2. Floor layout of supermarket

In order to match customers' movement data and record the trip of customer into database, the layout of supermarket containing 16 sections is reproduced into a picture from x and y coordinates on the scale of 15.7 pixels per meter (Fig. 2). While the customer passes a certain area of the supermarket with a shopping cart attached RFID tag, the information of customer position can be received by RFID receptor around the shelves and be transformed to a pixel point into database using the floor layout matching. RFID tag number attached to shopping cart, shopping date, time stamp, x and y coordinates of that time stamp, section of that coordinate and elapsed time are recorded, and Table 1 shows the sample data obtained using RFID system.

When the customers come to the checkout register and purchase, the POS data what he has bought are also recorded into database. The dataset is the shopping details as shown in Table 2. There are customer number, shopping date and time, name and category of the item, volume and amount, 7 columns in this table.

Actually, until the customer comes to purchase in registers, the tracking process seems to be competed. we define this process as a basic unit of customer's in-store behavior and give a unique ID to identify it. Also, the preprocessing system in the back-end server uses this ID to link the customer's purchase behavior obtained from POS data to her in-store behavior obtained from RFID data (Fig. 3).

Table 1. RFID data of the movement

RFID tag no	Date	Time	X	Y	Selling area	Elapsed time
T001	2009/05/20	18:15:17	91	542	Entrance	1
T001	2009/05/20	18:15:43	66	331	Vegetable	3
T001	2009/05/20	18:16:12	85	88	Fish	2
T001	2009/05/20	18:17:23	398	138	Coffee	2
T001	2009/05/20	18:17:57	487	160	Bread	3
T001	2009/05/20	18:18:21	556	361	Drinking	1
T001	2009/05/20	18:18:37	319	511	Register	1

Table 2. Detail of the POS data

Customer	Date	Time	Item name	Item category	Volume	Amount
Lucy	2009/05/20	18:18:52	Cabbage	Vegetable	1	158
Lucy	2009/05/20	18:18:52	Banana	Fruit	1	198
Lucy	2009/05/20	18:18:52	Sashimi	Fish	2	596
Lucy	2009/05/20	18:18:52	Loaf	Bread	1	112

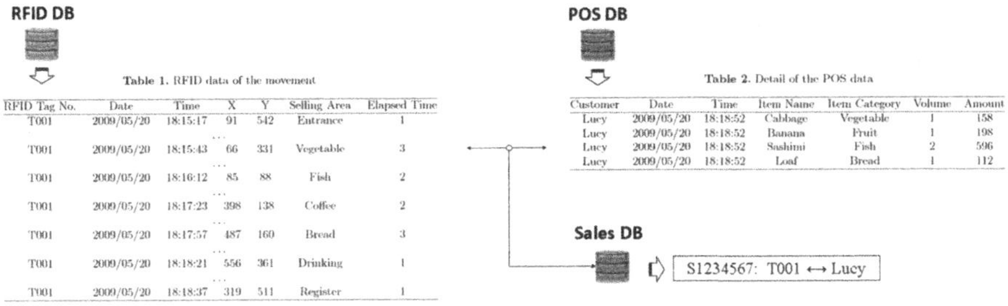

Fig. 3. Match purchase behavior to in-store behavior in the preprocessing system

2.2 Measuring Stay Time in Small Region

The experiment was carried out in a typical supermarket in Chubu region of Japan. Comparing with the previous studies, we focus on the customers' in-store behavior in a certain small area instead of the whole supermarket. Since bread is featured much more prominently than other kind of stable food on the Japanese plate, the bread selling area is selected as the experiment object. And, the measuring range is shown as the shadowy pattern in Fig. 4.

In this section, we also explain the definition of the stay time for the customers how they spent in bread selling area. For given a customer, her shopping trip which is tracked from her coming into the entrance until coming to the checkout register to purchase is tracked by the RFID tag. Here, the time spent

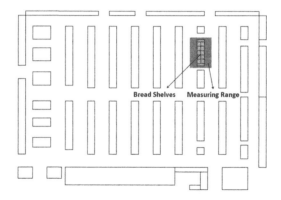

Fig. 4. Measuring area of products in bread category

by customer in the supermarket is expressed as follows:

$$Sum = \sum_{i=0}^{n} t_i \tag{1}$$

where the notation t_i denotes the "Elapsed Time" shown in the Table 1. And making an addition to Eq. (1), only if customer comes into the bread selling area (the shadow shown in Fig. 4), this in-store position of customer is adopted as the experiment target, and "Elapsed Time" t_i spent in this position is recognize as stay time for bread selling area. Therefore, the total stay time T of customer spent in bread selling area is defined as follow:

$$T = \sum_{i=0}^{n} t_i, \tag{2}$$

$$t_i = \begin{cases} \text{Elapsed Time, if position in bread selling area.} \\ 0, \qquad\qquad\qquad \text{otherwise.} \end{cases}$$

By using the Eq. (2), the stay time is calculated for the individual customer who has spent in the bread selling area.

3   Method

3.1   Bayesian Network

Bayesian network (BN) is a probabilistic graphical model that represents a set of random variables and their conditional dependencies via a directed acyclic graph. The probability theory of BN is based upon the bayes' rule, and if two observed events have the relation like $A \to B$ can be expressed as follows:

$$\Pr(A|B) = \frac{\Pr(B|A)\Pr(A)}{\Pr(B)}. \tag{3}$$

$\Pr(A)$ and $\Pr(A|B)$ denote the prior probability of event A and the posterior probability of event A when A according to the event B, respectively. $\Pr(B|A)$ denotes the likelihood function. The denominator $\Pr(B)$ is equal to $\Sigma_i \Pr(B|A = a_i) \Pr(A = a_i)$ which denotes the marginal distribution in all states of event A. By using Eq. (3), the prior probability $\Pr(A)$ is revised to $\Pr(A|B)$ by multiplying the likelihood $\Pr(B|A)$. Generally, for a set of variables, the method of maximum likelihood selects the set of values of the model parameters that maximizes the likelihood function.

3.2   Graph Structure and Probabilistic Reasoning

In our study, there are three variables - purchase behavior (P), purchase background (B) and stay time (T). Suppose that purchase behavior is depending on purchase background and stay time as $P \to B$ and $P \to T$, respectively. Also, suppose that stay time is also depending on purchase background as $T \to B$. As the purchase behavior is considered to be the response variable, purchase background and stay time to be the explanatory variables, bayesian network is constructed as Fig. 5.

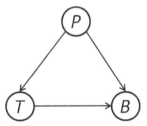

Fig. 5. Bayesian network for consumer behavior

According to the Fig. 5, the purchase probability based on purchase background and stay time is obtained as follows:

$$\Pr(P|B,T) = \frac{\Pr(P)\Pr(B|P)\Pr(T|B,P)}{\sum_P \Pr(P)\Pr(B|P)\Pr(T|B,P)} \tag{4}$$

where $\Pr(P)$ and $\Pr(P|B,T)$ denote the prior probability and posterior probability of purchase behavior, respectively. $\Pr(B|P)$ and $\Pr(T|B,P)$ are the likelihood function, also known as the conditional probability.

As shown in Eq. (4), $\Pr(P|B,T)$ denotes the posterior purchase probability in the condition of purchase background and stay time for given a customer. The notation P denotes the purchase event as a boolean variable and we define $P = 1$ if a purchase occurs and $P = 0$ otherwise. According to Bayes' rule, the purchase probability can be estimated by the Eq. (5) as follows. The denomination is the marginal distribution of the purchase in the purchased state $(P = 1)$ and unpurchased state $(P = 0)$.

$$\Pr(P = 1|B, T) = \frac{\Pr(P = 1)\Pr(B|P = 1)\Pr(T|B, P = 1)}{\sum\limits_{P \in \{0,1\}} \Pr(P)\Pr(B|P)\Pr(T|B, P)} \tag{5}$$

4   Experiment

4.1   Initialization of Variables

The experiment is carried out from May 11, 2009 to June 15, 2009 and total 1155 shopping paths are extracted for the bread category (The shopping paths having no position in measuring bread area are excluded from the experimental objects). As the hold-one-out validation is employed as the model estimation, we separate the data into 2 folds. One is from 11 May 2009 to 10 June 2009 used as the training data (containing 924 samples), the other one is the remained part from 11 June 2009 to 15 June 2009 used as the testing data (containing 231 samples).

– Response variable:
  • Purchase behavior is the response variable defined as a binary variable 0/1, which denotes unpurchased state and purchased state, respectively. As each shopping path is also linked to POS data to map the purchase behavior, the prior probability of observed purchase behavior $\Pr(P = 1)$ is 31.95 %.
– Explanatory variables:
  • Stay time is one of the explanatory variables used as the behavioral factor. Only if the customers come into the measuring range as shown in Fig. 4, the position point is accepted as the collecting target. The total elapsed time of each target point is recognized as the stay time (Eq. 2).
  • Purchase background is another explanatory variable used as the attitudinal factor. In contrast to purchase behavior and stay time, this variable has an independent experimental period just 3 months before the RFID experiment. Purchase behavior is generate from historical POS data during 11 February 2009 to 10 May 2009, which is also a cumulative factor to indicate the customers' attitude to the products. With a synergistic effect of stay time, decision-making process of purchase behavior can be demonstrated from both attitudinal and behavioral perspectives.

4.2   Optimization of Cluster Number

In this section, the effect of the cluster number to the predicting accuracy is discussed. For either of explanatory variables, the cluster number is taken from 2 to 10 by using k-meaning clustering. As shown in Eqs. (6) and (7), the notation T_l and B_m denote one cluster in stay time and purchase background respectively, and the notation L and M denote the total cluster number in stay time and purchase background respectively.

$$\{T_1, T_2, \cdots, T_L\} \tag{6}$$

$$\{B_1, B_2, \cdots, B_M\} \tag{7}$$

For the selection of cluster number, Bayesian information criterion (BIC) is adopted as the criterion to estimate and determine the optimal cluster number for bayesian network by the minimum value of BIC [17]. As we focus on the penalty term which denotes the number (k) of parameters in the model, k equals the number of total variables when the model is considered as 1-dimension linear model. In our case, due to both variable are in L-dimension and M-dimension respectively, the number of parameter in the penalty term can be estimated as $(L + M)$ for each combination, and BIC can be rewritten as follow:

$$\text{BIC} = n \cdot \ln(\sigma^2) + (L + M) \cdot \ln(n). \tag{8}$$

The notation σ^2 denotes the error variance. As purchase behavior is the binary variable, we use (1 - hit rate) as the residual term. The notation L and M denote the cluster number of purchase background and stay time, respectively. The notation n denotes the size of sample data. Eq. (8) is a decreasing function of hit rate and an increasing function of cluster number. By penalizing the cluster number, the lowest value of Eq. (8) is the one to be preferred.

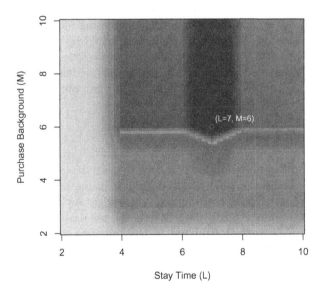

Fig. 6. Density estimation of BIC in all combinations of L and M

As shown in Fig. 6, the BIC results are drawn in density estimation. The X-axis and Y-axis denote cluster number of stay time (L) and purchase back-

Table 3. Accuracy comparison of predicting models

	LDA	LR	SVM	BN ($L = 7$, $M = 6$)
Training data				
Error	0.3019	0.2868	0.2078	0.1818
Testing data				
Accuracy	63.64 %	76.62 %	73.59 %	77.49 %

ground (M), respectively. This figure shows that the minimum value of BIC is obtained at $L = 7$ and $M = 6$.

4.3   Comparison of Accuracy

In the experiments, three prediction methods - linear discriminant analysis, logistic regression and support vector machine are employed as the comparing targets, which are denoted as LDA, LR and SVM in Table 3, respectively. With the optimal combination of cluster numbers $L = 7$ and $M = 6$ denoted as BN ($L = 7$, $M = 6$) in Table 3, our proposal is implemented by using the network as shown in Fig. 5 and Eq. (5). The comparison results are shown in Table 3, and our proposal shows the lowest Error in the training data and the highest Accuracy in the testing data.

In the linear models (LDA, LR and SVM with linear kernel), the stay time are considered as a positive effect and monotonic increasing relationship on the purchase behavior, however, actually the customers would be coming into a tedium emotion when they stay too long to make a purchase decision. This phenomenon which is disregarded by the traditional linear models, can be reproduced exactly by our proposal (see Table 6), therefore BN can obtain higher predicting performance than other models.

4.4   ROC Analysis

In most cases, decision functions are preferred to use a deterministic specification in the field of classification. With a discriminant value below or above 0, the target can be separated into two different categories. As a stochastic model is applied to this situation, 0.5 is widely used as the reference value. Nevertheless, when the positive (negative) data are in the minority group, it can cause a extreme skewness in results. Only maximizing the hit rate on sample data can increase the type I error (type II error). Here, we also introduce the multicriteria - sensitivity and specificity, which are denotes as Eq. (9) and Eq. (10), respectively.

$$Sensitivity = \frac{TP}{TP + FN}, \tag{9}$$

$$Specificity = \frac{TN}{FP + TN} \tag{10}$$

where TP, FN, TN and FP denote number of true positive data, false negative data, true negative data and false positive data, respectively. Therefore, the performance of models are assessed under the multicriteria - sensitivity and specificity, which denote the hit rate on positive data and negative data, respectively.

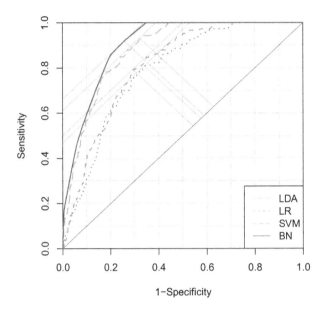

Fig. 7. ROC curve in adjustment of threshold

By using receiver operating characteristic (ROC) analysis, the results are drawn by plotting in adjustment of decision threshold in Fig. 7. The X-axis and Y-axis denote 1-Specificity (true negative rate) and Sensitivity (true positive rate), respectively. The dot line, dash line, long ash line and solid line denote the ROC curve of linear discriminant analysis, logistic regression, support vector machine and our proposal. As shown in Table 4, our proposal has shown a larger area under curve (AUC) than others.

Table 4. Comparison of area under ROC curve

	LDA	LR	SVM	BN ($L = 7$, $M = 6$)
AUC	0.7883	0.8094	0.8811	0.9023

Moreover, ROC curve can also show us the optimal decision threshold, which can separate data into two different categories that they should be belonged in as

Table 5. Comparison of multicriteria under optimal decision threshold

	LDA	LR	SVM	BN ($L = 7$, $M = 6$)
Training data				
Threshold	0.2423	0.2713	0.1574	0.3409
Testing data				
Accuracy	47.62 %	67.10 %	71.43 %	77.06 %
Sensitivity	24.23 %	78.26 %	51.81 %	88.41 %
Specificity	52.17 %	62.35 %	96.15 %	72.22 %

many as possible. As shown in Table 5, our proposal has shown a higher threshold than other models. Under individual optimal decision threshold, Table 5 also showed that our proposal has the highest performance in Accuracy and Sensitivity, however Specificity in our proposal is lower than it in SVM. Due to the Sensitivity in SVM is much worse, it seems SVM has an extreme skewness predicting in negative samples and increases the type II error, which also causes the low Accuracy. Therefore, under the muliticriteria our proposal not only obtains the best Accuracy, also obtain the optimal balance between Sensitivity and Specificity.

4.5   Discussion and Business Implication

Comparing with the previous studies, in which customers are always separated into typical 2 classes (above and below average) or 3 classes (low, medium, high), we investigated the classification of customers in a hierarchical process. As the cluster number of customers (from 2 to 10) is estimated and compared by the criterion - BIC, we can obtain the optimal clustering of customers for fitting the model in order to get the lowest training error. As shown in Sect. 4.2, the optimal cluster number of customers according to their purchase background is 6. In the meanwhile, the optimal cluster number of stay time is 7, which seems to separate the shopping process into 7 shopping sessions. Due to our proposal can provide a calibrated value of output in posterior probability, the change of purchase probability in each shopping session comparing with the previous shopping session is represented in Table 6. The column "Customer Group" shows the customers grouping by purchase background in ascending order, and The row "Shopping Session" shows the purchase intention separating by stay time in ascending order.

From the results, we can see that there is an nonmonotonic increasing phenomenon happened in all of customer groups, which supports the tedium effect on purchase behavior over time. For this issue, we explain the one of business implications in practice business situation basing on the results obtained by our proposal. As "the longer not means the better", an easily viewable layout arrangement is suggested to be possible to help the customers to decrease their wandering and make an efficient shopping. For the real instance in the coffee

Table 6. The change of purchase probability over shopping sessions in each purchase background groups

Customer Group	Shopping session						
	1	2	3	4	5	6	7
1	-	↗	↗	↗	↗	↘	↘
2	-	↗	↗	↗	↘	↘	↘
3	-	↗	↗	↗	↘	↗	↘
4	-	↗	↗	↘	↗	↘	↘
5	-	↗	↗	↗	↘	↗	↘
6	-	↗	↗	↘	↘	↘	↘

(a) Designing in coffee shelves

(b) Implementing in coffee shelves

Fig. 8. Implement business implication in real business situation

area (Fig. 8), we design the actual arrangement of coffee shelves like Fig. 8(a) to lay clearly by category-specific and also implement it into the real supermarket as Fig. 8(b). While the average stay time is decreasing from 10.81 s to 10.31 s, the purchase rate is increasing from 5.38 % to 6.44 %.

5   Conclusions

In this article, we present an operational improvement on tracking customers' in-store behavior. As the RFID tags attached to shopping carts can provide a direct observation of in-store behavior and generate a quantitative measure of in-store movement data, we introduce a new behavioral variable - stay time extracted from RFID data. Comparing with other customer factors (e.g., age, gender and income) used in previous studies, stay time is also a measurable and replicable variable without self-reported information, and this article applies it to a time-based prediction of purchase behavior not studied in existing literature. Therefore, this article also presents three methodological investigations on predicting customers' purchase behavior based on stay time. First, we propose an integrated predicting model constructed from purchase behavior, purchase background and stay time, and when bayesian network is employed to train the model, the predicting performance is much higher than other classic models widely used in this field. Second, due to bayesian network can treat the variables into discrete values, we examine the classification of customers in a comparable process. In contrast to typical 2 classes (above and below average) or 3 classes (low, medium, high), the clustering number is compared and estimated iteratively by the criterion - BIC. This improvement provides a subdivisible recognition on customer grouping than before, and also investigates the tedium effect on customers' purchase intention over time, when they are coming into long tail of shopping session. Third, we examine the ROC analysis to maximizing the accuracy under multicriteria - sensitivity and specificity. As the optimal decision threshold can be obtained by using ROC curve, even in the larger threshold, our proposal still shows higher sensitivity, specificity and accuracy than other models.

Furthermore, we also implemented managerial implications drawn from the results above. An easily viewable layout arrangement was proposed in the actual supermarket, which was suggested to lead the customers to decrease their wandering and make an efficient shopping. Due to the customers seemed to be able to defuse the tedium effect via decreased their stay time, supermarket gained an extra profit from customers accounting for their purchase rate increasing from 5.38 % to 6.44 %.

In future work, we plan to introduce a new variable to represent customers' brand switching behavior, and investigate its effect on stay time so as to affect purchase decision making when brand switching happened. We suggest this new variable of switching behavior can lead our proposal to reach higher accuracy level for predicting purchase behavior, and can also assist scholars and business persons related to this field to extract more business implications for actual applications.

Acknowledgment. This work was supported in part by MEXT Strategic Project to Support the Fomation of Research Bases at Private Universities (FY2014–2018) and MEXT Grant-in-Aid for Young Scientists (B) Grant Number 25780277.

References

1. Baker, R.G.V.: Impact of sales promotions on when, what, and how much to buy. Pap. Reg. Sci. **79**(4), 413–434 (2000)
2. Ben-Gal, I.E.: Bayesian networks. In: Ruggeri, F., Kenett, R., Faltin, F. (eds.) Encyclopedia of Statistics in Quality and Reliability. Wiley, Chichester (2007)

3. Chen, Z.Y., Fan, Z.P., Sun, M.: A hierarchical multiple kernel support vector machine for customer churn prediction using longitudinal behavioral data. Eur. J. Oper. Res. **223**(21), 461–472 (2012)
4. Frank, R.E., Massy, W.F., G.Morrison, D.: Bias in multiple discriminant analysis. J. Mark. Res. **2**(3), 250–258 (1965)
5. Gomez, B.G., Arranz, A.G., Cillan, J.G.: The role of loyalty programs in behavioral and affective loyalty. J. Consum. Mark. **23**(7), 387–396 (2006)
6. Guadagni, P.M., Little, J.D.C.: A logit model of brand choice calibrated on scanner data. Comput. Math. Organ. Theor. **2**(3), 203–238 (1983)
7. Gupta, S.: Impact of sales promotions on when, what, and how much to buy. J. Mark. Res. **25**(4), 342–355 (1988)
8. Hui, S.K., Bradlow, E.T., Fader, P.S.: Testing behavioral hypotheses using an integrated model of grocery store shopping path and purchase behavior. J. Consum. Res. **36**(3), 478–493 (2009)
9. Kim, G., Chae, B.K., Olson, D.L.: A support vector machine (SVM) approach to imbalanced datasets of customer responses: comparison with other customer response models. Serv. Bus. **7**(1), 167–182 (2013)
10. Larson, J.S., Bradlow, E.T., Fader, P.S.: An exploratory look at supermarket shopping paths. Int. J. Res. Mark. **22**(4), 395–414 (2005)
11. Little, J.D.C.: Feature article-aggregate advertising models: the state of the art. Oper. Res. **27**(4), 629–667 (1979)
12. Liu, Y.: The long-term impact of loyalty programs on consumer purchase behavior and loyalty. J. Mark. **71**(4), 19–35 (2007)
13. Nakahara, T., Yada, K.: Analyzing consumers' shopping behavior using RFID data and pattern mining. Adv. Data Anal. Classif. **6**(4), 355–365 (2012)
14. Pearl, J.: Fusion, propagation, and structuring in belief networks. Artif. Intell. **29**(3), 241–288 (1986)
15. Ravnik, R., Solina, F., Zabkar, V.: Modelling in-store consumer behaviour using machine learning and digital signage audience measurement data. In: Distante, C., Battiato, S., Cavallaro, A. (eds.) VAAM 2014. LNCS, vol. 8811, pp. 123–133. Springer, Heidelberg (2014)
16. Robertson, T.S., Kennedy, J.N.: Prediction of consumer innovators: application of multiple discriminant analysis. J. Mark. Res. **5**(1), 64–69 (1968)
17. Schwarz, G.: Estimating the dimension of a model. Ann. Stat. **6**(2), 461–464 (1978)
18. Sorensen, H.: The science of shopping. Mark. Res. **15**, 30–35 (2003)
19. Takai, K., Yada, K.: Relation between stay-time and purchase probability based on RFID data in a japanese supermarket. In: Setchi, R., Jordanov, I., Howlett, R.J., Jain, L.C. (eds.) KES 2010, Part III. LNCS, vol. 6278, pp. 254–263. Springer, Heidelberg (2010)
20. Yada, K.: String analysis technique for shopping path in a supermarket. J. Intell. Inf. Syst. **36**(3), 385–402 (2011)
21. Zuo, Y., Kita, E.: Stock price forecast using bayesian network. Expert Syst. Appl. **39**(8), 6729–6737 (2012)

Statistical Tests for Joint Analysis of Performance Measures

Alessio Benavoli[1,2] and Cassio P. de Campos[3(✉)]

[1] Istituto Dalle Molle di Studi Sull'Intelligenza Artificiale (IDSIA),
Scuola Universitaria Professionale Della Svizzera Italiana (SUPSI),
Manno, Switzerland
[2] Università Della Svizzera Italiana (USI), Lugano, Switzerland
[3] Queen's University Belfast, Belfast, UK
c.decampos@qub.ac.uk

Abstract. Recently there has been an increasing interest in the development of new methods using Pareto optimality to deal with multi-objective criteria (for example, accuracy and architectural complexity). Once one has learned a model based on their devised method, the problem is then how to compare it with the state of art. In machine learning, algorithms are typically evaluated by comparing their performance on different data sets by means of statistical tests. Unfortunately, the standard tests used for this purpose are not able to jointly consider performance measures. The aim of this paper is to resolve this issue by developing statistical procedures that are able to account for multiple competing measures at the same time. In particular, we develop two tests: a frequentist procedure based on the generalized likelihood-ratio test and a Bayesian procedure based on a multinomial-Dirichlet conjugate model. We further extend them by discovering conditional independences among measures to reduce the number of parameter of such models, as usually the number of studied cases is very reduced in such comparisons. Real data from a comparison among general purpose classifiers is used to show a practical application of our tests.

1 Introduction

In many real applications of machine learning, we often need to consider the trade-off between multiple conflicting objectives. For instance, measures like accuracy and architectural complexity are clearly two different (possibly conflicting) criteria. This issue can be tackled by considering a multi-objective decision making approach.

There are two main approaches to multi-objective decision making. The weighted-sum approach, which consists of transforming the original multi-objective problem into a single-objective problem by using a weighted formula; The Pareto approach, which considers directly the original multi-objective problem and searches for non-dominated solutions, that is, solutions that are not worse than any other solution with respect to all criteria.

© Springer International Publishing Switzerland 2015
J. Suzuki and M. Ueno (Eds.): AMBN 2015, LNAI 9505, pp. 76–92, 2015.
DOI: 10.1007/978-3-319-28379-1_6

Table 1. Architectural complexity and accuracy of two learning methods for PNN [3].

	New		State of art	
	Accuracy	Complexity	Accuracy	Complexity
IRIS	97.8	38.4	95.3	50.0
WINE	98.3	26.9	92.3	24.0
PIMA	72.1	28.6	65.3	37.7
BUPA	70.3	23.4	69.1	36.0

In a weighted-sum approach, a multi-objective problem is transformed into a single-objective problem by a numerical weight function that is assigned to objectives and then values of the weighted criteria are combined into a single value according to the weights. One of the reasons for its popularity is its simplicity. However, there are several drawbacks associated to it. First, the definition of weights in these formulas is often ad-hoc or requires great domain knowledge which might not be available. Second, the optimal solution strongly depends on that particular weight function, which misses the opportunity to find other models that might be actually more interesting to the user, for instance, representing a better trade-off between different criteria. Third, a weighted formula involving a linear combination of different criteria is meaningless in many scenarios, as the criteria may be non-commensurable (comparison of apples and oranges).

In the Pareto approach, instead of transforming a multi-objective problem into a single-objective problem and then solving it by using a single-objective decision making, a multi-objective algorithm is used to solve the original multi-objective problem. The advantage of the Pareto approach is that it can cope with any kind of non-commensurable criteria. Recently there has been an increasing interest in the development of new learning methods able to cope simultaneously with multi-objective criteria using Pareto optimality [1–4]. The disadvantage comes from the *power* of the Pareto approach in situations where a good weight function can be devised, as the Pareto approach is more conservative than using the weighted-sum idea. In this work we assume that a good weight function is not available. Consider for instance the work in [3], where it is proposed a multi-objective Pareto based optimization method for simultaneous optimization of architectural complexity and accuracy for Polynomial Neural Networks (PNN). By using multiple data sets, they compare their method with the state-of-art method for learning PNN, producing the results presented in Table 1.

Based on Table 1, [3] claims that a multi-objective approach (jointly optimizing architectural complexity and accuracy) is clearly beneficial. Can we say that their method is clearly better than the state of art for both criteria and also for each of them independently? For which criterion is it superior (respectively inferior)? To answer these questions we need a method that statistically assesses whether an algorithm is better than another in terms of all criteria. To the best knowledge of the authors, this method is lacking in machine learning and so it could not be used in [3].

Competing methods/algorithms are typically compared by means of a statistical test, whose aim is to assess whether an algorithm is significantly better than another (statistically comparing their performance on different data sets or problem instances). For comparing two algorithms over a collection of data sets, the most common approaches are the sign test or the Wilcoxon signed-rank test [5], however these tests are only able to cope with one performance measure (criterion) at a time, that is, they cannot consider a multi-objective approach without resorting to the weighted-sum approach described earlier. In this paper, we develop two tests that are able to cope jointly with multiple performance measures without having to somehow combine them: a frequentist procedure based on the generalized likelihood-ratio test and a Bayesian procedure based on a multinomial-Dirichlet conjugate model. We further extend them by discovering conditional independences among measures to reduce the number of parameters of such models, an important add-on since usually the number of data sets on which methods are compared is reduced. Applications of these new tests are numerous. Here we use real data from a comparison of general purpose classification methods to show a clear practical application of the tests.

2 Joint Analysis of Performance Criteria

Let $M_1, \ldots, M_m$ be a set of m performance measures (criteria) and assume that we are going to compare two algorithms A and B by jointly using these measures.

Definition 1. *We call a 'dominance statement' for B against A a sequence of m dominance conditions:*

$$D^{(BA)} = [\succ, \succ, \prec, \ldots, \succ],$$

where the comparison $\succ$ (or $\prec$) in the i-th entry of the vector $D^{(BA)}$ means that algorithm B is better than A (respectively, A is better than B) on measure M_i. □

Our goal is to make inferences on *dominance statements* by evaluating the m performance measures for the algorithms A and B on n different case studies (for instance, data sets, problem instances, etc.). In other words, we want to decide which $D^{(BA)}$ is the most appropriate for A and B given tables with values $M_{ij}^{(\text{Alg})}$ representing the j-th measure for the algorithm $\text{Alg} \in \{A, B\}$ in the i-th case study:

$$\mathbf{M}^{(\text{Alg})} = \begin{bmatrix} M_{11}^{(\text{Alg})} & M_{12}^{(\text{Alg})} & \cdots & M_{1m}^{(\text{Alg})} \\ M_{21}^{(\text{Alg})} & M_{22}^{(\text{Alg})} & \cdots & M_{2m}^{(\text{Alg})} \\ \vdots & \vdots & \vdots & \vdots \\ M_{n1}^{(\text{Alg})} & M_{n2}^{(\text{Alg})} & \cdots & M_{nm}^{(\text{Alg})} \end{bmatrix}. \tag{1}$$

Given the matrix of performances $\mathbf{M}^{(A)}$ and $\mathbf{M}^{(B)}$, we first build the binary matrix $\mathbf{X} = [\mathbf{M}^{(B)} \succ \mathbf{M}^{(A)}]$, whose entry x_{ij} is equal to one if algorithm B is better than algorithm A for the j-th measure in the i-th case study and zero otherwise. We assume that ties do not exist.[1] To each matrix $\mathbf{X}$ we associate a count vector $\mathbf{n}$, whose entries represent the counts for each one of the 2^m possible *dominance statements* (many of which might be zero).

Example 1. Consider the comparison of two algorithms in terms of accuracies M_1 (expressed in percent values in the first row) and time M_2 (in seconds, shown in the second row) on 12 data sets:

$$\mathbf{M}^A = \begin{bmatrix} 85 & 87 & 87 & 91 & 91 & 91 & 94 & 94 & 94 & 94 & 94 & 94 \\ 8 & 11 & 11 & 12 & 12 & 12 & 16 & 16 & 16 & 16 & 16 & 16 \end{bmatrix}^T,$$
$$\mathbf{M}^B = \begin{bmatrix} 84 & 86 & 86 & 92 & 92 & 92 & 95 & 95 & 95 & 95 & 95 & 95 \\ 9 & 10 & 10 & 13 & 13 & 13 & 15 & 15 & 15 & 15 & 15 & 15 \end{bmatrix}^T \tag{2}$$

where ^{T} denotes transpose.
    The matrix $\mathbf{X} = [\mathbf{M}^{(B)} \succ \mathbf{M}^{(A)}]$ is:[2]

$$\mathbf{X} = \begin{bmatrix} 0 & 0 & 0 & 1 & 1 & 1 & 1 & 1 & 1 & 1 & 1 & 1 \\ 0 & 1 & 1 & 0 & 0 & 0 & 1 & 1 & 1 & 1 & 1 & 1 \end{bmatrix}^T. \tag{3}$$

Hence, we derive that the dominance statement $[\prec, \prec]$ (or $[0,0]$), which means that B is worse than A on both measures, is observed $n_0 = 1$ time; the statement $[\prec, \succ]$ (or $[0,1]$), which means that B is worse than A on the first measure but better on the second, is observed $n_1 = 2$ times; the statement $[\succ, \prec]$ (or $[1,0]$) is observed $n_2 = 3$ times; the statement $[\succ, \succ]$ (or $[1,1]$) is observed $n_3 = 6$ times. Hence, we have that $\mathbf{n} = [1, 2, 3, 6]$ (a binary lexicographic order is used for the entries of $\mathbf{n}$). □

The matrix $\mathbf{X}$ or, equivalently, the vector $\mathbf{n}$, include all the information that we will use to derive our tests. While this approach might seem to lose information because we only account for the sign of each difference $M_{ij}^{(\text{Alg})} - M_{ij}^{(\text{Alg}')}$, there is no effective way of using the actual value of the difference across multiple measures if these measures are assumed to be expressed in *incomparable units*, as in this case no procedure could be used to compare the measures jointly or to collapse the measures into a single one in order to run standard tests (using some weighting function; we assume that normalizing the measures is not an option either, as it entails an additional assumption about the measures which might not hold). On the other hand, the sign of the difference is a proper comparable value among measures regardless of the particular meaning of each of them. In fact, we point out that the measures $M_{ij}^{(\text{Alg})}$ can themselves be obtained from

---

[1] If there are ties we treat a tie in a measure by a standard approach: we replicate the case with it into two and divide the weight of such case by two (this process might need to be performed multiple times until no ties are present in the data). Such approach preserves the sample size and fairly allocates ties between the algorithms being compared.

[2] An algorithm is better ($\succ$) than another when it has higher accuracy and lower computational time.

any arbitrary procedure (including statistical tests), as we only assume that the sign of the difference $M_{ij}^{(\text{Alg})} - M_{ij}^{(\text{Alg}')}$ is available (and we properly account for ties). This provides us with a very general setting, allowing for numerous applications.

3   Generalized Likelihood Ratio Test

We derive a simple null hypothesis significance test for the joint analysis of performance measures. We denote by θ_k, for $k = 0, \ldots, 2^m - 1$, the probability of obtaining one of the 2^m possible *dominance statements*. Hence, $\theta_k \geq 0$ and $\sum_{k=0}^{2^m-1} \theta_k = 1$. We have enumerated the *dominance statements* according to their "binary order", so that θ_0 is the probability of the statement $[\prec, \ldots, \prec, \prec]$, θ_1 is the probability of $[\prec, \ldots, \prec, \succ]$, θ_2 is the probability of $[\prec, \ldots, \prec, \succ, \prec]$, etc. Our goal is to find if there is a statement that is significantly more likely than all others based on the observation matrix $\mathbf{X}$. It is clear that $\mathbf{n}$ is a sufficient statistic for this test, since its k-th entry n_k corresponds to the counts for the k-th statement. Hence, to achieve our goal, we can perform a *Generalized Likelihood Ratio Test* (GLRT):

$$\lambda(\mathbf{n}) = \frac{\max_{\boldsymbol{\theta} \in \Theta^*} L(\boldsymbol{\theta}|\mathbf{n})}{\max_{\boldsymbol{\theta} \in \Theta} L(\boldsymbol{\theta}|\mathbf{n})}, \text{ where } L(\boldsymbol{\theta}|\mathbf{n}) = \prod_{k=0}^{2^m-1} \theta_k^{n_k}, \tag{4}$$

$\boldsymbol{\theta} = [\theta_0, \ldots, \theta_{2^m-1}]$, Θ is the simplex for $\boldsymbol{\theta}$, $\Theta^* = \{\boldsymbol{\theta} \in \Theta : \theta_{i^*} \leq \max(\boldsymbol{\theta} \setminus \theta_{i^*})\}$ (we abuse notation and indicate by $\boldsymbol{\theta} \setminus \theta_{i^*}$ all thetas apart from θ_{i^*}) and $i^* = \text{argmax}_{i=0,\ldots,2^m-1} n_i$. The rationality behind Eq.(4) is that we are testing two hypothesis: (H_0) $\theta_{i^*} \leq \max(\boldsymbol{\theta} \setminus \theta_{i^*})$ and (H_1) $\theta_{i^*} > \max(\boldsymbol{\theta} \setminus \theta_{i^*})$. Under H_0, the value of $\boldsymbol{\theta}$ which better explains the observations is the maximum likelihood estimate (MLE) subject to the constraint that $\boldsymbol{\theta} \in \Theta^*$. Its likelihood is the numerator of Eq. (4). The value of $\boldsymbol{\theta}$ which maximizes the likelihood is instead the MLE subject to $\boldsymbol{\theta} \in \Theta$. It is clear that $0 \leq \lambda(\mathbf{n}) \leq 1$. GLRT employs $\lambda(\mathbf{n})$ as a test statistic and rejects H_0 for small values of $\lambda(\mathbf{n})$, that is, when $\lambda(\mathbf{n}) \leq \rho$, where the value of ρ is determined by fixing the type-I error to be α. By Wilks' theorem, for large n, $-2\log(\lambda(\mathbf{n}))$ is chi-square distributed with one degree of freedom [6,7]. Hence, the rejection zone for the null hypothesis is approximately equal to

$$\mathcal{R} = \{\mathbf{n} : -2\log(\lambda(\mathbf{n})) > \chi_{1,\alpha}^2\}, \tag{5}$$

where α is the confidence level. Therefore, to apply GLRT, we must only compute $\lambda(\mathbf{n})$.

Theorem 1. *Given the count vector* $\mathbf{n}$, *it holds that*

$$\lambda(\mathbf{n}) = \frac{\left(\frac{n_a + n_b}{2}\right)^{n_a + n_b}}{n_a^{n_a} n_b^{n_b}}, \tag{6}$$

where n_a *is the greatest value among* $n_0, \ldots, n_{2^m-1}$ *and* n_b *the second greatest.* $\square$

Proof. The maximum likelihood estimate of $\boldsymbol{\theta}$ subject to the constraint $\boldsymbol{\theta} \in \Theta$ is

$$\left(\frac{n_0}{n}, \frac{n_1}{n}, \ldots, \frac{n_{2^m-1}}{n}\right),$$

in fact the only constraint on $\boldsymbol{\theta}$ in this case is that its elements sum up to 1. The maximum likelihood estimate of $\boldsymbol{\theta}$ subject to the constraint $\Theta^* = \{\boldsymbol{\theta} \in \Theta : \theta_{i*} \leq \max(\boldsymbol{\theta} \setminus \theta_{i*})\}$ can be computed using KKT conditions of optimality for optimization problems subject to inequality constraints [8]. To obtain this estimate let us assume without loss of generality that $n_0 \geq n_1 \geq n_2...$ Note that $i^* = \mathrm{argmax}_{i=0,...,2^m-1} n_i$ and so considering the equality constraint $\theta_{i*} = \max(\boldsymbol{\theta} \setminus \theta_{i*})$, we have that the maximum likelihood estimate of $\boldsymbol{\theta}$ is

$$\left(\frac{n_c}{n}, \frac{n_c}{n}, \frac{n_2}{n}, \ldots, \frac{n_{2^m-1}}{n}\right),$$

where $n_c = (n_0 + n_1)/2$. Then the likelihood ratio is

$$\frac{\left(\frac{n_c}{n}\right)^{n_0} \cdot \left(\frac{n_c}{n}\right)^{n_1} \cdots \left(\frac{n_{2^m-1}}{n}\right)^{n_0}}{\left(\frac{n_0}{n}\right)^{n_0} \cdot \left(\frac{n_1}{n}\right)^{n_1} \cdots \left(\frac{n_{2^m-1}}{n}\right)^{n_0}} = \frac{n_c^{n_0+n_1}}{n_0^{n_0} n_1^{n_1}},$$

which proves the theorem. $\square$

In case $n_a = n_b$, we have $\lambda(\mathbf{n}) = 1$ and $-2\log(\lambda(\mathbf{n})) = 0$, so that the null hypothesis can never be rejected. It can be shown that:

Theorem 2. *The GLRT (Eq. (5)) is (asymptotically) calibrated for a prescribed significance level α obtaining the maximum type-I error when $n_a + n_b = n$.* $\square$

This can be proven using an approach similar to that described in [9, Ex. 21.2].

Example 2. In Example 1, $m = 2$ and Eq.(2) yields $L(\boldsymbol{\theta}|\mathbf{n}) = \theta_0\theta_1^2\theta_2^3\theta_3^6$, where θ_0 is the probability of the statement $[\prec, \prec]$, θ_1 of $[\prec, \succ]$, θ_2 of $[\succ, \prec]$ and θ_3 of $[\succ, \succ]$. Hence, $n_a = 6$, $n_b = 3$, the statistic $\lambda(\mathbf{n}) = \frac{(\frac{9}{2})^9}{3^3 6^6} \approx 0.6$ and the p-value is 0.313. Given the value of the p-value, we cannot conclude that B is better than A on both measures. $\square$

GLRTs have the disadvantage that they do not provide the probability of the hypotheses, but only its p-value under H_0. This means that we do not have any information about the probability of the alternative hypothesis being true. To address this issue, in the next section we propose a Bayesian hypothesis test for testing a certain *dominance statement*.

4   Bayesian Test

We implement the Bayesian hypothesis test by following a Bayesian estimation approach, that is, by estimating the posterior probability of the vector of parameters $\boldsymbol{\theta}$. Given the count vector $\mathbf{n}$, the likelihood of $\boldsymbol{\theta}$ given the data is given by the right-hand side of Eq. (4), which is a multinomial distribution. As prior we then consider a Dirichlet distribution: $p(\boldsymbol{\theta}) \propto \prod\limits_{k=0}^{2^m-1} \theta_k^{\alpha_k - 1}$, where $\alpha_k > 0$ are the parameters of the Dirichlet distribution. In the rest of the paper, we will always use the symmetric prior $\alpha_k = 1/2^m$ (however, we can also use other priors such as the Jeffreys prior $\alpha_k = \frac{1}{2}$, or some robust prior model [10]). By conjugacy, the posterior is also a Dirichlet with updated parameters $n_k + \alpha_k$. In the Bayesian setting, to make inferences on a *dominance statement*, we have simply to compute the posterior probabilities $P(\theta_i > \max(\boldsymbol{\theta} \setminus \theta_i)|\mathbf{n})$, for $i = 0, \ldots, 2^m - 1$. This is the posterior probability that θ_i (associated to the i-statement) is greater than all other $\theta_{\neg i}$ values.

Proposition 1. *It holds that* $\sum\limits_{i=0}^{2^m-1} P(\theta_i > \max(\boldsymbol{\theta} \setminus \theta_i)|\mathbf{n}) = 1.$ $\qquad\square$

This result follows from the simple fact that $P(\theta_i = \theta_j|\mathbf{n}) = 0$ (i.e., since θ_i are continuous variables, it is clear that $P(\theta_i = \theta_j|\mathbf{n}) = 0$ since any probability density function on continuous variables assign probability zero to singletons). Hence, the above posterior probabilities enclose all the available information on the *dominance statements*. These probabilities can easily be computed by Monte Carlo sampling on the space of vectors $\boldsymbol{\theta}$ from the posterior Dirichlet distribution and then by counting the fraction of times we see $\theta_i > \max(\boldsymbol{\theta} \setminus \theta_i)$, for every i.

Example 3. Take again Example 1. We already know that $L(\boldsymbol{\theta}|\mathbf{n}) = \theta_0 \theta_1^2 \theta_2^3 \theta_3^6$, where θ_0 is the probability of the statement $[\prec, \prec]$, θ_1 of $[\prec, \succ]$, θ_2 of $[\succ, \prec]$ and θ_3 of $[\succ, \succ]$. The posterior probabilities of hypotheses are: $P(\theta_0 > \theta_{\neg 0}|\mathbf{n}) \approx 0.013$, $P(\theta_1 > \theta_{\neg 1}|\mathbf{n}) \approx 0.051$, $P(\theta_2 > \theta_{\neg 2}|\mathbf{n}) \approx 0.136$, and $P(\theta_3 > \theta_{\neg 3}|\mathbf{n}) \approx 0.80$. Hence the most probable dominance statement is $[\succ, \succ]$ and its probability is 0.8. These probabilities have been computed by Monte Carlo sampling as discussed above.

5   Bayesian Network

The columns of $\mathbf{X} = [\mathbf{M}^{(B)} \succ \mathbf{M}^{(A)}]$ can be seen as binary random variables $\mathcal{M} = \{M_1, \ldots, M_m\}$ representing which algorithm is better according to that measure. Because of possible stochastic conditional independences between these variables, the estimation of a joint probability $p(\mathcal{M})$ can be improved by using a Bayesian network (BN). A BN can be defined as a triple $(\mathcal{G}, \mathcal{M}, \mathcal{P})$, where $\mathcal{G} = (V_{\mathcal{G}}, E_{\mathcal{G}})$ is a directed acyclic graph (DAG) with $V_{\mathcal{G}}$ a collection of m nodes associated to the random variables $\mathcal{M}$ (a node per variable), and $E_{\mathcal{G}}$ a collection of arcs; $\mathcal{P}$ is a collection of conditional probabilities $p(M_i|PA_i)$ where PA_i denotes the parents of M_i in the graph (PA_i may be empty), corresponding to the relations of $E_{\mathcal{G}}$. In a Bayesian network, the Markov condition states that every variable is conditionally independent of its non-descendants given its parents. This structure induces a joint probability distribution by the factorization $p(M_1, \ldots, M_m) = \prod_i p(M_i|PA_i)$. Let $\boldsymbol{\theta}$ be the entire vector of parameters such that $\theta_{ijk} = p(M_i = k|PA_i = j)$, where $k \in \{0,1\}$, $j \in \{1, ..., 2^{|PA_i|}\}$ and $i \in \{1, \ldots, m\}$. Note that this represents a different parametrization with respect to the $\boldsymbol{\theta}$ of previous sections, but a simple transformation can be used to compute those values through the factorization expression. Given the table $\mathbf{X}$ with m measures and n case studies, the structure learning problem in Bayesian networks is to find a DAG $\mathcal{G}$ that maximizes its posterior probability, that is, $\mathcal{G}^* = \mathrm{argmax}_{\mathcal{G} \in \boldsymbol{\mathcal{G}}} p(\mathcal{G}|\mathbf{X})$, with $\boldsymbol{\mathcal{G}}$ the set of all DAGs over node set $\mathcal{M}$.

$$p(\mathcal{G}|\mathbf{X}) \propto p(\mathcal{G}) \cdot \int p(\mathbf{X}|\mathcal{G}, \boldsymbol{\theta}) \cdot p(\boldsymbol{\theta}|\mathcal{G}) d\boldsymbol{\theta},$$

where $p(\boldsymbol{\theta}|\mathcal{G})$ is the prior of $\boldsymbol{\theta}$ for a given graph $\mathcal{G}$, assumed to be a symmetric Dirichlet with positive hyper-parameter α^*:

$$p(\boldsymbol{\theta}|\mathcal{G}) = \prod_{i=1}^{m} \prod_{j=1}^{2^{|PA_i|}} \Gamma(\frac{\alpha^*}{2^{|PA_i|}}) \prod_{k=0}^{1} \frac{\theta_{ijk}^{\frac{\alpha^*}{2^{|PA_i|+1}} - 1}}{\Gamma(\frac{\alpha^*}{2^{|PA_i|+1}})}. \tag{7}$$

α^* is usually referred to as the Equivalent Sample Size (ESS). Such computation is known as the Bayesian Dirichlet Equivalent Uniform (BDeu) criterion [11,12], where we assume parameter independence and modularity [13]. We also assume $\alpha^* = 1$ and that there is no preference for any graph and set $p(\mathcal{G})$ as uniform.

In order to find the graph representing the best set of conditional independences over the space of all possible DAGs $\boldsymbol{\mathcal{G}}$, multiple approaches have been proposed in the literature. Because the number of measures is hardly above 15 to 20 and they are all binary, the combination of properties of the BDeu score [14] with a dynamic programming algorithm [15] usually suffices. Otherwise one might use more sophisticated ideas [16–18], which can deal with a greater number of variables. Given the optimal graph $\mathcal{G}$, we can employ the discovered conditional independences to write the joint distribution for $\mathcal{M}$ opportunely:

$$p(\mathbf{X}|\mathcal{G}, \boldsymbol{\theta}) = \prod_{i=1}^{m} \prod_{j=1}^{2^{|PA_i|}} \theta_{ij0}^{n_{ij0}} (1 - \theta_{ij0})^{n_{ij1}},$$

where n_{ijk} counts the number of times $(M_i = k \wedge PA_i = j)$ in the data. Combined with the prior $p(\boldsymbol{\theta}|\mathcal{G})$ of Eq. (7), this can be used to compute $P(\theta_i > \max(\boldsymbol{\theta} \setminus \theta_i)|\mathbf{X})$ by Monte Carlo sampling as before (even if different from previous sections, the parametrization of $\boldsymbol{\theta}$ used here also works for that). The advantages of using Bayesian networks are as follows. First, by using the $p(\mathcal{G}|\mathbf{X})$, the dependence model underlying the distribution is automatically adapted to what can be inferred from data, and so one usually needs fewer observations to learn a good model than when working with the full joint. Second, the graph can be used to identify relations between measures and how they are associated, which can be for instance used to ignore measures that are not able to help in discriminating the algorithms. Third, computations can be carried out efficiently (at least when we restrict ourselves to a couple of tens of variables). We will illustrate these benefits later on.

6   Experiments

In this section, we apply our tests to compare seven classifiers on 80 data sets (10 runs of 10-folds cross-validation) and using several performance measures. We have considered the following classifiers 'AODE' ($C1$), 'Bayes net' ($C2$), 'Bayes.NaiveBayes' ($C3$), 'trees.J48graft' ($C4$), 'trees.RandomForest' ($C5$), 'trees.bagging' ($C6$) and 'logistic' ($C7$). We have performed all the experiments using WEKA [19], which implements all such classifiers, and analyzed the results using simple scripts in R. We note that our purpose is not to conclude in favor or against any of the classifiers, but to illustrate the use of our new approaches to compare them. The measures used in the analysis are available at http://www.cs.qub.ac.uk/~c.decampos/benavoli-ambn2015.ods.

6.1   Accuracy and FPR-TPR

In this experiment, we have considered three measures. Accuracy is the percentage of correct predictions of a model, the most common measure to evaluate a classifier. For a binary classification problem, the true positive rate (*TPR*) defines how many correct positive results occur among all positive samples available during the test. The false positive rate (*FPR*), on the other hand, defines how many incorrect positive results occur among all negative samples available during the test. It is well known that accuracy is highly dependent on TPR and FPR (in the binary case it is just a convex combination of them). We compare the classifiers using (i) only accuracy and (ii) FPR-TPR jointly, and expect to see a great agreement between the results of (i) and (ii) because of the strong dependence between those measures. For (i) we use the Wilcoxon sign-rank test (which has more power than the sign test), and our tests for (ii). Matrix (8) (left) reports the statistical comparison of the seven classifiers performed by considering accuracy only. The numerical values in the matrix are the p-values of Wilcoxon sign-rank test computed on the direction ($\prec$ or $\succ$) corresponding to the highest value of the statistic (most likely direction to refute the null hypothesis). For instance, the meaning of the first matrix entry is as follows: C_1 has been found better than C_2 with p-value close to zero. Conversely, the first element in the second row means that C_2 has been found worse than C_3 (but non-significant with p-value 0.46). All pairwise comparisons with p-values less than $\alpha/2$ (e.g., $\alpha = 0.1$ or 0.05) are significant. To control the family-wise type-I error of many pairwise comparisons, the significance level should be adjusted by the Bonferroni correction (or other more efficient approaches) [5]. Hereafter, we report the p-values of the frequentist tests, so the implementation of such corrections is straightforward.

$$
\begin{array}{c}
\begin{array}{ccccccc}
 & C_2 & C_3 & C_4 & C_5 & C_6 & C_7 \\
C_1 & \succ 0 & \succ 0 & \prec & \prec.17 & \succ 0 & \succ 0 \\
C_2 & & \prec.46 & \prec 0 & \prec.046 & \succ 0 & \succ 0 \\
C_3 & & & \prec 0 & \prec.048 & \succ 0 & \succ 0 \\
C_4 & & & & \succ.026 & \succ 0 & \succ 0 \\
C_5 & & & & & \succ 0 & \succ 0 \\
C_6 & & & & & & \prec 0
\end{array}
\quad
\begin{array}{ccccccc}
 & C_2 & C_3 & C_4 & C_5 & C_6 & C_7 \\
C_1 & \succ.99 & \succ.99 & \prec.99 & \prec.85 & \prec 1 & \prec 1 \\
C_2 & & \succ.27 & \prec 1 & \prec.92 & \succ 1 & \succ 1 \\
C_3 & & & \prec 1 & \prec.90 & \succ 1 & \succ 1 \\
C_4 & & & & \succ.93 & \succ 1 & \succ 1 \\
C_5 & & & & & \succ 1 & \succ 1 \\
C_6 & & & & & & \prec 1
\end{array}
\end{array}
\tag{8}
$$

Matrix (8) (right) reports the comparison performed with the Bayesian test considering jointly TPR and -FPR (negative FPR so that as higher as better). In this case, each entry of the matrix represents the most probable joint dominance and the numerical value the relative probability. The first element says that C_1 is jointly better than C_2 because has higher -FPR (so lower FPR) and higher TPR and this statement holds with posterior probability 0.99. The test using the Bayesian network model achieved almost equal results for the probabilities (variations only because of Monte Carlo, data not shown), because the two measures are well correlated (so the Bayesian network inferred the joint model as the most probable, which reduced it to the standard Bayesian test without the Bayesian network). Also the GLRT is consistent with the results obtained by the Bayesian test. For instance, its p-values relative to the C_1 row are $0.014, 0.024, 0, 0.29, 0, 0$. Apart from 0.29 all the p-values are significant for $\alpha = 0.05$. A reason to prefer GLRT to the Bayesian test is that we have shown that it is calibrated to type-I error. On the other hand the probabilities returned by the Bayesian test have a more direct interpretation. For this reason, in the following we will just show the results of the Bayesian test. Comparing the two matrices is clear that the results are quite in agreement (smaller p-values correspond to higher probabilities and vice versa). The advantage of the new approach accounting for multiple measures altogether is that it is able to jointly consider them and thus its conclusions have additional meaning.

6.2   Accuracy, F-Measure and Weighted-AUC

In this section we compare the classifiers using accuracy, F-measure and weighted-AUC: (i) separately; (ii) considering pairwise combinations of these measures; (iii) considering the three measures together.

For the case of Accuracy and Weighted-AUC, Matrix (9) (on the left) reports the results of the comparison obtained considering separately each of these measures (each cell contains the result for Accuracy on top and Weighted-AUC below it), while Matrix (9) (on the right) is the result of the Bayesian joint test. For performing the separate tests, we have used the Wilcoxon sign-rank test [5]. The numerical values in the Matrix (9) (on the left) are the p-values of Wilcoxon sign-rank test computed on the direction ($\prec$ or $\succ$) corresponding to the highest value of the statistic (most likely direction to refute the null hypothesis). For instance, the meaning of the comparison C_1 versus C_5, is as follows: C_1 has been found worse than C_5 in accuracy (with p-value 0.17) and better in Weighted-AUC (with p-value 0.14). All pairwise comparisons with p-values less than $\alpha/2$ (e.g., $\alpha = 0.1$ or 0.05) are significant.[3] Matrix (9) (on the right) reports the comparison performed with the Bayesian test considering jointly Accuracy and Weighted-AUC. In this case, each entry of the matrix represents the most probable joint dominance statement and the numerical value is the relative probability. Comparing the two matrices, there are two cases where the tests are in

---

[3] To control the family-wise type-I error of many pairwise comparisons, the significance level should be adjusted, as previously described.

clear contradiction (in bold) and a case (C_4 vs. C_7) where the joint comparison gives an evident advantage in power. This means that C_4 is better than C_7 jointly on both accuracy and Weighted-AUC, while this is not true when the two performance measures are considered separately. Therefore, it is evident that decisions derived by a joint test can be very different from the decisions carried out using a separate test for each performance measure. If the goal is to compare algorithms considering jointly the measures, then it is more appropriate to use the new methods proposed here. The GLRT is overall consistent with the results obtained by the Bayesian test (results not shown). For instance, its p-value for "C_4 better than C_7 on both the performance measures" is almost zero (so "very" significant). The choice between GLRT and the Bayesian test depends on the user's needs.

$$(9)$$

Now we consider Weighted-AUC and F-measure together. Matrix (10) (on the left) reports the results of the comparison based on separate tests (each cell contains the result for Weighted-AUC on top and F-measure below it), while Matrix (10) (on the right) regards the Bayesian joint test. There are five cases where the tests are in contradiction (in bold). In particular, in the comparisons C_2 vs. C_5 and C_3 vs. C_5, the Bayesian test asserts that C_5 is jointly better with probability 0.91, while the separate tests do not find a significant dominance. Again for C_4 vs. C_7, it is evident that the joint comparison gives an advantage in power.

$$(10)$$

Finally we consider the three performance measures together. Matrix (11) reports the result of the Bayesian joint test.

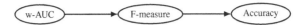

Fig. 1. Three measures used to compare C_4 and C_5 and their (in)dependences.

$$
\begin{array}{c}
\\
C_1 \\
C_2 \\
C_3 \\
C_4 \\
C_5 \\
C_6
\end{array}
\begin{array}{cccccc}
C_2 & C_3 & C_4 & C_5 & C_6 & C_7 \\
\left(\begin{array}{cccccc}
.99 & .99 & 1 & .81 & 1 & 1 \\
 & .31 & 1 & .91 & 1 & 1 \\
 & & 1 & .91 & 1 & 1 \\
 & & & .55 & 1 & 1 \\
 & & & & 1 & 1 \\
 & & & & & 1
\end{array}\right)
\end{array}. \qquad (11)
$$

We can then assert that C_1 is better than C_2 and C_3 jointly on all performance measures. Overall, C_5 appears to be jointly the best classifier followed by C_4. By using the Bayesian network inference to compare C_4 and C_5, we achieve the very same conclusions (results not shown). The interesting outcome of that inference is that we can graphically see the relation between measures in Fig. 1, which is automatically learned from the matrix of measures, and not surprisingly, all three measures of classification accuracy are dependent.

6.3   Comparison Using Six Measures

In this section we compare the same seven classifiers but now using six performance measures jointly: Accuracy, F-measure, weighted-AUC, Kappa statistics, root mean squared error (RMSE), and mean absolute error (MAE). In order to illustrate the capabilities of the proposed approach, let us take on the task of comparing the classifiers C_1 and C_2. By using the BN and the learned conditional (in)dependences displayed in Fig. 2, we obtain the probability of C_1 to be better than C_2 jointly in all six measures to be 0.5, while the value reaches 0.9 without using the BN, which suggests that an unreliable decision could be taken because independent measures where assumed to be dependent (the model without the Bayesian network was learned with very few data, about 80 cases for a parameter space of dimension 63, which is clearly insufficient). From Fig. 2 we see that RMSE and MAE are independent measures with respect to the others and each other. With such information, we can look to their importance separately. Using the Bayesian test for MAE we get a very low probability of 0.54 towards C_2, while RMSE achieves 0.99 towards C_2. The other four connected measures in Fig. 2 achieve probability 0.9999 towards C_1. Hence we are able to identify the source of this difference between the result with the Bayesian network and without it, which clarifies the measures under which one classifier is better than the other. Further applications are numerous, but they go beyond the scope of this work.

Fig. 2. Six measures used to compare classifiers C_1 and C_2 and their dependences.

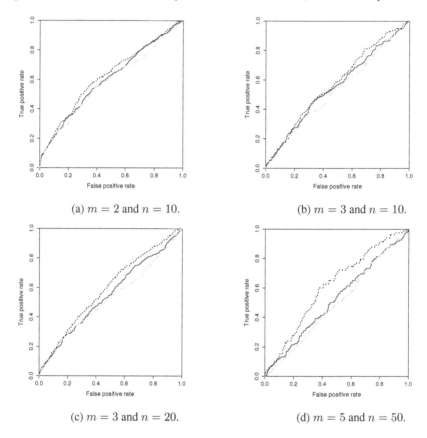

Fig. 3. ROC curves for the GLRT (gray dashed-dotted) and the Bayesian test with (black dashed) and without (black contiguous) the Bayesian network use during learning. Distributions and data (n samples) are generated for a domain with m measures.

6.4   Simulation Study

Finally, we perform a simulated study to understand the benefit of using the Bayesian networks. We study scenarios with m equal to 2, 3 and 5 measures from which we uniformly draw at random the multinomial parameters, that is, $2^2 - 1 = 3$, $2^3 - 1 = 7$ and $2^5 - 1 = 31$ independent parameters, respectively. We label each test case as follows: if the maximum θ is greater than the second greatest plus 0.1%, then this is labeled as a case where there is a difference

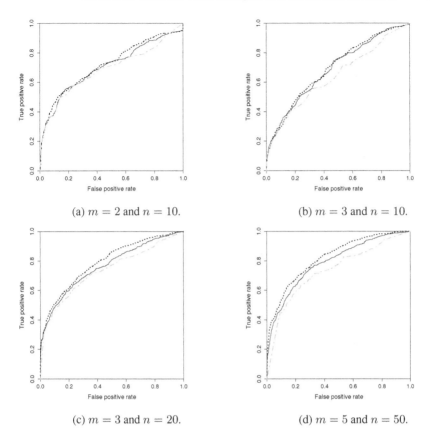

(a) $m = 2$ and $n = 10$.

(b) $m = 3$ and $n = 10$.

(c) $m = 3$ and $n = 20$.

(d) $m = 5$ and $n = 50$.

Fig. 4. ROC curves for the GLRT (gray dashed-dotted) and the Bayesian test with (black dashed) and without (black contiguous) the Bayesian network use during learning. Distributions and data (n samples) are generated for a domain with m measures uniformly at random assuming that all measures are independent from each other.

between the maximum and the others. Otherwise we say the maximum is not greater than the others (and we force the maximum and second greatest to be equal to each other). Then we randomly generate n samples (n =10, 20 or 50) from the distribution and run the GLRT and the Bayesian test with and without the support of the Bayesian network to learn the underlying distribution from data. For each test case, we record the probability that the maximum parameter is greater than the others (or the p-value in the case of the GLRT). This procedure is repeated one thousand times for cases where the maximum is greater (so *positive* cases) and one thousand times with the maximum equal to the second greatest value (*negative* cases). The results over these two thousand test cases are used to build a receiver operating characteristic (ROC) curve according to the usual procedure: True/false positive/negative are defined by varying the threshold for the probability (or respectively the p-value) such that the method takes a decision of whether it is a positive or negative case. In this way, we obtain the percentage (over two thousand test cases) of true positive, true negative, false positive and false negative for each method for each threshold. The curves with the GLRT (gray dashed-dotted) and the Bayesian test with the Bayesian network (black dashed) and without it (black contiguous) are shown in Fig. 3 for different values of m and n. In all cases, the GLRT is equal or inferior to the Bayesian test, and the Bayes test with the Bayesian network version is always equal or superior to the Bayesian test alone. We notice that the curves are barely superior than random guess (which would correspond to the identity function in the ROC graph). This happens because there are too many parameters to learn in the multinomial with respect to the amount of data. We see that with the increase of data ($n = 50$) and use of the Bayesian network to better fit a model, the curve begins to improve with respect to random guesses.

We repeat the experiment but we now assume that the five measures are independent from each other. In this scenario we expect the method with the Bayesian network to be superior than using the full joint distribution, as it can estimate a more appropriate distribution (given the limited amount of data). The idea is that the Bayesian network can learn the fact that the measures are independent (this fact is not disclosed to the methods, as in practice we usually would not know it beforehand). Again we uniformly draw at random the parameters of the multinomial (respecting the independence assumption among all measures), then we draw the data and we label the cases as before. The ROC curves for this scenario are shown in Fig. 4. Again, the Bayesian test with the Bayesian network achieves the best curves.

Table 2 shows the area under the curves for each method and scenario. The values obtained by GLRT are inferior to those of the Bayesian test. The latter has consistently produced better results with the support of the Bayesian network

Table 2. Area under the ROC curve for each method in each scenario. m is the number of measures, n the number of data points over which the measures are compared, and Type describes whether the simulation sampled the parameters without restriction (full) or with the forced assumption that each measure is independent of each other (indep).

m	n	Type	GLRT	Bayesian test	Bayesian test + BN
2	10	indep	0.686	0.703	0.715
2	10	full	0.583	0.601	0.622
3	10	indep	0.641	0.688	0.694
3	10	full	0.530	0.555	0.577
3	20	indep	0.735	0.764	0.791
3	20	full	0.524	0.549	0.590
5	50	indep	0.735	0.790	0.822
5	50	full	0.500	0.522	0.613

for learning the distribution. The superiority of the method with the Bayesian network is justified by the better estimation of the joint distribution with its underlying independence assessments.

7   Conclusions

In machine learning and artificial intelligence, a very important task is to compare the performance of algorithms on different case studies and to use multiple different performance measures. This is typically performed using statistical tests. In this paper, we have developed new statistical tests that are able to compare the algorithms considering all the performance measures jointly. This allows for example to make statements such as a classifier is jointly better than another on multiple measures as well as on particular subsets of measures, which can be identified with the use of a Bayesian network modeling the (in)dependences among measures. With artificial and real-data examples we have shown that the decisions derived by a joint test can be very different from the decisions carried out using a separate test for each performance measure. We argue that the ideas developed here can offer a new way for comparing algorithms using multiple performance measures. Future work includes the exploration of applications and the further use of the Bayesian network structure to understand the relations between performance measures and their importance for the evaluation of algorithms. Moreover, we plan to extend this approach to be able to compare multiple measures on multiple algorithms at the same time.

References

1. Dehuri, S., Cho, S.-B.: Multi-criterion pareto based particle swarm optimized polynomial neural network for classification: a review and state-of-the-art. Comput. Sci. Rev. 3(1), 19–40 (2009)
2. Cai, W., Chen, S., Zhang, D.: A multiobjective simultaneous learning framework for clustering and classification. IEEE Trans. Neural Netw. 21(2), 185–200 (2010)
3. Shi, C., Kong, X., Philip, S.Y., Wang, B.: Multi-objective multi-label classification. In: SIAM International Conference on Data Mining, pp. 355–366. SIAM (2012)
4. Hsiao, K.J., Xu, K., Calder, J., Hero, A.O.: Multi-criteria anomaly detection using pareto depth analysis. In: Advances in Neural Information Processing Systems, vol. 25, pp. 845–853. Curran Associates Inc (2012)
5. Demšar, J.: Statistical comparisons of classifiers over multiple data sets. J. Mach. Learn. Res. 7, 1–30 (2006)
6. Wilks, S.S.: The large-sample distribution of the likelihood ratio for testing composite hypotheses. Ann. Math. Stat. 9, 60–62 (1938)
7. Rice, J.: Mathematical Statistics and Data Analysis. Cengage Learning, Belmont (2006)
8. de Campos, Cassio P., Tong, Yan, Ji, Qiang: Constrained maximum likelihood learning of bayesian networks for facial action recognition. In: Forsyth, David, Torr, Philip, Zisserman, Andrew (eds.) ECCV 2008, Part III. LNCS, vol. 5304, pp. 168–181. Springer, Heidelberg (2008)

9. DasGupta, A.: Asymptotic Theory of Statistics and Probability, 1st edn. Springer, New York (2008)
10. Walley, P.: Inferences from multinomial data: learning about a bag of marbles. J. R. Statist. Soc. B **58**(1), 3–57 (1996)
11. Buntine, W.: Theory refinement on Bayesian networks. In: Conference on Uncertainty in Artificial Intelligence, pp. 52–60. Morgan Kaufmann (1991)
12. Cooper, G.F., Herskovits, E.: A bayesian method for the induction of probabilistic networks from data. Mach. Learn. **9**, 309–347 (1992)
13. Heckerman, D., Geiger, D., Chickering, D.M.: Learning Bayesian networks: the combination of knowledge and statistical data. Mach. Learn. **20**, 197–243 (1995)
14. de Campos, C.P., Ji, Q.: Roperties of Bayesian Dirichlet scores to learn Bayesian network structures. In: AAAI Conference on Artificial Intelligence, pp. 431–436. AAAI Press, 2010
15. Silander, T., Myllymaki, P.: A simple approach for finding the globally optimal bayesian network structure. In: Conference on Uncertainty in Artificial Intelligence, pp. 445–452. AUAI (2006)
16. Barlett, M., Cussens, J.: Advances in Bayesian network learning using integer programming. In: Conference on Uncertainty in Artificial Intelligence, pp. 182–191. AUAI (2013)
17. de Campos, C.P., Ji, Q.: Efficient structure learning of Bayesian networks using constraints. J. Mach. Learn. Res. **12**, 663–689 (2011)
18. Yuan, C., Malone, B.: Learning optimal Bayesian networks: a shortest path perspective. J. Artif. Intell. Res. **48**, 23–65 (2013)
19. Hall, M., Frank, E., Holmes, G., Pfahringer, B., Reutemann, P., Witten, I.H.: The weka data mining software: an update. ACM SIGKDD Explorations Newsletter **11**(1), 10–18 (2009)

Extending Naive Bayes Classifier with Hierarchy Feature Level Information for Record Linkage

Yun Zhou$^{(\boxtimes)}$, John Howroyd, Sebastian Danicic, and J. Mark Bishop

Tungsten Centre for Intelligent Data Analytics (TCIDA),
Goldsmiths, University of London, London, UK
{y.zhou,j.howroyd,s.danicic,m.bishop}@gold.ac.uk

Abstract. Probabilistic record linkage has been well investigated in recent years. The Fellegi-Sunter probabilistic record linkage and its enhanced version are commonly used methods, which calculate match and non-match weights for each pair of corresponding fields of record-pairs. Bayesian network classifiers – naive Bayes classifier and TAN have also been successfully used here. Very recently, an extended version of TAN (called ETAN) has been developed and proved superior in classification accuracy to conventional TAN. However, no previous work has applied ETAN in record linkage and investigated the benefits of using a naturally existing hierarchy feature level information. In this work, we extend the naive Bayes classifier with such information. Finally we apply all the methods to four datasets and estimate the F_1 scores.

Keywords: Probabilistic record linkage · Naive Bayes classifier · TAN and ETAN · Hierarchy feature level information

1 Introduction

Record linkage (RL) [1] proposed by Halbert L. Dunn (1946) refers to the task of finding records that refer to the same entity across different data sources. These records contain identifier fields (e.g. name, address, time, postcode etc.). The simplest kind of record linkage, called deterministic or rules-based record linkage, requires all or some identifiers are identical giving a deterministic record linkage procedure. This method works well when there exists a common/key identifier in the dataset. However, in *real world* applications, deterministic record linkage is problematic because of the incompleteness and privacy protection [2] of the key identifier field.

To mitigate against this problem, probabilistic record linkage (also called fuzzy matching) is developed, which takes a different approach to the record linkage problem by taking into account a wider range of potential identifiers. This method computes weights for each identifier based on its estimated ability to correctly identify a match or a non-match, and uses these weights to calculate

Y. Zhou—The authors would like to thank the Tungsten Network for their financial support.

J. Suzuki and M. Ueno (Eds.): AMBN 2015, LNAI 9505, pp. 93–104, 2015.
DOI: 10.1007/978-3-319-28379-1_7

a score (usually *log-likelihood* ratio) that two given records refer to the same entity.

Record-pairs with scores above a certain threshold are considered to be matches, while pairs with scores below another threshold are considered to be non-matches; pairs that fall between these two thresholds are considered to be "possible matches" and can be dealt with accordingly (e.g., human reviewed, linked, or not linked, depending on the requirements). Whereas deterministic record linkage requires a series of potentially complex rules to be programmed ahead of time, probabilistic record linkage methods can be *trained* to perform well with much less human intervention.

The Fellegi-Sunter probabilistic record linkage (PRL-FS) [3] is one of the most commonly used methods. It assigns the match/non-match weight for each corresponding field of record-pairs based on *log-likelihood* ratios. For each record-pair, a composite weight is computed by summing each field's match or non-match weight. When a field agrees (the contents of the field are the same), the field match weight is used for computing the composite weight; otherwise the non-match weight is used. The resulting composite weight is then compared to the aforementioned thresholds to determine whether the record-pair is classified as a match, possible match (hold for clerical review) or non-match. Determining where to set the match/non-match thresholds is a balancing act between obtaining an acceptable sensitivity (or recall, the proportion of truly matching records that are classified match by the algorithm) and positive predictive value (or precision, the proportion of records classified match by the algorithm that truly do match).

In PRL-FS method, a match weight will only be used when two strings exactly agree in the field. However, in many *real world* problems, even two strings describing the same field may not exactly (character-by-character) agree with each other because of typographical error (mis-spelling). For example, the field (first name) comparisons such as (*Andy, Andrew*) and (*Andy, John*) are both treated as non-match in PRL-FS even though the terms *Andy* and *Andrew* are more likely to refer to the same person. Moreover, such mis-spellings are not uncommon according to the research results [4] of US Census Bureau, which show that 25 % of first names did not agree character-by-character among medical record-pairs that are from the same person. To obtain a better performance in *real world* usage, Winkler proposed an enhanced PRL-FS method (PRL-W) [5] that takes into account field similarity (similarity of two strings for a field within a record-pair) in the calculation of field weights, and showed better performance of PRL-W compared to PRL-FS [6].

Probabilistic graphical models for classification such as naive Bayes (NBC) and tree augmented naive Bayes (TAN) are also used for record linkage [7], where the single class variable contains two states: match and non-match. These models can be easily improved with domain knowledge. For example, monotonicity constraints (i.e. a higher field similarity value indicating a higher degree of 'match') can be incorporated to help reduce overfitting in classification [8]. Recently, a state-of-the-art Bayesian network classifier called ETAN [9,10] has been proposed and shown outperform the NBC and TAN in many cases. ETAN relaxes the assumption about independence of features, and does not require features to be connected to the class.

In this paper we will apply ETAN to the probabilistic record linkage problem. Also we will extend the naive Bayes classifier (referred to as HR-NBC) by introducing hierarchy restrictions between features. As discussed in previous work [11,12], these hierarchy restrictions are very useful to avoid unnecessary computation of field comparison, and to help refine the Bayesian network structure.

In our model, such hierarchy restrictions are mined from the semantic relationships between features, which widely exist in *real world* record matching problems. An example of this occurs especially in address matching. For example, two restaurants with the same name located in two cities are more likely to be recognized as two different restaurants. Because they might be two different branches in two cities. In this case, the city locations have higher importance than the restaurant names. And we can introduce a connection between these two features.

To deal with mis-spellings in records, we use the Jaro-Winkler similarity function to measure the differences between fields of two records. These field difference values and known record linkage labels are used to train the classifier. Finally, we compare all the methods – PRL-W, TAN, ETAN, NBC and HR-NBC in four datasets. The results show the benefits of using different methods under different settings.

2   Probabilistic Record Linkage

2.1   PRL-FS and PRL-W

Let us assume that there are two datasets A and B of n-tuples of elements from some set F. (In practice F will normally be a set of a strings.) Given an n-tuple a we write a_i for the i-th component (or field) of a.

Matching. If an element of $a \in A$ is the representation of the same object as represented by an element of $b \in B$ we say a *matches* b and write $a \sim b$. Some elements of A and B match and others do not. If a and b do not match we write $a \nsim b$. We write $M = \{(a, b) \in A \times B | a \sim b\}$ and $U = \{(a, b) \in A \times B | a \nsim b\}$. The problem is then, given an element x in $A \times B$ to define an algorithm for deciding whether or not $x \in M$.

Comparison Functions on Fields. We assume the existence of a function:

$$cf : F \times F \to [0, 1].$$

With the property that $\forall h \in F, cf(h, h) = 1$. We think of cf as a measure of *how similar* two elements of F are. Many such functions exist on strings including the normalised Levenshtein distance or Jaro-Winkler. In conventional PRL-FS method, its output is either 0 (non-match) or 1 (match). In PRL-W method, a field similarity score (Jaro-Winkler distance [5, 13]) is calculated, and normalized between from 0 and 1 to show the degree of match.

Discretisation of Comparison Function. Same as previous work [6], rather than concern ourselves with the *exact* value of $cf(a_i, b_i)$ we consider a set of $I_1, \cdots I_s$ of disjoint ascending intervals exactly covering the closed interval $[0, 1]$. These intervals are called *states*. We say $cf(a_i, b_i)$ is in state k to mean $cf(a_i, b_i) \in I_k$.

Given an interval I_k and a record-pair (a, b) we define two values[1]:

- $m_{k,i}$ is the probability that $cf(a_i, b_i) \in I_k$ given that $a \sim b$.
- $u_{k,i}$ is the probability that $cf(a_i, b_i) \in I_k$ given that $a \not\sim b$.

Given a pair (a, b), the *weight* $w_i(a, b)$ of their i-th field is defined as:

$$w_i(a, b) = \sum_{k=1}^{s} w_{k,i}(a, b)$$

where

$$w_{k,i}(a, b) = \begin{cases} \ln(\frac{m_{k,i}}{u_{k,i}}) & \text{if } cf(a_i, b_i) \in I_k \\ \ln(\frac{1-m_{k,i}}{1-u_{k,i}}) & \text{otherwise.} \end{cases}$$

The *composite weight* $w(a, b)$ for a given pair (a, b) is then defined as

$$w(a, b) = \sum_{i=1}^{n} w_i(a, b).$$

2.2   The E-M Estimation of Parameters

In practice, the set M, the set of matched pairs, is unknown. Therefore, the values $m_{k,i}$, and $u_{k,i}$, defined above, are also unknown. To accurately estimate these parameters, we applied the expectation maximization (EM) algorithm with randomly sampled initial values for all these parameters.

The Algorithm

1. Choose a value for p, the probability that an arbitrary pair in $A \times B$ is a match.
2. Choose values for each of the $m_{k,i}$ and $u_{k,i}$, defined above.
3. *E-step*: For each pair (a, b) in $A \times B$ compute

---

[1] Note in conventional PRL-FS method [3], two fields are either matched or unmatched. Thus the k of $m_{k,i}$ can be omitted in this case.

$$g(a,b) = \frac{p \prod\limits_{(a,b)\in A\times B} \prod\limits_{k=1}^{s} m'_{k,i}(a,b)}{p \prod\limits_{(a,b)\in A\times B} \prod\limits_{k=1}^{s} m'_{k,i}(a,b) \quad + \quad (1-p) \prod\limits_{(a,b)\in A\times B} \prod\limits_{k=1}^{s} u'_{k,i}(a,b)} \quad (1)$$

where

$$m'_{k,i}(a,b) = \begin{cases} m_{k,i} & \text{if } cf(a_i,b_i) \in I_k \\ 1 & \text{otherwise.} \end{cases}$$

and

$$u'_{k,i}(a,b) = \begin{cases} u_{k,i} & \text{if } cf(a_i,b_i) \in I_k \\ 1 & \text{otherwise.} \end{cases}$$

4. *M-step*: Then recompute $m_{k,i}$, $u_{k,i}$, and p as follows:

$$m_{k,i} = \frac{\sum\limits_{(a,b)\in A\times B} g'_{k,i}(a,b)}{\sum\limits_{(a,b)\in A\times B} g(a,b)}, \quad u_{k,i} = \frac{\sum\limits_{(a,b)\in A\times B} \tilde{g}'_{k,i}(a,b)}{\sum\limits_{(a,b)\in A\times B} 1 - g(a,b)}, \quad p = \frac{\sum\limits_{(a,b)\in A\times B} g(a,b)}{|A \times B|}$$

$$(2)$$

where

$$g'_{k,i}(a,b) = \begin{cases} g(a,b) & \text{if } cf(a_i,b_i) \in I_k \\ 0 & \text{otherwise.} \end{cases}$$

and

$$\tilde{g}'_{k,i}(a,b) = \begin{cases} 1 - g(a,b) & \text{if } cf(a_i,b_i) \in I_k \\ 0 & \text{otherwise.} \end{cases}$$

In usage, we iteratively run the E-step and M-step until the convergence criteria are satisfied: $\sum(|\Delta m_{k,i}|) \le 1 \times 10^{-8}$, $\sum(|\Delta u_{k,i}|) \le 1 \times 10^{-8}$, and $|\Delta p| \le 1 \times 10^{-8}$. Having obtained values for $m_{k,i}$ and $u_{k,i}$. We can then compute the composite weight (the natural logarithm of $g(a,b)$) for each pair defined earlier.

In our implementation, we set the decision threshold as 0.5, and do not consider possible matches. Because using a domain expert to manually examine these possible matches is expensive. Thus, the record-pair (a,b) is recognized as a match if $g(a,b) > 0.5$; otherwise it is a non-match.

3   Bayesian Network Classifiers for Record Linkage

In this section we discuss different Bayesian network classifiers (NBC, TAN and ETAN) for record linkage. After that, we discuss the hierarchy structure between features, and the proposed hierarchy-restricted naive Bayes classifier (HR-NBC).

3.1   The Naive Bayes Classifier

Let record-pair feature vector $\overrightarrow{f}$ be an input vector[2] to the classifier, and C_k be a possible class of the binary variable C, where $C_1 = 0$ indicates a non-match and $C_2 = 1$ indicates a match. The model calculates the probability of C_k given the feature values (distance for each field-pair). This can be formulated as:

$$P(C_k|\overrightarrow{f}) = P(C_k) \times \frac{P(\overrightarrow{f}|C_k)}{P(\overrightarrow{f})} \tag{3}$$

In the naive Bayes classifier (Fig. 1(a)), we assume the conditional independence of features, $P(\overrightarrow{f}|C_k)$ can be decomposed as $P(\overrightarrow{f}|C_k) = \prod_{i=1}^{n} P(f_i|C_k)$. Thus, Eq. (3) becomes:

$$P(C_k|\overrightarrow{f}) = P(C_k) \times \frac{\prod_{i=1}^{n} P(f_i|C_k)}{P(\overrightarrow{f})} \tag{4}$$

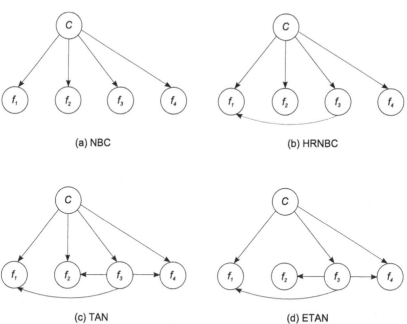

Fig. 1. The graphical representation of NBC, HR-NBC, TAN, ETAN. The blue arrow represents the dependency introduced by hierarchy feature level information.

---

[2] Here $\overrightarrow{f} = \{f_i | f_i = I_k, i = 1, ..., n\}$ contains n elements, whose values indicate the distances between two records on specific fields, I_k is the state/interval discretised from $cf(a_i, b_i)$.

With this equation, we can calculate $P(C_k|\vec{f})$ to classify $\vec{f}$ into the class (match/non-match) with the highest $P(C_k|\vec{f})$. This approach is one of the baseline methods we compare our model to.

Like the probabilistic record linkage, one of the often-admitted weaknesses of this approach is that it depends upon the assumption that each of its fields is independent from the others. The tree augmented naive Bayes classier (TAN) and its improved version ETAN relax this assumption by allowing interactions between feature fields.

3.2   The Tree Augmented Naive Bayes Classifier

TAN [14] can be seen as an extension of the naive Bayes classifier by allowing a feature as a parent (Fig. 1(c)). In NBC, the network structure is naive, where each feature has the class as the only parent. In TAN, the dependencies between features are learnt from the data. Given a complete data set $D = \{D_1, ..., D_L\}$ with L labelled instances, where each instance is an instantiation of all the variables. Conventional score-based algorithms for structure learning make use of certain heuristics to find the optimal DAG that best describes the observed data D over the entire space. We define:

$$\hat{G} = \arg\max_{G \in \Omega} \ell(G, D) \qquad (5)$$

where $\ell(G, D)$ is the *log-likelihood* score, which is the logarithm of the likelihood function of the data that measures the fitness of a DAG G to the data D. Ω is a set of all DAGs.

Assume that the score (i.e. BDeu score [15]) is decomposable and respects likelihood equivalence, we can devise an efficient structure learning algorithm for TAN. Because every feature f_i has C as a parent, the structure (f_i has f_j and C as parents, $i \neq j$) has the same score with the structure, where f_j has f_i and C as parents:

$$\ell(f_i, \{f_j, C\}, D) + \ell(f_j, C, D) = \ell(f_j, \{f_i, C\}, D) + \ell(f_i, C, D) \qquad (6)$$

Beside the naive Bayes structure, in the TAN, features are only allowed to have at most one other feature as a parent. Thus, we have a tree structure between the features. Based on the symmetry property (Eq. (6)), we can have an efficient algorithm to find the optimal TAN structure by converting the original problem (Eq. (5)) into a minimum spanning tree construction problem. More details could be found in [9].

3.3   The Extended TAN Classifier

As discussed in the previous section, the TAN encodes a tree structure over all the attributes. And it has been shown to outperform the naive Bayes classifier in a range of experiments [14]. However, when the training data are scarce or a feature and the class are conditionally independent given another feature, we might not get a TAN structure. Therefore, people have proposed the Extended TAN (ETAN) classifier [9,10] to allow more structure flexibility.

ETAN is a generalization of TAN and NBC. It does not force a tree to cover all the attributes, and a feature to connect with the class. As shown in Fig. 1(d), ETAN could disconnect a feature if such a feature is not important to predict C. Thus, ETAN's search space of structures includes that of TAN and NBC, and we have:

$$\ell(\hat{G}_{ETAN}, D) \geq \ell(\hat{G}_{TAN}, D) \quad and \quad \ell(\hat{G}_{TAN}, D) \geq \ell(\hat{G}_{NBC}, D) \qquad (7)$$

which means the score of the optimal ETAN structure is superior or equal to that of the optimal TAN and NBC (Lemma 2 in [9]).

In the ETAN, the symmetry property (Eq. (6)) does not hold, because a feature (e.g. f_2 in Fig. 1(d)) is allowed to be disconnected from the class. Thus, the undirected version of minimum spanning tree algorithm cannot be directly applied here. Based on Edmonds' algorithm for finding minimum spanning trees in directed graphs, people developed the structure learning algorithm of ETAN, whose computational complexity is quadratic in the number of features (as is TAN). For detailed discussions we direct the reader to the papers [9,10].

3.4 Hierarchy Restrictions Between Features

To utilize the benefits of existing domain knowledge, we extend the NBC method by allowing hierarchy restrictions between features (HR-NBC). These restrictions are modelled as dependencies between features in HR-NBC.

Hierarchy restrictions between features commonly occur in *real world* problems. For example, Table 1 shows four address records, which refer to two restaurants (there are two duplicates). The correct linkage for these four records is: (1) record 1 and 2 refer to one restaurant in Southwark, and (2) record 3 and 4 refer to another restaurant in Blackheath. As we can see, even record 1 and 3 exactly match with each other in the field of restaurant name, they cannot be linked with each other because they are located in a different borough.

Based on the description of the example Table 1, we can see there is a hierarchy restriction between the *name* and *borough* fields, where the *borough* field has higher feature level than *name* field. Thus, intuitively, it is recommended to compare the *borough* field first to filter record linkage pairs. To let our classifier

Table 1. Four restaurant records with name, address, borough/town and type information.

Index	Name (f_1)	Address (f_2)	Borough (f_3)	Type (f_4)
1	Strada	Unit 6, RFH Belvedere Rd	Southwark	Roman
2	Strada at Belvedere	Royal Festival Hall	Southwark	Italian
3	Strada	5 Lee Rd	Blackheath	Italian
4	Strada at BH	5 Lee Road	BLACKHEATH	Italian

capture such hierarchy restriction, we introduce a dependency between these two fields ($f_3 \rightarrow f_1$) to form our HR-NBC model (Fig. 1(b)). Thus, Eq. (4) now becomes:

$$P(C_k|\overrightarrow{f}) = P(C_k) \times \frac{P(f_1|f_3, C_k)\prod_{i=2}^{n} P(f_i|C_k)}{P(\overrightarrow{f})} \qquad (8)$$

Parameter Estimation. Let θ denote the parameters that need to be learned in the classifier and let r be a set of fully observable record-pairs. The classical maximum likelihood estimation (MLE) finds the set of parameters that maximize the data *log-likelihood* $\ell(\theta|r) = \log P(r|\theta)$.

However, for several cases in the unified model, a certain parent-child state combination would seldom appear, and the MLE learning fails in this situation. Hence, maximum a posteriori algorithm (MAP) is used to mediate this problem via the *Dirichlet* prior: $\hat{\theta} = \arg\max_\theta \log P(r|\theta)P(\theta)$. Because there is no informative prior, in this work we use the BDeu prior [15] with equivalent sample size (ESS) equal to 1.

4   Experiments

This section compares PRL-W to different Bayesian network classifiers. The goal of the experiments is to do an empirical comparison of the different methods, and show the advantages/disadvantages of using different methods in different settings. Also, it is of interest to investigate how such hierarchy feature level information could improve the classifier's performance.

4.1   Settings

Our experiments are performed on four different datasets[3], two synthetic datasets [12] (*Country* and *Company*) with sampled spelling errors and two real datasets (*Restaurant* and *Tungsten*). The *Country* and *Company* datasets contain 9 and 11 fields/features respectively. All the field similarities are calculated by the Jaro-Winkler similarity function.

Restaurant is a standard dataset for record linkage study [8]. It was created by merging the information of some restaurants from two websites. In this dataset, each record contains 5 fields: name, address, city, phone and restaurant-type[4].

Tungsten is a commercial dataset from an e-invoicing company named Tungsten Corporation. In this dataset, there are 2744 duplicates introduced by user entry errors. Each record contains 5 fields: company name, country code, address line 1, address line 4 and address line 6.

---

[3] These datasets can be found at http://yzhou.github.io/.

[4] Because the phone number is unique for each restaurant, it, on its own, can be used to identify duplicates without the need to resort to probabilistic record linkage techniques. Thus, this field is not used in our experiments.

The experiment platform is based on the Weka system [16]. Since TAN and ETAN can not deal with continuous field similarity values, these values are discretised with the same routine as described in PRL-W. To simulate *real world* situation, we use an affordable number (10, 50 and 100) of labelled records as our training data. The reason is clear that it would be very expensive to manually label hundreds of records. The experiments are repeated 100 times in each setting, and the results are reported with the mean.

To evaluate the performance of different methods, we compare their ability to reduce the number of *false decisions*. False decisions include **false matches** (the record-pair classified as a match for two different records) and **false non-matches** (the record-pair classified as a non-match for two records that are originally same). Thus these methods are expected to get high *precision* and *recall*, where *precision* is the number of correct matches divided by the number of all classified matches, and *recall* is the number of correct matches divided by the number of all original matches.

To consider both the *precision* and *recall* of the test, in this experiment, we use F_1 score as our evaluation criteria. This score reaches its best value at 1 and worst at 0, and is computed as follows:

$$F_1 = 2 \times \frac{precision \times recall}{precision + recall} \tag{9}$$

4.2 Results

The F_1 score of all five methods in different scenarios are shown in Table 2, where the highest average score in each setting is marked bold. Statistically significant improvements of the best result over competitors are indicated with asterisks * ($p = 0.05$).

As we can see, the PRL-W gets the best result in *Company* and *Restaurant* datasets. And its performance does not depends on the number of labelled data. The reason is the record linkage weights were computed with an EM-algorithm as described in Eqs. (1) and (2) over the whole dataset (labelled and unlabelled data). When two classes are easy to distinguish, it is not surprising that the PRL-W could get good performance with limited labelled data.

Because the scarce labelled data and large number of features, TAN and the state-of-the-art ETAN methods have relatively bad performances in *Country* and *Company* datasets. Although it is proven that ETAN provides higher fit to the data (Eq. (7)) than TAN, it receives lower classification accuracies in most settings due to overfitting. In the *Tungsten* dataset, TAN gets the best performance.

According to the results, both NBC and HR-NBC get high F_1 scores in all settings. This demonstrates the benefits of using these two methods when the labelled data is scarce. Moreover, the performance of our HR-NBC[5] is equal or superior to that of NBC in all settings.

---

[5] In each dataset, we only introduce one hierarchy restriction between the *name* and *address* fields.

Table 2. The F_1 score of five record linkage methods in different datasets.

Dataset	L	PRL-W	TAN	ETAN	NBC	HR-NBC
Country	10	**0.974**	0.920*	0.899*	0.938*	0.941*
	50	0.971*	0.970*	0.967*	**0.976**	**0.976**
	100	0.967*	0.977*	0.978	0.980	**0.981**
Company	10	**0.999**	0.969*	0.965*	0.987*	0.988*
	50	**0.999**	0.995*	0.992*	0.997*	0.997*
	100	**0.999**	0.997*	0.996*	0.998	**0.999**
Restaurant	10	**0.996**	0.874*	0.863*	0.884*	0.897*
	50	**0.996**	0.950*	0.952*	0.957*	0.958*
	100	**0.995**	0.957*	0.958*	0.959*	0.960*
Tungsten	10	0.872	**0.878**	0.877	**0.878**	0.877
	50	0.873*	**0.904**	0.900	**0.904**	**0.904**
	100	0.873*	**0.914**	0.911	0.911*	0.912

*$p = 0.05$

5   Conclusions

In this paper, we discussed the hierarchy restrictions between features, and exploited the classification performance of different methods for record linkage on both synthetic and real datasets.

Results demonstrate that, in settings of limited labelled data, PRL-W works well and its performance is independent of the number of labelled data, and show that TAN, NBC and HR-NBC have better performances than ETAN even though the latter method provides theoretically better fit to the data. Compared with NBC, HR-NBC achieves equal or superior performances in all settings, which show the benefits of introducing hierarchy restrictions between features in these datasets.

We note, however, that our method might not be preferable in all cases. For example, in a medical dataset, a patient could move her or his address and have multiple records. In this case, two records with different addresses refer to the same person. Thus, the hierarchy restrictions used in this paper will introduce extra false non-matches.

In future work we will investigate other sources of domain knowledge to enhance the performance of the resultant classifier, such as improving accuracy by using specific parameter constraints [17] elicited from experts.

References

1. Dunn, H.L.: Record linkage*. Am. J. Public Health Nations Health **36**(12), 1412–1416 (1946)
2. Tromp, M., Ravelli, A.C., Bonsel, G.J., Hasman, A., Reitsma, J.B.: Results from simulated data sets: probabilistic record linkage outperforms deterministic record linkage. J. Clin. Epidemiol. **64**(5), 565–572 (2011)

3. Fellegi, I.P., Sunter, A.B.: A theory for record linkage. J. Am. Stat. Assoc. **64**(328), 1183–1210 (1969)
4. Winkler, W.E.: The state of record linkage and current research problems. In: Statistical Research Division, US Census Bureau, Citeseer (1999)
5. Winkler, W.E.: String comparator metrics and enhanced decision rules in the Fellegi-Sunter model of record linkage. In: Proceedings of the Section on Survey Research, pp. 354–359 (1990)
6. Li, X., Guttmann, A., Cipiere, S., Maigne, L., Demongeot, J., Boire, J.Y., Ouchchane, L.: Implementation of an extended Fellegi-Sunter probabilistic record linkage method using the Jaro-Winkler string comparator. In: 2014 IEEE-EMBS International Conference on Biomedical and Health Informatics (BHI), pp. 375–379. IEEE (2014)
7. Elmagarmid, A.K., Ipeirotis, P.G., Verykios, V.S.: Duplicate record detection: a survey. IEEE Trans. Knowl. Data Eng. **19**(1), 1–16 (2007)
8. Ravikumar, P., Cohen, W.W.: A hierarchical graphical model for record linkage. In: Proceedings of the 20th Conference on Uncertainty in Artificial Intelligence, pp. 454–461. AUAI Press (2004)
9. de Campos, C.P., Zaffalon, M., Corani, G., Cuccu, M.: Extended tree augmented naive classifier. In: van der Gaag, L.C., Feelders, A.J. (eds.) PGM 2014. LNCS, vol. 8754, pp. 176–189. Springer, Heidelberg (2014)
10. de Campos, C.P., Corani, G., Scanagatta, M., Cuccu, M., Zaffalon, M.: Learning extended tree augmented naive structures. Int. J. Approximate Reasoning **68**, 153–163 (2016)
11. Ananthakrishna, R., Chaudhuri, S., Ganti, V.: Eliminating fuzzy duplicates in data warehouses. In: Proceedings of the 28th International Conference on Very Large Data Bases, pp. 586–597. VLDB Endowment (2002)
12. Leitao, L., Calado, P., Herschel, M.: Efficient and effective duplicate detection in hierarchical data. IEEE Trans. Knowl. Data Eng. **25**(5), 1028–1041 (2013)
13. Jaro, M.A.: Advances in record-linkage methodology as applied to matching the 1985 census of Tampa, Florida. J. Am. Stat. Assoc. **84**(406), 414–420 (1989)
14. Friedman, N., Geiger, D., Goldszmidt, M.: Bayesian network classifiers. Mach. Learn. **29**(2–3), 131–163 (1997)
15. Heckerman, D., Geiger, D., Chickering, D.M.: Learning Bayesian networks: the combination of knowledge and statistical data. Mach. Learn. **20**(3), 197–243 (1995)
16. Hall, M., Frank, E., Holmes, G., Pfahringer, B., Reutemann, P., Witten, I.H.: The weka data mining software: an update. ACM SIGKDD Explor. Newslett. **11**(1), 10–18 (2009)
17. Zhou, Y., Fenton, N., Neil, M.: Bayesian network approach to multinomial parameter learning using data and expert judgments. Int. J. Approximate Reasoning **55**(5), 1252–1268 (2014)

Empirical Behavior of Bayesian Network Structure Learning Algorithms

Brandon Malone[(✉)]

Max Planck Institute for the Biology of Ageing, Cologne, Germany
brandon.malone@age.mpg.de

Abstract. Bayesian network structure learning (BNSL) is the problem of finding a BN structure which best explains a dataset. Score-based learning assigns a score to each network structure. The goal is to find the structure which optimizes the score. We review two recent studies of empirical behavior of BNSL algorithms.

The score typically reflects fit to a training dataset; however, models which fit training data well may generalize poorly. Thus, it is not clear that finding an optimal network is worthwhile. We review a comparison of exact and approximate search techniques. Sometimes, approximate algorithms suffice; for complex datasets, the optimal algorithms produce better networks.

BNSL is known to be NP-hard, so exact solvers prune the search space using heuristics. We next review problem-dependent characteristics which affect their efficacy. Empirical results show that machine learning techniques based on these characteristics can often be used to accurately predict the algorithms' running times.

Keywords: Bayesian networks · Structure learning · Algorithm selection · Empirical hardness

1 Introduction

Bayesian networks (BNs) (Pearl 1988) are a widely-used formalism for representing uncertain relationships among variables in a domain of interest. In some cases, domain experts can specify these relationships as a BN structure; however, when they are unknown, we must learn the structure from data.

In the commonly-used *score-based* framework (Heckerman et al. 1995), a score is assigned to each structure. The score is typically a penalized log-likelihood which trades off the fit of a BN to the data with the complexity of the structure. The BN structure learning problem (BNSL) is then cast as an optimization problem in which the goal is to find a BN structure with an optimal score.

BNSL is known to be NP-hard (Chickering 1996), so early optimization algorithms (such as Cooper and Herskovits (1992), Heckerman et al. (1995),

---

B. Malone—This paper is based on Malone et al. (2014, 2015), with co-authors Matti Järvisalo, Petri Myllymäki, Kusta Kangas and Mikko Koiviso from HIIT and the Department of Computer Science at the University of Helsinki.

J. Suzuki and M. Ueno (Eds.): AMBN 2015, LNAI 9505, pp. 105–121, 2015.
DOI: 10.1007/978-3-319-28379-1_8

Friedman et al. (1999), Chickering (2002), Moore and Wong (2003), Teyssier and Koller (2005), Tsamardinos et al. (2006)) used local search techniques. However, these algorithms suffer from the same problem faced by all local search techniques: the quality of the found solution relative to an optimal one is unknown. Consequently, a variety of algorithms have been proposed which solve the problem exactly (Ott et al. 2004; Koivisto and Sood 2004; Silander and Myllymäki 2006; Parviainen and Koivisto 2009; de Campos and Ji 2011; Yuan and Malone 2013; Bartlett and Cussens 2015; van Beek and Hoffmann 2015).

Since BNSL is NP-hard, the exact algorithms have exponential worst-case behavior. Nevertheless, many of the algorithms employ sophisticated heuristics, such as branch-and-bound techniques, to provably rule out many possible structures. In practice, these algorithms can learn provably optimal networks for modestly-sized datasets; in general, optimal networks on the order of 50 variables can be learned with reasonable resources (Malone et al. 2014).

The score of a BN structure is ideally a reflection of how well it models a training dataset. The general assumption has been that networks which model the training data well also accurately reflect new data. However, it is well-known that a model can describe a training set very well, yet generalize poorly to new data (Mitchell 1997). Thus, there is no guarantee that a network which optimizes a score for a training set will generalize well to new data.

Until a recent study (Malone et al. 2015), there was no clear empirical evidence on whether the increased computational efforts required by exact approaches to BNSL are justifiable in terms of generalization to unseen testing data. As the first half of this paper, we review that work, which shows that for some datasets, simple strategies such as greedy hill climbing can provide good generalization. However, the simple strategies fail to generalize well on other datasets. Predictive likelihood results show that the optimal algorithms consistently generalize well.

Because of their guarantees, all of the exact algorithms find optimal, equivalent networks. So, in terms of generalization, these algorithms are equivalent. As previously mentioned, though, the algorithms use sophisticated, and very different, heuristics to find the optimal network and prove its optimality. In terms of resource requirements, then, specific implementations of these algorithms, *solvers*, are very different (van Beek and Hoffmann 2015).

For the second half of this work, we review a study (Malone et al. 2014) which shows that machine learning techniques can learn a simple, yet nontrivial, model that accurately predicts the fastest solver for a given instance. Additional features are shown to capture the hardness of an instance more accurately. Models with the additional features significantly improve prediction accuracy.

The rest of this paper is structured as follows. In Sect. 2, we formally introduce Bayesian networks and BNSL. Section 3 provides an overview of the specific solvers used in this work, while Sect. 4 outlines the datasets used. Generalization of learned networks is reviewed in Sect. 5, and Sect. 6 reviews results on exact solver behavior. Finally, Sect. 7 concludes the paper.

2   Background

A Bayesian network (Pearl 1988) is a compact representation of a joint probability distribution over the random variables $\mathbf{V} = \{X_1, \ldots, X_n\}$. It consists of a directed acyclic graph (DAG) in which each vertex corresponds to one of the random variables; a directed edge indicate direct dependence between two variables. Additionally, each variable X_i has an associated probability distribution, conditioned on its parents in the DAG, PA_i. The joint probability distribution given by the network is

$$P(\mathbf{V}) = \prod_{i=1}^{n} P(X_i | PA_i). \tag{1}$$

Given a dataset $\mathcal{D} = \{D_1, \ldots D_N\}$, where each D_i is a complete instantiation of $\mathbf{V}$, the goal of structure learning is to find a Bayesian network $\mathcal{N}$ which best fits $\mathcal{D}$. The fit of $\mathcal{N}$ to $\mathcal{D}$ is quantified by a scoring function s. Many scoring functions have been proposed in the literature, including Bayesian scores (Cooper and Herskovits 1992; Heckerman et al. 1995), MDL-based scores (Suzuki 1999; Silander et al. 2008), and independence-based scores (de Campos and Huete 2000), among others. The scoring functions can typically be interpretted as penalized log-likelihood functions. All commonly used scoring functions are *decomposable* (Heckerman et al. 1995); that is, they decompose into a sum of *local scores* for each variable, its parents, and the data,

$$s(\mathcal{N}; \mathcal{D}) = \sum_{i=1}^{n} s_i(PA_i; \mathcal{D}), \tag{2}$$

where $s_i(PA_i)$ gives the score of X_i using PA_i as its parents and is non-negative. We omit $\mathcal{D}$ when it is clear from context.

A variety of pruning rules (Suzuki 1999; Tian 2000; Teyssier and Koller 2005; de Campos and Ji 2011) can be used to demonstrate that some parent sets are never optimal for some variables. Additionally, in practice, large parent sets are often pruned *a priori*. We refer to parent sets remaining after all pruning as *candidate parent sets* and denote all candidate parent sets of X_i as $\mathcal{P}_i$.

The *Bayesian network structure learning* problem (BNSL) is defined as follows.

The BNSL Problem

Input: A set $\mathbf{V} = \{X_1, \ldots, X_n\}$ of variables and a local score $s_i(PA_i)$ for each $PA_i \in \mathcal{P}_i$ for each X_i.

Task: Find a DAG N^* such that

$$N^* \in \arg\min_{N} \sum_{i=1}^{n} s_i(PA_i),$$

where PA_i is the parent set of X_i in N and $PA_i \in \mathcal{P}_i$.

---

3   Solvers

This section describes all *solvers* (algorithm implementations) used in this work.
Hill climbing with a tabu list and random restarts (TABU, http://www.
bnlearn.com). Hill climbing is a widely-used local search technique in discrete
optimization (Russell and Norvig 2003) that typically finds local optima for
an objective function f by maintaining a *current* solution and applying *search
operators*. At each step, all search operators are tentatively applied to the current solution to find its *neighborhood*. The member of the neighborhood which
results in the biggest improvement to f is selected as the new current solution. This process is repeated until a local optimum is found. *Random restarting*
is a strategy to escape from a local optimum by randomly changing a locally
optimal solution and restarting the search from the new random solution. The
tabu list strategy (Glover 1990) augments random restarts by keeping track of
recently visited solutions; solutions in the tabu list are ignored when considering
new neighborhoods. Even with random restarts and a tabu list, the algorithm
provides no guarantees on the proximity of local optima to globally optimal
solutions.

In the context of BNs, each solution corresponds to a network; the search
operators considered here are edge addition, deletion and reversal (as long as the
resulting structure is a DAG). The objective function f is exactly the scoring
function s.

Max-min hill climbing (MMHC, http://www.bnlearn.com). Max-min hill
climbing (Tsamardinos et al. 2006) is a two-phase hybrid learning algorithm.
During the first phase, it uses a set of statistical independence tests to identify arcs that are forbidden from appearing in the learned network. The second
phase uses TABU to find local optima within this restricted space. Here we use
a mutual information statistical test during the first phase. The first phase of
MMHC is similar to constraint-based methods such as PC (Spirtes et al. 2000).
Empirically, MMHC has been shown to outperform several other state-of-the-
art algorithms, including PC, sparse candidate, three phase dependency analysis, optimal reinsertion and greedy equivalence search (Tsamardinos et al. 2006).
While MMHC does guarantee to recover BN structures when the data are faithful to a DAG in the large sample limit (Tsamardinos et al. 2006), it does not
offer any non-trivial guarantees about the generalization quality of the learned
network for unfaithful, finite datasets.

Chow-Liu (CL). The Chow-Liu algorithm (Chow and Liu 1968) is an exact,
polynomial-time algorithm for finding an optimal tree-structured BN. The algorithm calculates the mutual information between all pairs of variables to form a
weighted graph. The maximum spanning tree through the graph corresponds to
the optimal tree-structured BN.

A* (A*, http://www.urlearning.org). State space search using A* (Yuan and
Malone 2013) is a provably optimal algorithm which is guaranteed to optimize
s. It is based on casting BNSL as a shortest-path finding problem; A* is then

used to solve the shortest path problem, which gives the optimal network for the given local scores. For the exact solver comparisons, we refer to a variant of A* which uses multiple pattern databases as A*EC.

Integer linear programming (ILP, http://www.cs.york.ac.uk/aig/sw/ gobnilp/). Another approach to solving BNSL optimally is based on integer linear programming (ILP) (Bartlett and Cussens 2015). In ILP, BNs are defined as vertices on a particular polytope, and a cutting plane approach is used to find the vertex corresponding to the optimal BN.

Branch and Bound (BNB, http://www.ecse.rpi.edu/~cvrl/structlearning. html). The branch-and-bound search algorithm (de Campos and Ji 2011) searches for optimal networks in a relaxed space of directed graphs that may contain cycles. Found cyclic solutions are iteratively ruled out by removing one arc in it and branching over the possible choices of the arc to remove.

Provably optimal (OPT). All optimal algorithms (including A*, ILP, BNB, and their variants) find equivalent networks[1]. Thus, in the context of the generalization analysis, they are equivalent and only one of the optimal algorithms is used for each dataset.

Solver resource constraints. For running the experiments we used a cluster of Dell PowerEdge M610 computing nodes equipped with two 2.53-GHz Intel Xeon E5540 CPUs and 32-GB RAM. For each individual run, we used a timeout of 2 h and a 28-GB memory limit. We treat the runtime of any instance as 2 h if a solver exceeds either the time or memory limit.

4 Datasets

We used a similar set of benchmark datasets for both studies; in total, we used 48 distinct datasets[2]:

- Datasets sampled from benchmark Bayesian networks. 19 datasets, SAMPLED.
- Datasets from the UCI repository. 19 datasets, UCI.
- Datasets sampled from random Bayesian networks. 7 datasets, SYN.
- Datasets we compiled by processing log files. 3 datasets, LOG.

We preprocessed each dataset by removing all continuous variables, variables with very large domains (e.g., unique identifiers), and variables that take on only one value. Other than preprocessing, the datasets were used slightly differently in the generalization study compared to the exact solver analysis; the relevant sections discuss exactly how the datasets were used.

---

[1] This work assumes s is score-equivalent (Heckerman et al. 1995).

[2] The datasets are available at http://bnportfolio.cs.helsinki.fi/.

5 Generalization of Learned Networks

Our aim in the first part of this work is to shed light on the relationship of different learning strategies, based on the solvers discussed in Sect. 3, and the unknown discrepancy between training set scores and generalization. In particular, we address the following research questions for different fixed learning algorithms and training sets.

Q1 How do hard constraints on the number of parents in learned structures affect their generalization?

Q2 How does the amount of training data affect the generalization of learned structures?

Q3 Which learning strategies result in networks with the best generalization?

Our main findings, based on a rigorous experimental setup, are the following. With respect to Q1, we show that for small datasets, hard constraints limiting the maximum number of parents to 2 improves generalization on a few datasets for local search algorithms; however, optimal algorithms usually benefit from a higher limit. We answer Q2 by using increasingly large subsets of available training data. Regardless of the algorithms' guarantees, more training data results in more accurate predictions on testing data. Finally, we address Q3 by considering all of the data collected during the evaluation. For some datasets, simple strategies such as the tractable Chow-Liu algorithm can provide good generalization. However, the simple strategies fail to generalize well on other datasets. Predictive likelihood results show that OPT consistently generalizes well.

5.1 Experimental Setup

Datasets. We used 29 datasets from the UCI and SAMPLED categories; the number of variables in the datasets ranges from 17 to 60, and the number of records ranges from about 30 to 20 000. We used standard 10-fold cross-validation in order to evaluate the learning strategies.

Parent limit. For all algorithms except CL, we used hard limits of 2 and 8 on the number of parents. When discussing algorithms, we use a subscript to indicate the maximum number of parents, such as OPT_8.

Scoring function. We selected the commonly-used Bayesian Dirichlet with score equivalence and uniform structure prior (BDeu) scoring function (Heckerman et al. 1995) with an equivalent sample size (ESS) of 1 as the scoring function.

Inference. For all learned structures, parameter values were set using a symmetric Dirichlet prior with a concentration parameter of 1 (which is equivalent to Laplacian smoothing). All testing likelihood calculations were performed by multiplying relevant family factors.

Evaluation. In order to address our research questions, we use the predictive likelihood to evaluate the generalization capability of the learned networks. In particular, for a dataset d and learning strategy l, we calculate the per-prediction-likelihood, $\ell_{pp}^{d,l}$, which is the likelihood of each prediction on the test set,

$$\ell_i^{d,l} = \sum_{r=1}^{N} \log P(d_r|\mathcal{N}) = \sum_{r=1}^{N} \sum_{i=1}^{n} \log P(X_i^r|PA_i^r) \tag{3}$$

$$\ell_{pp}^{d,l} = -\frac{\sum_{i=1}^{10} \ell_i^{d,l}}{N_d \cdot n_d}, \tag{4}$$

summing over the folds $i = 1..10$, where $\ell_i^{d,l}$ is the predictive likehood on the test set for fold i using learning strategy l, N_d is the number of records in the test set, and n_d is the number of variables in the dataset.

The numerator of Eq. 4 is the sum over all of the test set predictive likelihoods for learning strategy l and dataset d. Each $\ell_i^{d,l}$ term comprises $\frac{N_d}{10} \cdot n_d$ terms. In total, the sum in the numerator includes $N_d \cdot n_d$ terms, each of which corresponds to the log probability of one variable of one record from the test set. Consequently, the denominator serves as a normalizing constant, and $\ell_{pp}^{d,l}$ is the average log probability of each prediction.

In order to compare learning strategies, we normalize the $\ell_{pp}^{d,l}$ values for each dataset between 0 and 1 to obtain

$$\hat{\ell}_{pp}^{d,l} = 1 - \frac{\ell_{pp}^{d,l} - \min_{l'}\{\ell_{pp}^{d,l'}\}}{\max_{l'}\{\ell_{pp}^{d,l'}\} - \min_{l'}\{\ell_{pp}^{d,l'}\}} \tag{5}$$

where l' ranges over all learning strategies. Note that, after normalization, the learning strategy with the best $\ell_{pp}^{d,l}$ has $\hat{\ell}_{pp}^{d,l} = 0$ while the worst learning strategy has $\hat{\ell}_{pp}^{d,l} = 1$.

It is important to note that $\ell_{pp}^{d,l}$ and $\hat{\ell}_{pp}^{d,l}$ consider all variables equally. In particular, they do not consider a special "class" variable.

5.2   Impact of Restricting Parent Set Size

We study question Q1 by comparing the $\hat{\ell}_{pp}^{d,l}$ among datasets when using $k = 2$ and $k = 8$ as the maximum number of parents for each learning algorithm. The BDeu score implicitly restricts the maximum number of selected parents as a soft constraint by integrating over all parameterizations of parent instantiations. Other scores, such as MDL, explicitly incorporate a complexity penalty to discourage large parent sets. In both cases, though, this restriction is a soft constraint. Here, we consider the maximum number of parents as a *hard constraint*.

Optimal. Figure 1 (left) shows the performance (in terms of $\hat{\ell}_{pp}^{d,l}$) of generalization using OPT$_k$ for parent limits $k = 2, 8$. The (left, top) and (left, bottom) plots show distinctly different patterns. Figure 1 (left, top) clearly shows that OPT$_2$ results in better generalization for SAMPLED datasets with 100 records. However, as the number of records increases, OPT$_8$ yields better performance. In contrast, for UCI datasets, OPT$_8$ is almost always better.

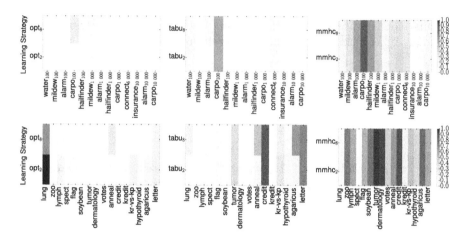

Fig. 1. The $\hat{\ell}_{pp}^{d,l}$ values for OPT (left), TABU (center) and MMHC (right) with a hard limit of $k = 2$ and $k = 8$ for SAMPLED (top) and UCI (bottom) datasets. The datasets are sorted in ascending number of records. Lighter colors indicate better performance. Close inspection of the MMHC strategies show some slight difference; however, they are difficult to discern in the scaled image.

Tabu. In contrast to the results for OPT, Fig. 1 (center, bottom) shows that TABU$_2$ generalizes better than TABU$_8$ for UCI datasets. One possible explanation for this difference is that the greedy strategy of TABU$_8$ favors structures which improve the likelihood while increasing the complexity of the learned structures. Thus, the learned structure overfits the training data and does not generalize well to testing data. In contrast, as OPT is guaranteed to find the best-scoring structure, it finds structures which better balance training set likelihood and complexity. The hard constraints on the number of parents for TABU$_2$ forbid it from selecting the complex structures. Both TABU$_2$ and TABU$_8$ typically generalize well on SAMPLED datasets.

MMHC. Figure 1 (right) shows that the hard parent limit has little effect on $\hat{\ell}_{pp}^{d,l}$ for MMHC. The first phase of MMHC uses a set of statistical independence tests to restrict the learned network structures. For many of the datasets, the relatively small number of records restricts the power of these tests and leads to a very small search space in the second phase, despite initially allowing many more structures for the 8-parent space.

In summary, the answer to Q1 clearly depends both on the training datasets and learning algorithm; the global guarantees of OPT allow it to fully take advantage of the larger $k = 8$ search space, but the local search strategy of TABU performs better in the more restricted $k = 2$ space.

More data is required to accurately estimate the conditional probability distributions for complex structures (with more parameters). This may explain why OPT$_2$ generalizes better than OPT$_8$ for datasets with a small number of records.

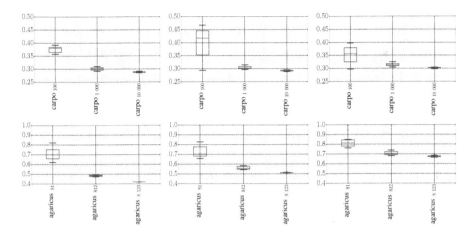

Fig. 2. The $\hat{\ell}_{pp}^{d,l}$ values for using the OPT8 (left), TABU2 (center), and CL (right) learning strategies as the number of records increases. The top row is for the *carpo* dataset (SAMPLED); the bottom row is for the *agaricus* dataset (UCI). Note the different y-axes for the plots. Lower values and smaller boxes are better.

5.3   Impact of Amount of Training Data

To investigate the impact of the amount of available training data, to answer Q2 we compared how $\ell_{pp}^{d,l}$ of OPT8, TABU2 and CL behave as the number of records available for training increases. Figure 2 shows that for all algorithms on both SAMPLED and UCI datasets, more records lead to better $\ell_{pp}^{d,l}$. Furthermore, the plots also show that with more records, the variance of $\ell_{pp}^{d,l}$ decreases. Interestingly, the plot also shows that CL performs better than OPT8 and TABU2 on *carpo*, a SAMPLED dataset, when only 100 records are available. This again highlights that restricted model classes can generalize better than those which allow more parameters, especially when little data is available to estimate the parameter values. Despite the differences in guarantees, OPT8, TABU2 and *cl* perform similarly for *carpo*₁ ₀₀₀ and *carpo*₁₀ ₀₀₀.

As with *carpo*, for the UCI *agaricus* dataset, the likelihood improves and variance decreases as the number of records increases. However, OPT8 improves from $\ell_{pp}^{d,l} \approx 0.7$ for 81 records to $\ell_{pp}^{d,l} \approx 0.48$ with 812 records. In contrast, TABU2 only improves from $\ell_{pp}^{d,l} \approx 0.7$ to about $\ell_{pp}^{d,l} \approx 0.55$, and CL exhibits even less improvement. For *agaricus*, OPT8 using only 812 records results in better generalization than TABU2 or CL with all 8 123 records.

We observed similar behavior on other SAMPLED and UCI datasets as the amount of training data was varied. The same general trends hold for all algorithms and datasets with respect to Q3. Namely, the predictive likelihood improves and variance decreases as the size of the training set increases.

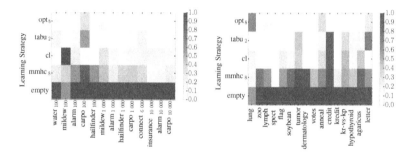

Fig. 3. The $\hat{\ell}_{pp}^{d,l}$ values for the best learning strategies. The empty network is included as a baseline. The SAMPLED datasets are shown in the left heatmap, and UCI datasets are in the right. The datasets are sorted in ascending number of records. Lighter colors indicate better performance (Color figure online).

5.4   Comparison Across Learning Strategies

Finally, based on the previous results, we studied Q3 by choosing the best learning strategies and comparing their $\hat{\ell}_{pp}^{d,l}$ across all of the datasets. In essence, we fix the training set while varying the learning strategy. Additionally, EMPTY (with no edges) was included as a baseline. The results in Fig. 3 show several expected trends and a few surprises. As expected, EMPTY is the worst on almost all of the datasets. For the reasons mentioned in Sect. 5.2, MMHC$_8$ was typically worse than the other strategies. These trends are consistent for both SAMPLED and UCI datasets. For SAMPLED datasets, TABU$_2$ and OPT$_8$ have very similar $\hat{\ell}_{pp}^{d,l}$ for most datasets; the $\hat{\ell}_{pp}^{d,l}$ of CL is also surprisingly similar to that of the two more "sophisticated" strategies.

For UCI datasets, OPT$_8$ continues to consistently have good $\hat{\ell}_{pp}^{d,l}$. On the other hand, CL and TABU$_2$ exhibit much more inconsistency in their generalization relative to OPT$_8$. For some datasets, such as *dermatology* and *kredit*, they match OPT$_8$; on others, such as *credit* and *tumor*, CL and TABU$_2$ do not generalize well. Surprisingly, CL exhibits the best $\hat{\ell}_{pp}^{d,l}$ for *letter*, the UCI dataset with the most records.

For Q3, OPT guarantees consistently translate into networks with good generalization. Algorithms with weaker guarantees produce networks with inconsistent generalization.

Comments on Datasets. Besides the behavior of the learning algorithms, these results also suggest differences in the datasets themselves. In particular, it seems that SAMPLED datasets are "easier," in the sense that many learning strategies find networks which generalize well. On the other hand, only the strategy with strong guarantees consistently generalizes well on UCI datasets. In some sense, this result is not surprising. The SAMPLED data is by construction accurately modeled by a BN, while it is very unlikely that UCI datasets are faithful to any BN. These caveats are important for future evaluations.

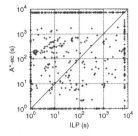

Fig. 4. Comparison of two state-of-the-art algorithms for finding an optimal Bayesian network. Runtimes below 1 or above 7200 s are rounded to 1 and 7200, respectively. See Sect. 5 for descriptions of the solvers and the datasets.

6   Exact Solver Empirical Hardness Models

As shown in Sect. 5, exact algorithms often lead to networks which generalize better than those found with approximation algorithms. Due to the intrinsic differences between the algorithmic approaches underlying the solvers, it is not surprising that their relative efficiency varies on a per-instance basis. To exemplify this, a comparison of the runtimes of ILP and A^*EC is illustrated in Fig. 4 using typical benchmark datasets. Evidently, neither of these two solvers dominates the other, as there clearly are instances on which one solver is much more efficient than the other.

To explain the observed orthogonal performance characteristics shown in Fig. 4, it has been suggested, roughly, that typical instances can be solved to optimum by A^* if the *number of variables* n is at most around 50 (Fan et al. 2014), and by ILP if the number of *candidate parent sets* m is not very large (Bartlett and Cussens 2015).

Unfortunately, beyond this rough characterization, the practical time complexity of the fastest algorithms is currently poorly understood. The gap between the analytic worst-case and best-case bounds is very large, and typical instances fall somewhere in between. Moreover, the sophisticated search heuristics employed by the algorithms are quite sensitive to small variations in the instances, which results in somewhat chaotic looking behavior of runtimes. Even the following basic questions are open:

Q4 Are the simple features, the number of variables n and the number of candidate parent sets m, sufficient for determining which of the available solvers is fastest for a given instance?

Q5 Are there other efficiently computable features that capture the hardness of the problem significantly more accurately than n and m alone?

In this section, we answer both these questions in the affirmative. We answer Q4 by learning a simple, yet nontrivial, model that accurately predicts the fastest solver for a given instance based on n and m only. We show how this yields an algorithm portfolio that almost always runs as fast as the fastest algorithm, thus significantly outperforming any fixed algorithm on a large collection of instances. To address this issue and answer Q5, we introduce and study several additional features that potentially capture the hardness of the problem more accurately for a given solver. In particular, we show that learning models with a much wider variety of features yields significant improvement in the prediction accuracy.

Related Work. The idea of learning to predict an algorithm's runtime from empirical data is not new. Rice (1976) proposed feature-based modeling to facilitate the selection of the best-performing algorithm for a given problem instance, considering various example problems. More recently, machine learning and empirical hardness models (Leyton-Brown et al. 2002) have been used for solver portfolios in several domains.

6.1   Capturing Hardness

The *hardness* of a BNSL instance, relative to a given solver, is the runtime of the solver on the instance. We aim to find a *model* that approximates the hardness and is efficient to evaluate for any given instance from a small set of efficiently computable *features* of BNSL instances. We can then learn the model by computing the feature values and collecting empirical runtime data from a set of BNSL instances. We first introduce several candidate features that are potentially informative about the hardness of BNSL instances for one or more solvers. We then explain how we learn a hardness model and estimate its prediction accuracy.

Features for BNSL. We consider several features which naturally fall into four categories, explained next, based on the strategy used to compute them: **Basic**, **Basic extended**, **Lower bounding**, and **Probing**. Due to space constraints, please refer to the original paper for the complete list of features.

The **Basic** features include the number of variables n and the mean number of candidate parent sets per variable, m/n. The features in **Basic extended** summarize the size distribution of the collections $\mathcal{P}_i$ and the parent sets PA_i in each $\mathcal{P}_i$. In the **Lower bounding** category, the features reflect statistics from a directed graph that is an optimal solution to a relaxation of the original BNSL problem. In the **Simple LB** subcategory, a graph is obtained by letting each variable select its best parent set according to the scores. Many solvers use this lower bounding technique. In the **Pattern database LB** subcategory, the features are the same but the graph is obtained from a more sophisticated relaxation using pattern databases (Yuan and Malone 2013).

Probing refers to running a solver for several seconds and collecting statistics about its behavior during the run. We consider three probing strategies: TABU, an anytime variant of A* (Malone and Yuan 2013), and ILP (Cussens et al. 2013). Probing is implemented by running each algorithm for 5 s and collecting several features of the learned structure.

Model Training and Evaluation. Based on the features discussed in the previous section, we trained reduced error pruning trees (REP trees) (Quinlan 1987) to predict the runtime of an instance of BNSL for each solver. We chose these decision tree models because of their interpretability, compared to techniques such as neural networks or support vector machines, and because of their flexibility, compared to linear regression and less expressive model classes.

6.2 Experiment Setup

We used all of the datasets mentioned in Sect. 4. We considered 5 different scoring functions[3]: BDeu with the Equivalent Sample Size selected from $\{0.1, 1, 10, 100\}$ and BIC. For each dataset and scoring function, we generated scores with parent limits ranging from 2 to up to 6. The size of the datasets ranged from about 100 records to over $60,000$ records. For portfolio construction we removed very easy instances (solved within 5 s by all solvers) as uninteresting, and instances on which all solvers failed, leaving 586 instances. We evaluated the portfolios using 10-fold cross-validation.

6.3 Portfolios for BNSL

This section focuses on the construction of practical BNSL solver portfolios in order to address question Q4. Optimal portfolio behavior is to always select the best-performing solver for a given instance. As the main results, we will show that, perhaps somewhat surprisingly, it is possible to construct a practical BNSL solver portfolio that is close-to-optimal using only the **Basic** features.

As the basis of this work, we ran all the solvers and their parameterizations on all the benchmark instances. Figure 5 (left) shows the number of instances for which each solver was the fastest. The performance of BNB is in general inferior to the other solvers; in the following we will focus on ILP and A*EC. However, recall Fig. 4: while ILP is clearly best measured in the number of instances solved the fastest, the performance of ILP on a per-instance basis is very much orthogonal to that of A*. We now show that a simple BNSL solver portfolio can capture the best-case performance of *both* of these approaches.

A Very Simple Solver Portfolio. We found that using only the **Basic** features are enough to construct a highly efficient BNSL solver portfolio. While on an intuitive level the importance of these two features may be to some extent unsurprising, such intuition does not directly translate into an actual predictor that would close-to-optimally predict the best-performing solver.

Figure 5 (right) shows the performance of each individual solver variant, as well as the Virtual Best Solver (VBS), which is the theoretically optimal portfolio which always selects the best algorithm, constructed by selecting *a posteriori* the fastest solver for each input instance. "portfolio" is our simple portfolio which uses only the **Basic** features. As the figure shows, the performance of our simple portfolio is very close to the theoretically optimal performance of VBS and greatly outperforms the individual solvers.

---

[3] Our results were not very sensitive to the scoring function, except its effect on the number of CPSs, so our results generalize to other decomposable scores.

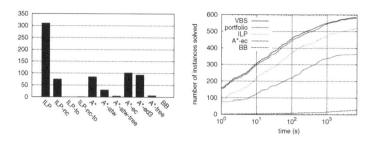

Fig. 5. (left) VBS contributions of each solver, i.e., the number of instances for which a solver was fastest. Several variants of ILP and A* were used. Please see the original paper for more details. (right) Solver performance: VBS, our simple portfolio, ILP, A*EC, and BNB.

6.4 Predicting Runtimes

To address Q5, we investigate the effect of the feature sets on prediction accuracy.

As just shown, the **Basic** features can effectively distinguish between solvers to use on a particular instance of BNSL. However, notable improvements in runtime prediction accuracy are gained by employing a much wider range of features. Figure 6 (left) compares the actual runtimes for A*EC to the predictions made by the REP tree model trained using only the **Basic** features. The model clearly splits the instances into only a few bins and predicts the same runtime for all instances in the bin. The actual runtime ranges are quite wide within each bin. For example, for the bin with predictions near 80 s, the actual runtimes span from around 5 s to about an hour. Even though these predictions allow for good portfolio behavior, they are not useful to estimate actual runtime.

On the other hand, Fig. 6 (right) shows the same comparison for models learned using **A* probing** features (1–38, 51–62). Many more of the predictions fall along or near the main diagonal. That is, the larger, more sophisticated feature set results in more accurate runtime predictions. We observed similar, though less pronounced, trends for ILP.

6.5 REP Tree Characteristics

For additional insight, we considered how often specific features were selected (Table 1). A feature is rarely selected for predicting both solvers. This further confirms that the solver runtimes are influenced by different structural properties of instances. Nevertheless, **Simple LB** features were helpful for both algorithms. Somewhat surprisingly, the **Pattern database LB** features were more useful for ILP, even though A*EC directly uses the pattern database in its search. For all of the graph-based features (node degree and non-trivial SCCs), the standard deviation was always selected over the maximum and mean. This suggests that systematic variations between nodes are important for determining the hardness of an instance. The table also shows that a small number of features were consistently selected for most of the cross-validation folds for any particular solver.

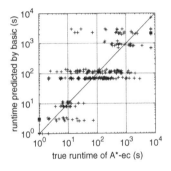 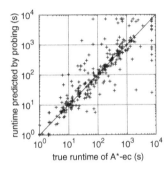

Fig. 6. Predicted vs. actual runtimes for A*EC using **Basic** features only (left) and all features up to A* probing (right).

Qualitatively, this implies that most of the trees were based on the same small set of features. Developing a more in-depth understanding of these instance characteristics in light of solver performance is an important aspect of future work.

Table 1. Features used for A*EC and ILP in more than 5 of the 10 cross-validation folds. For each solver, the set of possible features consisted of non-probing features (1–38) and the relevant probing features.

	Feature	A*-ec	ILP
(1)	Number of variables, n	10	0
(2)	Number of CPS, mean	0	7
(3)	Number of CPS, sum, m	2	10
(4)	Number of CPS, max	0	7
(8)	CPS cardinalities, sd	0	8
(11)	Simple LB, Node in-degree, sd	0	7
(14)	Simple LB, Node out-degree, sd	8	0
(17)	Simple LB, Node degree, sd	10	0
(26)	Pd LB, Node in-degree, sd	1	9
(38)	Pd LB, Size of non-trivial SCCs, sd	0	8
(62)	A* probing, Error bound	10	0
(68)	ILP probing, Node out-degree, sd	0	10
(74)	ILP probing, Error bound	0	10

7   Discussion

Bayesian network structure learning (BNSL) continues to be an area of very active research. In this review, we have presented two orthogonal studies of

BNSL algorithms. The first demonstrated that, whenever possible, exact learning algorithms should be used for finding structures. The second study showed that it is typically possible to not only select the best exact learning algorithm for a given dataset but also predict how long it will take to find the optimal structure.

These studies suggest a variety of future investigations. For example, the most "Bayesian" approach to generalization should be a model averaging strategy, but the current work considers only a single structure. In light of the generalization results, empirical hardness models could be built for different dataset categories.

References

Bartlett, M., Cussens, J.: Integer linear programming for the Bayesian network structure learning problem. In: Artificial Intelligence (2015)

Chickering, D.M.: Learning Bayesian networks is NP-complete. In: Fisher, D., Lenz, H.-J. (eds.) Learning from Data: Artificial Intelligence and Statistics V. Lecture Notes in Statistics, vol. 112, pp. 121–130. Springer, New York (1996)

Chickering, D.M.: Learning equivalence classes of Bayesian-network structures. J. Mach. Learn. Res. **2**, 445–498 (2002)

Chow, C., Liu, C.: Approximating discrete probability distributions with dependence trees. IEEE Trans. Inf. Theory **14**(3), 462–467 (1968)

Cooper, G.F., Herskovits, E.: A Bayesian method for the induction of probabilistic networks from data. Mach. Learn. **9**, 309–347 (1992)

Cussens, J., Bartlett, M., Jones, E.M., Sheehan, N.A.: Maximum likelihood pedigree reconstruction using integer linear programming. Genet. Epidemiol. **37**(1), 69–83 (2013)

de Campos, C.P., Ji, Q.: Efficient learning of Bayesian networks using constraints. J. Mach. Learn. Res. **12**, 663–689 (2011)

de Campos, L.M., Huete, J.F.: A new approach for learning belief networks using independence criteria. Int. J. Approximate Reasoning **24**(1), 11–37 (2000)

Fan, X., Yuan, C., Malone, B.: Tightening bounds for Bayesian network structure learning. In: Proceedings of the 28th AAAI Conference on Artificial Intelligence (2014)

Friedman, N., Nachman, I., Peer, D.: Learning Bayesian network structure from massive datasets: the "sparse candidate" algorithm. In: Proceedings 13th Conference on Uncertainty in Artificial Intelligence (1999)

Glover, F.: Tabu search: a tutorial. Interfaces **20**(4), 74–94 (1990)

Heckerman, D., Geiger, D., Chickering, D.M.: Learning Bayesian networks: the combination of knowledge and statistical data. Mach. Learn. **20**, 197–243 (1995)

Koivisto, M., Sood, K.: Exact Bayesian structure discovery in Bayesian networks. J. Mach. Learn. Res. **5**, 549–573 (2004)

Leyton-Brown, K., Nudelman, E., Shoham, Y.: Learning the empirical hardness of optimization problems: the case of combinatorial auctions. In: Hentenryck, P. (ed.) CP 2002. LNCS, vol. 2470, pp. 556–572. Springer, Heidelberg (2002)

Malone, B., Järvisalo, M., Myllymäki, P.: Impact of learning strategies on the qualpacking Bayesian networks: an empirical evaluation. In: Proceedings of the 31st Conference on Uncertainty in Artificial Intelligence (2015)

Malone, B., Kangas, K., Järvisalo, M., Koivisto, M., Myllymäki, P.: Predicting the hardness of learning Bayesian networks. In: Proceedings of the 28th AAAI Conference on Artificial Intelligence (2014)

Malone, B., Yuan, C.: Evaluating anytime algorithms for learning optimal Bayesian networks. In: Proceedings of the 29th Conference on Uncertainty in Artificial Intelligence (2013)

Mitchell, T.: Machine Learning. McGraw-Hill, New York (1997)

Moore, A., Wong, W.-K.: Optimal reinsertion: a new search operator for accelerated and more accurate Bayesian network structure learning. In: Proceedings of the 20th International Conference on Machine Learning, pp. 552–559 (2003)

Ott, S., Imoto, S., Miyano, S.: Finding optimal models for small gene networks. In: Proceedings of the Pacific Symposium on Biocomputing (2004)

Parviainen, P., Koivisto, M.: Exact structure discovery in Bayesian networks with less space. In: Proceedings of the Twenty-Fifth Conference on Uncertainty in Artificial Intelligence, Montreal, Quebec, Canada. AUAI Press (2009)

Pearl, J.: Probabilistic Reasoning in Intelligent Systems: Networks of Plausible Inference. Morgan Kaufmann Publishers Inc., San Mateo (1988)

Quinlan, J.R.: Simplifying decision trees. Int. J. Man Mach. Stud. **27**, 221–234 (1987)

Rice, J.R.: The algorithm selection problem. Adv. Comput. **15**, 65–118 (1976)

Russell, S.J., Norvig, P.: Artificial Intelligence: A Modern Approach. Pearson Education, Upper Saddle River (2003)

Silander, T., Myllymäki, P.: A simple approach for finding the globally optimal Bayesian network structure. In: Proceedings of the 22nd Conference on Uncertainty in Artificial Intelligence (2006)

Silander, T., Roos, T., Kontkanen, P., Myllymäki, P.: Factorized normalized maximum likelihood criterion for learning Bayesian network structures. In: Proceedings of the 4th European Workshop on Probabilistic Graphical Models (2008)

Spirtes, P., Glymour, C., Schemes, R.: Causation, Prediction, and Search, 2nd edn. MIT Press, Cambridge (2000)

Suzuki, J.: Learning Bayesian belief networks based on the MDL principle: an efficient algorithm using the branch and bound technique. IEICE Trans. Inf. Syst. **E82–D**(2), 356–367 (1999)

Teyssier, M., Koller, D.: Ordering-based search: a simple and effective algorithm for learning Bayesian networks. In: Proceedings of the 21st Conference on Uncertainty in Artificial Intelligence (2005)

Tian, J.: A branch-and-bound algorithm for MDL learning Bayesian networks. In: Proceedings of the 16th Conference on Uncertainty in Artificial Intelligence (2000)

Tsamardinos, I., Brown, L., Aliferis, C.: The max-min hill-climbing Bayesian network structure learning algorithm. Mach. Learn. **65**, 31–78 (2006)

van Beek, P., Hoffmann, H.-F.: Machine learning of Bayesian networks using constraint programming. In: Pesant, G. (ed.) CP 2015. LNCS, vol. 9255, pp. 429–445. Springer, Heidelberg (2015)

Yuan, C., Malone, B.: Learning optimal Bayesian networks: a shortest path perspective. J. Artif. Intell. Res. **48**, 23–65 (2013)

On Model Selection, Bayesian Networks, and the Fisher Information Integral

Yuan Zou and Teemu Roos[✉]

Helsinki Institute for Information Technology HIIT,
Gustaf Hällströmin katu 2b, 00014 Helsinki, Finland
{yuan.zou,teemu.roos}@hiit.fi
http://www.hiit.fi/cosco/promo

Abstract. We study BIC-like model selection criteria and in particular, their refinements that include a constant term involving the Fisher information matrix. We observe that for complex Bayesian network models, the constant term is a negative number with a very large absolute value that dominates the other terms for small and moderate sample sizes. We show that including the constant term degrades model selection accuracy dramatically compared to the standard BIC criterion where the term is omitted. On the other hand, we demonstrate that exact formulas such as Bayes factors or the normalized maximum likelihood (NML), or their approximations that are not based on Taylor expansions, perform well. A conclusion is that in lack of an exact formula, one should use either BIC, which is a very rough approximation, or a very close approximation but not an approximation that is truncated after the constant term.

1 Introduction

A Bayesian network encodes joint probability distributions of a set of random variables via a directed acyclic graph (DAG). Since Bayesian networks with different network topologies form a lattice-like hierarchy with both nested and non-nested relations, it becomes imperative to regularize model complexity when learning the structure from finite data. In this paper we study BIC-like model selection criteria that can be derived via a Laplace approximation, and their properties in the case of Bayesian networks. Our main focus is on complexity regularization and in particular, the lower-order terms such as the constant term, $\log \int_\Theta \sqrt{\det I(\theta)} \, d\theta$, involving the Fisher information, $I(\theta)$, which are omitted in the standard BIC formula.

An approximation of the Bayes factor (or the marginal likelihood) [5] under Jeffreys' prior, where the constant term is retained, results in a so called Fisher information approximation (FIA). We show that contrary to what might be expected, namely that a more refined approximation such as FIA should be better than a rough approximation such as BIC, FIA tends to be extremely inaccurate for small and moderate sample sizes. In particular, we observe that for complex Bayesian network models (with thousands or tens of thousands of independent parameters), the constant term is a negative number with a very

© Springer International Publishing Switzerland 2015
J. Suzuki and M. Ueno (Eds.): AMBN 2015, LNAI 9505, pp. 122–135, 2015.
DOI: 10.1007/978-3-319-28379-1_9

large absolute value that dominates all the other terms in FIA unless the sample size is greater than the number of parameters. The absolute value of the term grows rapidly with increasing model order, which makes the FIA criterion favor complex models unless the sample size is extremely large. Similar results have been reported for other model families such as the exponential model [9] and Markov sources [15].[1]

In this paper, we first review the FIA approximation and discuss its relation to certain other model selection criteria. Because there is no closed form formula for the Fisher information integral under most model families, including Bayesian networks, we illustrate how to estimate it with arbitrarily fine precision using Monte Carlo techniques. We carry out model selection experiments where we highlight the complexity regularization performance by the various criteria in order to determine which of the criteria are safe and which should be avoided under given conditions.

2 The Fisher Information Approximation

In this section, we discuss what we call the Fisher information approximation (FIA), and relate it to other model selection criteria. First, let's consider the Bayes factor criterion before investigating asymptotic approximations. The Bayes factor measures the ratio of marginal likelihoods between competing models.

$$\mathrm{BF}_{12} = \frac{p(x^n \, ; \, \mathcal{M}_1)}{p(x^n \, ; \, \mathcal{M}_2)} = \frac{\int_{\Theta_{\mathcal{M}_1}} p(x^n \, ; \, \theta_1, \mathcal{M}_1) \, p(\theta_1) \, d\theta_1}{\int_{\Theta_{\mathcal{M}_2}} p(x^n \, ; \, \theta_2, \mathcal{M}_2) \, p(\theta_2) \, d\theta_2}, \tag{1}$$

where $p(\theta_1)$ and $p(\theta_2)$ denote the parameter priors under the two models, $\mathcal{M}_1$ and $\mathcal{M}_2$, respectively.

The marginal likelihood has a built-in, implicit penalty for model complexity, see [10]. A closed form solution for the marginal likelihood is only available for a limited set of model families when conjugate priors exist. For other model families, we usually need to resort to sampling methods such as MCMC methods [3]. Furthermore, even when an efficient formula for calculating Bayes factors is available, like in the case of Bayesian networks discussed in this work, model selection performance may be highly sensitive to the choice of the associated parameter priors [18].

2.1 Approximation of Marginal Likelihood

To avoid the selection of a specific prior and to obtain a more objective method for model selection, we can use asymptotic (large-sample) approximations of the

---

[1] Our earlier paper on this topic appeared as an invited paper at the ITA-2013 workshop. Hence, no prior peer-reviewed publication of this material exists beyond the basic Monte Carlo approximation proposed in [13]. In particular, this is the first study where the lower-order terms of information criteria are discussed in conjunction with a model class for which model selection criteria are being intensively developed.

Bayes factor or the marginal likelihood such as the classic BIC criterion [16]. The BIC can be obtained via Laplace approximation, which involves a Taylor expansion of the log-likelihood function around its maximum. For instance, if we have a model $\mathcal{M}$ with $d_{\mathcal{M}}$ free parameters, jointly denoted by $\theta \in \Theta_{\mathcal{M}}$, and a data set x^n with sample size n, the Laplace approximation of the log-marginal likelihood is given by

$$
\begin{aligned}
\log p(x^n \,;\, \mathcal{M}) &= \log \int_{\Theta_{\mathcal{M}}} p(x^n \,;\, \theta, \mathcal{M})\, p(\theta)\, d\theta \\
&= \log p(x^n \,;\, \hat{\theta}(x^n)) + \log p(\hat{\theta}(x^n)) \\
&\quad + \frac{d_{\mathcal{M}}}{2} \log(2\pi) - \frac{1}{2} \log \det \hat{I}(\hat{\theta}(x^n)) + o(1),
\end{aligned}
\tag{2}
$$

where $p(\theta)$ is the parameter prior, the maximum likelihood parameters are denoted by $\hat{\theta}(x^n)$, and $\hat{I}(\theta)$ is the empirical Fisher information matrix at θ. If the distributions of model $\mathcal{M}$ are independent and identically distributed (i.i.d.), by the law of large numbers, we have the average per-symbol empirical Fisher information converging to its expectation $I(\hat{\theta}(x))$:

$$
n^{-1}\hat{I}(\hat{\theta}(x^n)) \to I(\hat{\theta}(x^n)), \text{ where } I(\theta) = \mathbb{E}_\theta\, \hat{I}(\theta).
\tag{3}
$$

Then by simple manipulation, the fourth term in Eq. (2) can be approximated as

$$
\frac{1}{2} \log \det \hat{I}(\hat{\theta}(x^n)) = \frac{d_{\mathcal{M}}}{2} \log n + \frac{1}{2} \log \det I(\hat{\theta}(x^n)) + o(1).
\tag{4}
$$

Finally, we can obtain the approximation of log marginal likelihood as

$$
\begin{aligned}
\log p(x^n \,;\, \mathcal{M}) &= \log p(x^n \,;\, \hat{\theta}(x^n)) - \frac{d_{\mathcal{M}}}{2} \log n \\
&\quad + \log p(\hat{\theta}(x^n)) + \frac{d_{\mathcal{M}}}{2} \log(2\pi) - \frac{1}{2} \log \det I(\hat{\theta}(x^n)) + o(1).
\end{aligned}
\tag{5}
$$

When the sample size n increases, lower order terms that are independent of n will eventually be dominated by the terms that grow with n. Therefore, for very large sample sizes, we can omit the last four terms in Eq. (5) and change the sign to obtain the familiar BIC criterion:

$$
\mathrm{BIC}(x^n \,;\, \mathcal{M}) = -\log p(x^n \,;\, \hat{\theta}_{\mathcal{M}}(x^n)) + \frac{d_{\mathcal{M}}}{2} \log n,
\tag{6}
$$

To get a more precise approximation, we would need to include the lower-order terms as well. However, they depend on the chosen prior. An often quoted objective choice is the Jeffreys prior. The Jeffreys prior was initially proposed to acquire an invariance property under reparameterization [4]. Later, studies have shown that the Jeffreys prior also has several minimax properties [1,11]. For example, it achieves the asymptotic minimax risk for model families with smooth finite-dimensional parameters. This requirement is met in most of the cases for Bayesian networks. However, when the maximum likelihood parameters lie on the boundary of the parameter space, Jeffreys prior may fail to achieve the asymptotic minimax property. In this work, for the sake of simplicity, we assume that the necessary conditions are satisfied and ignore the boundary issues.

The Jeffreys prior is proportional to the square root of the determinant of the Fisher information matrix:

$$p(\theta) = \text{FII}(\mathcal{M})^{-1} \sqrt{\det I(\theta)}. \tag{7}$$

The normalizing term, which we call the *Fisher information integral* (FII), is given by

$$\text{FII}(\mathcal{M}) = \int_{\Theta_{\mathcal{M}}} \sqrt{\det I(\theta)} \, d\theta.$$

Plugging Eq. (7) in Eq. (5), we get the Fisher information approximation:

$$\text{FIA}(x^n \, ; \, \mathcal{M}) = \log p(x^n \, ; \, \hat{\theta}_{\mathcal{M}}(x^n)) - \frac{d_{\mathcal{M}}}{2} \log \frac{n}{2\pi} - \log \text{FII}(\mathcal{M}) + o(1). \tag{8}$$

For Bayesian networks, which is the model class studied in this work, the Jeffreys prior has been derived in [7]. Unfortunately, as the authors showed, evaluating it is NP-hard. Therefore, it is unlikely that an efficient formula for FII could be obtained for Bayesian networks. To get around this difficulty, we introduce a way to approximate FII by first linking the marginal likelihood to another model selection criterion via the FIA formula.

2.2   Approximation of Normalized Maximum Likelihood

The FIA formula is important not only because it approximates the Bayesian marginal likelihood. It also coincides with the asymptotic form of the normalized maximum likelihood (NML) model selection criterion [17]. NML is a modern form of the minimum description length (MDL) principle, which is an information theoretic approach to select the model that has the shortest code length for describing the information in the data [2,12].

The NML model is defined as:

$$\text{NML}(x^n \, ; \, \mathcal{M}) = \frac{p(x^n \, ; \, \hat{\theta}_{\mathcal{M}}(x^n))}{C_n^{\mathcal{M}}}, \tag{9}$$

where the normalizing factor $C_n^{\mathcal{M}}$ is the sum of the maximum likelihoods over all potential data sets:

$$C_n^{\mathcal{M}} = \sum_{x^n} p(x^n \, ; \, \hat{\theta}_{\mathcal{M}}(x^n)). \tag{10}$$

NML provides a unique solution to minimize the *worst case regret* under log loss for all possible distributions, and the constant $\log C_n^{\mathcal{M}}$ is the minimax and maximin regret, see [17,20].

As stated above, the logarithm of the NML probability shares the same asymptotic expansion as the marginal likelihood under Jeffreys prior, given by

FIA. The regularity conditions required for this to hold are discussed in [11]. Therefore, we can combine Eq. (8) with Eq. (9) and obtain an estimate of $\log \mathrm{FII}(\mathcal{M})$ by:

$$\log \mathrm{FII}(\mathcal{M}) = \log C_n^{\mathcal{M}} - \frac{d_{\mathcal{M}}}{2} \log \frac{n}{2\pi} + o(1), \tag{11}$$

However, the normalizing constant, $C_n^{\mathcal{M}}$ also lacks a closed form solution for most of model families and therefore, its value can be calculated efficiently only for a restricted set of model families such as the Bernoulli and multinomial models [6]. For other cases, one possible solution is to use factorized variants of NML [14], which approximate the formula by factorizing it as a product of locally minimax optimal models. The study in [19] proves that for Bayesian networks, the factorized NML (fNML) is asymptotically equivalent to BIC but leads to improved model selection accuracy for finite samples. In this work, we provide further evidence about the behavior of fNML.

However, instead of resorting to factorized NML variants, where no numerical guarantees about the approximation error are known, we estimate NML by Monte Carlo sampling in the same fashion as in [13]. The obtained estimates can be shown to be consistent as the number of simulated samples is increased. Hence they provide a sound approach for approximating NML and thereby also the FII constant: once we have obtained an estimate of the NML normalizing term, we deduct other terms as in Eq. (8) to approximate $\log \mathrm{FII}(\mathcal{M})$. After that, by plugging in the approximated value of $\log \mathrm{FII}(\mathcal{M})$ in Eq. (11), we can calculate FIA for any sample size without having to repeat the sampling procedure.

3   Monte Carlo Approximation of NML

For Bayesian networks, there is no efficient way to compute the exact value of $\log C_n^{\mathcal{M}}$. We need to consider other approximate methods such as the Monte Carlo sampling method introduced in [13]. Based on the law of large numbers, the sample average is guaranteed to converge to the mean if the sampling size is large. By sampling m data sets $\{x_1^n, \ldots, x_m^n\}$ from distribution $q(\cdot)$, we have a consistent *importance sampling estimator* for $C_n^{\mathcal{M}}$ as:

$$\frac{1}{m} \sum_{t=1}^{m} \frac{p(x_t^n \,;\, \hat{\theta}_{\mathcal{M}}(x_t^n))}{q(x_t^n)} \xrightarrow{a.s.} C_n^{\mathcal{M}} \quad \text{as } m \to \infty. \tag{12}$$

Ideally, any proposal distribution q with full support will guarantee convergence.

However, the shape of q significantly affects the rate of convergence and the variance of the estimator. We need to choose a sampling distribution q that is similar to the target distribution. Following [13], we use the sampling distribution by drawing each set of the parameters independently from the Dirichlet distribution $\mathrm{Dir}(\frac{1}{2}, \frac{1}{2}, \ldots, \frac{1}{2})$, which results in the Krichevsky-Trofimov universal model (K-T model) [8]. It has been proved that the K-T model is asymptotically equivalent to NML as long as the parameters are not on the boundary.

4   Numerical Results Concerning the Lower-Order Terms

In this section we present some properties of $\log \mathrm{FII}(\mathcal{M})$ that are important to the model selection behavior of the FIA formula.

4.1   Numerical Values of $\log C_n^{\mathcal{M}}$ and $\log \mathrm{FII}(\mathcal{M})$

Firstly, for each combination of maximum indegree, number of nodes, and alphabet size, which together determine the number of parameters, we generate 100 Bayesian networks randomly. We estimate the $\log C_n^{\mathcal{M}}$ under different sample sizes to show how the $\log C_n^{\mathcal{M}}$ curve relates to the BIC curve and its upper bound. Note that while the main determinant of the model complexity, as measured by $\log C_n^{\mathcal{M}}$, is the number of parameters, these different Bayesian network models usually have somewhat different complexities. As we will see, however, the variance among networks with a fixed number of parameters is relatively small compared to the differences between networks with a different number of parameters.[2]

As an example, we show the results of Bayesian networks with $l = 20$ nodes, alphabet size $|\mathcal{X}| = 4$, and indegree (number of parents) of each node $k = 5, \ldots, 8$ subject to the acyclicity condition. All estimates of $\log C_n^{\mathcal{M}}$ under each sample size are calculated separately for 100 different Bayesian networks to obtain the mean and the standard deviation. (The variance is due to both the aforementioned differences between different model structures as well as the noise inherent to the Monte Carlo technique.)

Because $C_n^{\mathcal{M}}$ is defined as the sum of maximized likelihoods over all possible data sets, and because in the discrete case the likelihood is always at most one, a trivial upper bound for $\log C_n^{\mathcal{M}}$ is defined as

$$\log C_n^{\mathcal{M}} \leq nl \log |\mathcal{X}|. \tag{13}$$

Figure 1 shows that for small sample sizes, this upper bound tightly squeezes $\log C_n^{\mathcal{M}}$ towards zero. On the other hand, up to constant terms, $\log C_n^{\mathcal{M}}$ shares the same asymptotic form with the BIC (Eqs. (6) and (11)). As the sample size increases, the slope of the $\log C_n^{\mathcal{M}}$ curve will tend to the slope of $\frac{d_{\mathcal{M}}}{2} \log n$. In terms of the graph, where the sample size is shown on a logarithmic scale, the $\log C_n^{\mathcal{M}}$ curve becomes a straight line that is parallel to the corresponding BIC curve. The difference between the curves tends to the constant $\log \mathrm{FII}(\mathcal{M}) - \frac{d_{\mathcal{M}}}{2} \log 2\pi$. The figure suggests that the constant grows rapidly as the model order is increased.

If the sample size is small, the sum of lower-order terms may be a very important part that should not be ignored. For example, Fig. 1 shows that for Bayesian networks with 20 nodes, alphabet size $|\mathcal{X}| = 4$ and maximum indegree $k = 6$, when the sample size is $n = 1000$, the sum of lower terms amount to a

---

[2] An interesting line of future research will be to zoom in into the differences in model complexity within classes of networks with a fixed number of parameters by the techniques we use here.

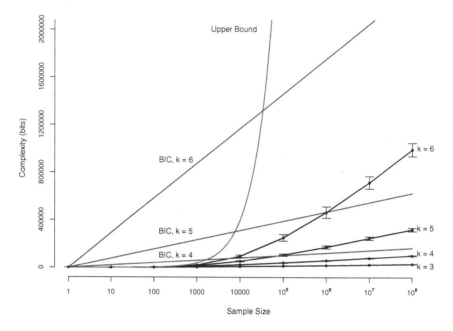

Fig. 1. Estimates of $\log C_n^{\mathcal{M}}$ by Monte Carlo sampling for Bayesian networks with $l = 20$ nodes, $|\mathcal{X}| = 4$ and $k = \{3, \ldots, 6\}$ with increasing sample sizes n from 1 to 10^8 (shown in log-scale). The red line shows the upper bound $nl \log |\mathcal{X}|$. The blue curves are the BIC complexity penalties over different k respectively. The black lines link the means of $\log C_n^{\mathcal{M}}$ at increasing sample sizes with the same k. The error bars showing the standard deviations for each estimates (Color figure online).

number less than $-800,000$. This is because $\log C_n^{\mathcal{M}}$ is restricted by its upper bound to almost zero but the term $\frac{d_{\mathcal{M}}}{2} \log n$ is larger than $800,000$.

4.2   Accuracy of FIA for Small Sample Sizes

Secondly, we look into the accuracy of FIA as an approximation of $\log C_n^{\mathcal{M}}$ when the sample size is small. Here we estimate $\log C_n^{\mathcal{M}}$ by the Monte Carlo sampling method for both small and large sample sizes. We show the estimated values for a set of nested Bayesian networks of 20 nodes. The models are nested in the sense that simpler (less edges) Bayesian networks are obtained by removing edges from a complex ($k = 8$), randomly generated Bayesian network. We simulate $m = 100$ data sets in each case and take the average to estimate the $\log C_n^{\mathcal{M}}$ value. On the other hand, we also estimate the constant term $\log \mathrm{FII}(\mathcal{M})$ (by Eq. (11)) for the same networks using a sample size of 10^9 to make sure that the term $o(1)$ becomes negligible, and plug in the resulting constant into the FIA formula for the smaller sample sizes. Table 1 lists related quantities for Bayesian networks with 20 nodes and alphabet size $|\mathcal{X}| \in \{2, 4\}$, when sample sizes are 10^3 or 10^5 and maximum indegrees are from one to eight.

Based on Table 1, a significant observation is that when the model is very complex, for instance, when $|\mathcal{X}| = 4$ and $k \geq 6$, the $\log \text{FII}(\mathcal{M})$ is a negative number with very large absolute value (less than -10^6). However, the absolute values of the term $\frac{d_\mathcal{M}}{2} \log \frac{n}{2\pi}$, as shown in the third row of Table 1 are much smaller than $\log \text{FII}(\mathcal{M})$ for small sample sizes. Therefore, the term $\frac{d_\mathcal{M}}{2} \log \frac{n}{2\pi}$ is dominated by $\log \text{FII}(\mathcal{M})$, which results in negative values of the sum. For example, as shown in the fourth row of Table 1, for sample size $n = 10^3$, this is the case for alphabet size $|\mathcal{X}| = 4$, with maximum indegree $k \geq 4$; and for alphabet size $|\mathcal{X}| = 2$, with maximum indegree $k = 8$. When the sample size increases to $n = 10^5$, for some simpler networks like $|\mathcal{X}| = 2$, and $k \leq 5$, the values of $\log C_n^\mathcal{M}$ and the sum are fairly close to each other. But for the most complex networks when $|\mathcal{X}| = 4$ and $k \geq 7$, sample sizes as large as 10^5 are still far from enough to even make the sum positive. The more complex the model, the larger sample size that we need to get sensible complexity penalties.

Due to the properties discussed above, the model selection by FIA fails under several conditions. For example, with $|\mathcal{X}| = 2$ and sample size $n = 10^3$, the FIA penalty for Bayesian networks with maximum indegree $k = 6$ is larger than for $k = 7$. Because the simpler network is a subset of the more complex one, the maximum likelihood value for the network with $k = 7$ is always higher or equal to that for the model with $k = 7$. Therefore, the FIA criterion will select the Bayesian network with $k = 7$ rather than the one with $k = 6$, *no matter what the data are*. For sample size $n = 10^5$ the problem does not occur when the alphabet size of $|\mathcal{X}| = 2$ but with $|\mathcal{X}| = 4$, the same problem occurs for $k \geq 7$ even with sample size $n = 10^5$. The rule of thumb that one should have more samples than there are free parameters in the model seems to hold quite well in these situations.

The above observations underline the importance of paying attention to the potential problems due to the $o(1)$ terms involved in the approximations for small and moderate sample sizes. Curiously enough, the BIC formula, which is based on omitting all $O(1)$ terms does not have a similar problem; we will return to this issue below.

5   Model Selection Simulations

In the above, we already made some remarks on the likely consequences of the identified properties of FIA to model selection performance. In this section, we perform a set of simulation experiments to investigate them in detail. We focus in particular on complexity regularization in Bayesian networks. We consider networks with $l = 20$ and $l = 40$ discrete-valued nodes. The alphabet size of each node is varied to be $|\mathcal{X}| = 2$ or $|\mathcal{X}| = 4$.

In each simulation, we restrict the model comparison to a set of eight network topologies that are obtained by constructing a random DAG with each node's indegree $k = 8$ (subject to the acyclicity condition) and removing edges from it to obtain DAGs with maximum indegrees $k = 7, \ldots, 1$. Such a comparison is

Table 1. The $\log C_n^{\mathcal{M}}$ estimates based on FIA (the fourth row) or Monte Carlo sampling (the fifth row), the Fisher information integral $\log \mathrm{FII}$ and the higher order term $\frac{d}{2} \log \frac{n}{2\pi}$ for Bayesian networks of $k = \{1, \ldots, 8\}$, alphabet size $|\mathcal{X}| = \{2, 4\}$ with number of nodes $l = 20$ and sample size $n \in \{10^3, 10^5\}$. Values that are based on Monte Carlo approximation are reported with four significant digits

$$|\mathcal{X}| = 2, \mathbf{n} = 10^3$$

k	1	2	3	4	5	6	7	8
$\log \mathrm{FII}$	-22.88	-37.57	-96.27	-349.9	-1004	-2565	-6488	-14330
$d_{\mathcal{M}}$	39	75	143	271	511	959	1791	3327
$\frac{d_{\mathcal{M}}}{2} \log \frac{n}{2\pi}$	142.6	274.3	523.0	991.1	1869	3507	6550	12167
sum	119.8	236.7	426.7	641.2	864.1	941.7	61.45**	-2163*
$\log C_n$	179.5	298.9	481.2	711.0	1092	1565	2056	2698

$$|\mathcal{X}| = 2, \ \mathbf{n} = 10^5$$

k	1	2	3	4	5	6	7	8
$\log \mathrm{FII}$	-22.88	-37.57	-96.27	-349.9	-1004	-2565	-6488	-14330
$d_{\mathcal{M}}$	39	75	143	271	511	959	1791	3327
$\frac{d_{\mathcal{M}}}{2} \log \frac{n}{2\pi}$	272.2	523.4	998.0	1891	3566	6693	12500	23219
sum	249.3	485.9	901.7	1541	2562	4128	6011	8889
$\log C_n$	308.0	542.4	941.8	1545	2608	4204	6390	10270

$$|\mathcal{X}| = 4, \ \mathbf{n} = 10^3$$

k	1	2	3	4	5	6	7	8
$\log \mathrm{FII}$	-86.96	-1123	-8211	-48710	-239000	-1135000	-5105000	-21230000
$d_{\mathcal{M}}$	231	879	3327	12543	47103	176127	655359	2424831
$\frac{d_{\mathcal{M}}}{2} \log \frac{n}{2\pi}$	844.8	3215	12167	45872	172263	644122	2396742	8867956
sum	757.8	2092	3956	-2840*	-66720*	-490700*	-2709000*	-12360000*
$\log C_n$	832.4	2289	5522	10300	16880	21070	23050	24500

$$|\mathcal{X}| = 4, \ \mathbf{n} = 10^5$$

k	1	2	3	4	5	6	7	8
$\log \mathrm{FII}$	-86.96	-1123	-8211	-48710	-239000	-1135000	-5105000	-21230000
$d_{\mathcal{M}}$	231	879	3327	12543	47103	176127	655359	2424831
$\frac{d_{\mathcal{M}}}{2} \log \frac{n}{2\pi}$	1612	6135	23219	87539	328735	1229203	4573798	16923071
sum	1525	5012	15010	38830	89750	94330	-531500*	-4308000*
$\log C_n$	1582	5059	15310	4137	112500	261100	494000	858900

*) $\log C_n^{\mathcal{M}}$ approximations by FIA with negative values
**) $\log C_n^{\mathcal{M}}$ approximations by FIA with a changing order

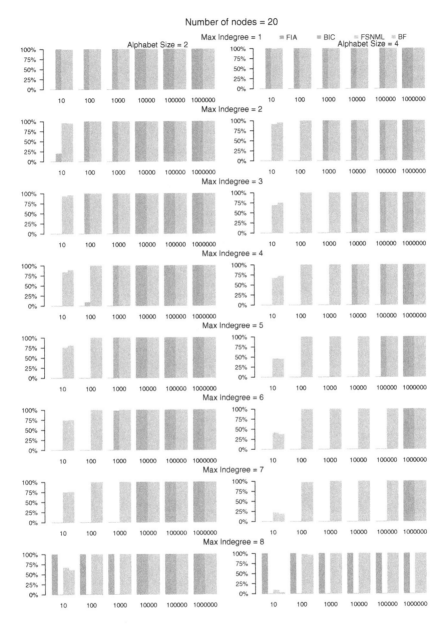

Fig. 2. Model selection experiments for selecting Bayesian networks with 20 nodes and maximum indegree $k = \{1, \ldots, 8\}$. Bars show percentages of correctly identified models by four different criteria as a function of sample size $n = \{10, 10^2, \ldots, 10^6\}$. For the left plots, we have alphabet size $|\mathcal{X}| = 2$, and for the right ones we have $|\mathcal{X}| = 4$. Four criteria are FIA (Fisher information approximation) by Eq. (8), BIC by Eq. (6), fsNML (factorized sequential NML) [19], and BF (Bayes factor with "true" prior) (Color figure online).

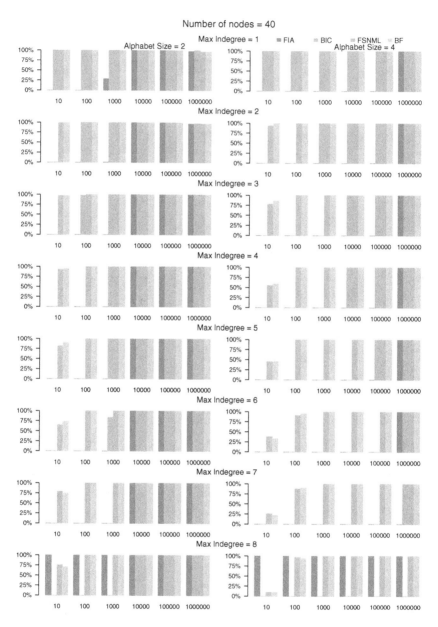

Fig. 3. Model selection experiments with the same settings for Bayesian networks with 40 nodes. (cont'd from Fig. 2) (Color figure online)

admittedly atypical since most practical scenarios involve several possible network topologies with the same maximum indegree, whereas we only consider one topology for each value of k. We adopt the present methodology for the purpose of highlighting the complexity regularization aspect and in order to be able to estimate the FII term accurately for each individual Bayesian network model.[3]

Within each group of Bayesian networks, we compare FIA with other model selection criteria of varying levels of approximation, including BIC [16], and fNML [19]. To obtain a measure of the ideal performance, we also include the Bayes factor based on the "true" prior. In practice, the true prior is obviously not known in advance, and therefore, the Bayes factor criterion should be taken simply as a yardstick against which to compare the other methods. The effect of using different priors in Bayes factors has been studied in [18].

We perform the comparison for sample sizes $10, 100, \ldots, 10^6$. For each sample size we draw 100 random data sets from the true network, and apply the different criteria to select one of the eight possible network structures. We show the results as percentages of correctly identified models in Figs. 2 and 3. For the Bayesian networks with alphabet size $|\mathcal{X}| = 2$ (for both $l = 20$ and $l = 40$), sample size 10^4 is enough for FIA to achieve nearly 100 % accuracy. But for the cases when $|\mathcal{X}| = 4$, FIA needs $n \geq 10^6$ to achieve good performance. Most of the failures are caused by selecting the most complex models with maximum indegree $k = 8$: see the bottom panels of each figure to verify that when the true model is $k = 8$, FIA achieves 100 % accuracy just because it always favors the most complex model available unless the sample size is large enough to avoid the reversed complexity penalty phenomenon discussed in the previous section.

On the contrary, the BIC criterion works better than FIA except when the true model is the most complex one. Its accuracy decreases when the maximum indegree of the true model increases. For networks with $|\mathcal{X}| = 4$ and $k = 8$, the BIC criterion fails even when the sample size reaches 10^6. Based on Table 1, we can see that BIC puts unnecessary large penalties to complex models. Therefore, it tends to select simple models. On the other hand, we note that the fNML criterion performs almost as well as the Bayes factor criterion with the true prior.

6   Conclusions

The simulation experiment verifies that whenever the sample size is not sufficient, the FIA model selection criterion is unreliable for Bayesian network model selection. We emphasize that none of the above suggests that NML or Bayes factors have similar issues for small sample sizes. Indeed, the experiments also show that another kind of (non-asymptotic) approximation of NML, the fNML criterion, behaves almost as well as Bayes factor with the true prior. A remarkable fact is that a very rough approximation (of the Bayes factor as well as the NML), namely the classic BIC criterion where all $O(1)$ terms are ignored, was in our experiments actually never worse and often much better than the FIA criterion where the asymptotic formula is truncated only at the $o(1)$ term.

---

[3] Unlike in the numerical studies in the previous section, here we want to take into account the fine-grained differences between FII values between different Bayesian network models with a fixed number of parameters.

Comparing FIA penalties with $\log C_n^{\mathcal{M}}$ makes it clear that the $o(1)$ term in Eq. (8) is also an essential part when the sample size is small, which leads to huge differences between the FIA penalty and $\log C_n^{\mathcal{M}}$. Similar results are also reported in the early work in [9] for an exponential model and in [15] for Markov sources. Based on the simulation experiment, we suggest that including the constant term alone may actually be dangerous, and in case useful asymptotic formulas are sought after, one should consider more refined approximations that also include $o(1)$ terms.

It is important to note that our goal in this study was not the evaluate the model selection performance of a criterion where the constant FII term is obtained by Monte Carlo techniques. Such a criterion may not be very practical since for complex networks, the sample size at which the $o(1)$ term becomes negligible can be enormous and drawing a sufficient number of random data sets would be time consuming. Instead, we wanted to illustrate the performance of the FIA criterion, independently of the method by which the FII term is obtained. In other words, we wanted to find out whether evaluating the FII term via an approximate analytic formula, for example, would lead to a useful model selection criterion. The answer turns out to be negative. Hence, studying analytic approximations without paying close attention to the $o(1)$ terms is likely to be of limited interest.

In the future, we can extend the study to other model classes such as generalized linear models with continuous parameters to see if the problem of FIA for small sample sizes also applies to them. To address the small sample issues related to FIA, we may also try to analytically break down the $o(1)$ term to obtain more reliable approximations. A closer study for the performance of FIA in general can then be done in these two directions.

References

1. Clarke, B.S., Barron, A.R.: Jeffreys prior is asymptotically least favorable under entropy risk. J. Stat. Plan. Infer. **41**(1), 37–61 (1994)
2. Grünwald, P.D.: The Minimum Description Length Principle. MIT Press, Cambridge (2007)
3. Han, C., Carlin, B.P.: Markov chain Monte Carlo methods for computing Bayes factors. J. Am. Stat. Assoc. **96**(455), 1122–1132 (2001)
4. Jeffreys, H.: An invariant form for the prior probability in estimation problems. J. Roy. Stat. Soc. A **186**(1007), 453–461 (1946)
5. Kass, R.E., Raftery, A.E.: Bayes factors. J. Am. Stat. Assoc. **90**(430), 773–795 (1995)
6. Kontkanen, P., Myllymäki, P.: A linear-time algorithm for computing the multinomial stochastic complexity. Inf. Process. Lett. **103**(6), 227–233 (2007)
7. Kontkanen, P., Myllymäki, P., Silander, T., Tirri, H., Grünwald, P.: On predictive distributions and Bayesian networks. Stat. Comput. **10**, 39–54 (2000)
8. Krichevsky, R., Trofimov, V.: The performance of universal coding. IEEE Trans. Inf. Theor. **27**(2), 199–207 (1981)
9. Navarro, D.: A note on the applied use of MDL approximations. Neural Comput. **16**(9), 1763–1768 (2004)

10. Rasmussen, C. E., Ghahramani, Z.: Occam's razor. In: Leen, T., Dietterich, T., Tresp, V. (eds.) Advances in Neural Information Processing Systems, pp. 294–300 (2001)
11. Rissanen, J.: Fisher information and stochastic complexity. IEEE Trans. Inf. Theor. **42**(1), 40–47 (1996)
12. Rissanen, J.: Information and Complexity in Statistical Modeling. Springer, New York (2007)
13. Roos, T.: Monte Carlo estimation of minimax regret with an application to MDL model selection. In: Proceedings of the IEEE Information Theory Workshop, pp. 284–288. IEEE Press (2008)
14. Roos, T., Rissanen, J.: On sequentially normalized maximum likelihood models. In: Rissanen, J., Liski, E., Tabus, I., Myllymäki, P., Kontoyiannis, I., Heikkonen, J. (eds.) Proceedings of the Workshop on Information Theoretic Methods in Science and Engineering (WITMSE 2008), Tampere, Finland (2008)
15. Roos, T., Zou, Y.: Keep it simple stupid – on the effect of lower-order terms in BIC-like criteria. In: Information Theory and Applications Workshop (ITA), pp. 1–7. IEEE Press (2013)
16. Schwarz, G.: Estimating the dimension of a model. Ann. Stat. **6**, 461–464 (1978)
17. Shtarkov, Y.M.: Universal sequential coding of single messages. Probl. Inf. Transm. **23**(3), 3–17 (1987)
18. Silander, T., Roos, T., Kontkanen, P., Myllymäki, P.: Factorized normalized maximum likelihood criterion for learning Bayesian network structures. In: Jaeger, M., Nielsen, T. D. (eds.) Proceedings of the 4th European Workshop on Probabilistic Graphical Models (PGM 2008), pp. 257–272 (2008)
19. Silander, T., Roos, T., Myllymäki, P.: Learning locally minimax optimal Bayesian networks. Int. J. Approx. Reason. **51**(5), 544–557 (2010)
20. Xie, Q., Barron, A.R.: Asymptotic minimax regret for data compression, gambling, and prediction. IEEE Trans. Inf. Theor. **46**(2), 431–445 (2000)

Unsupervised Evolutionary Algorithm for Dynamic Bayesian Network Structure Learning

Jingguo Dai and Jia Ren[✉]

College of Information Science and Technology, Hainan University,
Haikou 570228, China
renjia@hainu.edu.cn

Abstract. The introduction of temporal dimension makes it difficult and complex to learn dynamic Bayesian network (DBN) structure for huge search space, hence many studies focus on some particular types of DBN, such as dynamic Naive Bayesian Classifier (DNBC). In order to overcome the limited applicability of DBN structure learning methods, this paper proposes an unsupervised evolutionary algorithm in which the selection of initial population has been implemented by means of mutual information to reduce the search space. Furthermore, in view of the poor performance of traditional encoding scheme and the recount of Bayesian information criterion (BIC) score when calculating the individual fitness, we provide a new structure representation without a necessity of the acyclicity test and an updating algorithm for BIC scores with the help of family inheritance to improve the efficiency. Simulations on synthetic data demonstrate that the proposed unsupervised evolutionary algorithm is effective in DBN structure learning.

Keywords: Dynamic Bayesian networks · Structure learning · Genetic algorithm · Mutual information · Family BIC score

1   Introduction

Bayesian network (BN) is a multivariate statistical model based on probability and graph theory to describe the uncertain relationships among a given set of variables. Because BN uses a directed acyclic graph (DAG) to visually complete the acquisition and representation of knowledge, it has become an important tool in the fields of data mining and artificial intelligence [1–3]. Structure learning lays the groundwork for BN studies, especially when there is no prior model. However, it is known that BN structure learning from data is an NP-hard problem [4]. Many methods have been proposed to deal with this difficult situation, leading to three different schemes: model selection by search and score [5,6], and by conditional independency tests [7,8], and hybrid approaches combining the above two schemes [9,10].

© Springer International Publishing Switzerland 2015
J. Suzuki and M. Ueno (Eds.): AMBN 2015, LNAI 9505, pp. 136–151, 2015.
DOI: 10.1007/978-3-319-28379-1_10

As an extension of BN, dynamic Bayesian network (DBN) that takes the temporal characteristic into account has attracted the attention of many researchers in recent years [11–13]. This kind of model has shown better performance than static BN when the stochastic time-varying evolution of a set of variables should be considered [14]. Because of adding the temporal dimension, it becomes more difficult and complex to explore DBN structure learning approaches. Moreover, many mature structure learning methods for static BN cannot be directly applied to dynamic systems. Therefore, some researchers have dedicated to DBN structure learning techniques. Migual et al. [14] presented an evolutionary optimization algorithm to learn the structure of dynamic naive Bayesian classifier (DNBC), this approach effectively acquired the dependencies and eliminated irrelevant attributes to get higher classification accuracy. Another interesting method proposed by Wu et al. [15] used Gaussian assumption to explore a new model, termed Gaussian DBN. It included temporal information and did not require discrete data. However, all of these above structure learning methods only aimed at some certain types of DBNs, i.e., they cannot be applied to other models since much stronger assumptions were required to reduce the solution space, resulting from the exponentially growing dimensionality of the search space with the increasing number of variables.

There are many score-based heuristic algorithms used in the current structure learning researches, including K2 algorithm [16], genetic algorithm [17], ant colony optimization [18], and simulated annealing algorithm [19]. And K2 algorithm is one of the most popular and typical structure learning methods, but its effectiveness depends on the order of nodes, which is the necessary input when researchers search for the parents set. However, it is always difficult to acquire correct priori information in many practical applications, which probably lead to wrong node ordering, and as a result, there may be an incorrect network structure. Compared with K2 algorithm that relies on the prior knowledge, genetic algorithm only uses the fitness function based on data to evaluate each individual, and it has a strong ability of universal search [20] because of many advantages, such as large search range and high search efficiency. However, we found that most of the existing genetic methods have several obvious drawbacks. On one hand, in the procedure of individual coding, it always uses adjacency matrixes that cause low generating probability of feasible solutions and decrease the optimization efficiency since it needs to repeatedly examine the new individual and ensure its acyclicity [19,21], thus increasing the time cost. On the other hand, in the previous studies, it had to recalculate the score of the whole network structure when the researchers attempted to obtain the fitness of the new individual [22]. In fact, there is usually no change for the local structure in the evolutionary process, hence in order to further improve the learning efficiency, it is necessary to consider the local stability of the network. Therefore, aiming at these above problems, this paper proposes an unsupervised evolutionary method inspired from genetic algorithm to complete DBN structure learning. Firstly, in order to reduce the search space, we introduce mutual information to the population initialization for dependency analysis; Secondly,

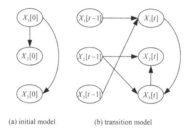

(a) initial model   (b) transition model

Fig. 1. An example of a three-node DBN. (a) gives an initial model, and (b) shows a transition model, including connections both within and across time slots.

we provide a special encoding scheme and new corresponding genetic operators to generate new individual without acyclicity test so as to decrease the time cost; Finally, with respect to the inheritability of Bayesian information criterion (BIC) score for new individual, we concentrate on the estimate whether the scores of the offspring can be inherited from the parent generation.

The rest of this paper is organized as follows. Section 2 gives a brief introduction of DBN and presents the problems of DBN structure learning using genetic algorithms in the previous studies. Section 3 describes the improvement of the previous approaches, including the new coding scheme and corresponding genetic operators. In Sect. 4, the experimental results are reported. Finally, we conclude the paper and discuss the potential extension for future work in Sect. 5.

2   The Problem

At the beginning of this section, some basic concepts need to be introduced first, and then we give the description of shortcomings for DBN structure learning approaches based on genetic algorithm in the previous researches.

2.1   Dynamic Bayesian Network

A DBN is a probabilistic graphical model that combines BN with time information to deal with the temporal data, it is actually made of BN as base structure which is extended to temporal domain. Therefore, we define a group of n random processes as $X = \{X_i\}_{i=1,2,...,n}$ and $X_i[t]$ is the random variable of process X_i at discrete time t. An example of DBN is illustrated graphically in Fig. 1. We can see that a DBN is consisted of two kinds of networks, i.e., $G = (G_0, G_{\rightarrow})$, where G_0, which is a static BN, defines the probability distribution at the initial time; $G_{\rightarrow}$ denotes the connections within and across time slots. Because of the Markovian property, the parents of nodes in $G_{\rightarrow}$ are only allowed to be from the same or previous time slot, i.e., $pa(X_i[t]) \subset \{X[t-1], X[t]\}$.

The process of building a DBN can be separated into two phases: structure learning and parameter learning. Due to the fact that parameter learning is based on the known network structure, it is essential to explore the techniques

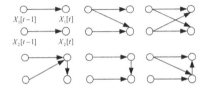

Fig. 2. Several possible transition structures of a two-node DBN.

Table 1. Number of structures for BN and DBN with n nodes

n	BN	DBN
1	1	1
2	3	33
3	25	10591

for structure learning. And usually, the initial structure G_0 of DBN has been already given by the prior knowledge from the start, so in this article, we devote to developing an unsupervised evolutionary algorithm without training data to learn the transition structure $G_\rightarrow$ of a DBN.

2.2   The Solution Space

BN structure learning is a modeling process to find the causal dependencies between variables from sample data. Robinson [23] has proved that the size of the solution space with n nodes for BN is given by the following equation:

$$f(n) = \begin{cases} 1, & n = 0; \\ \sum_{i=1}^{n}(-1)^{i+1}\binom{n}{i}2^{i(n-i)}f(n-i), & n > 0. \end{cases} \tag{1}$$

It is no hard to see that the number of BN structures grows exponentially with the increasing number of nodes. Furthermore, because of the addition of temporal dimension, it is more intractable to learn DBN structure directly and precisely from a huge solution space. More vividly, we plot part of all possible transition structures for DBN with two nodes in Fig. 2, and it can be seen that though they are two-node DBN models, we have to take relationships among four nodes into consideration, including two adjacent time slots, thus it greatly increases the number of possible networks.

Table 1 shows the number of BN and DBN structures, respectively. And it is clear that the number of DBN structures is significantly more than that of BN structures in case of the same number of nodes. Therefore, in order to improve the search efficiency of the proposed evolutionary method, we need to take effective measures to reduce the search space.

Table 2. Rate of randomly generated adjacent matrixes without loop circuit

n	$f(n)$	$F(n)$	$r(n)$
1	1	2	50.000 %
2	3	16	18.750 %
3	25	512	4.883 %
4	543	65536	0.829 %

2.3　The Encoding Scheme

For BN, it usually uses 0 and 1 to represent whether there is an arc between two nodes. For example, when a BN has n nodes, there will be an adjacent matrix $(e_{ij})_{n \times n}$. If X_i points to X_j, $e_{ij} = 1$, i.e., X_i is a parent of X_j; otherwise, $e_{ij} = 0$. Then, this above matrix will be rewritten as a vector in the previous coding scheme, i.e., $(e_{11}, e_{12}, \ldots, e_{1n}, e_{21}, e_{22}, \ldots, e_{2n}, \ldots, e_{n1}, e_{n2}, \ldots, e_{nn})$, which is regarded as the individual coding for BN. As an extension of BN, DBN can also employ this encoding scheme for the nodes in the same time slot as a part of its individual coding. However, the possibility of obtaining a directed acyclic graph is not high by using the above traditional coding scheme. And taking static BN as an example, in Table 2, $F(n) = 2^{n \times n}$ is the total number of two-dimensional $n \times n$ matrixes, $f(n)$ is given in Sect. 2.2, and $r(n) = \frac{f(n)}{F(n)}$ denotes the rate of randomly generated adjacent matrixes without loop circuit.

We can see from Table 2 that the rate of randomly generated adjacent matrixes which satisfy the acyclicity feature decreases promptly with the growth of the number of nodes, that is, when employing an arbitrary adjacent matrix to act as an individual, we have to examine its acyclicity, and its possibility of containing loop circuit grows fast, leading to high time complexity. Hence, it is necessary to provide an alternative encoding scheme to improve code efficiency of the evolutionary algorithm.

2.4　BIC Scoring Function

In regard to the evaluation of the relative merits for different network structures, there are many metrics. And BIC score-based approach is one of the most popular strategies [15] to pick out the optimal solution from the space of possible network structures, according to the rankings. The formula for BIC is given by the following expression:

$$score_{BIC} = \sum_{i=1}^{n} \sum_{j=1}^{q_i} \sum_{k=1}^{r_i} m_{ijk} \log \frac{m_{ijk}}{m_{ij*}} - \sum_{i=1}^{n} \frac{q_i(r_i - 1)}{2} \log m. \tag{2}$$

where n is the number of nodes, q_i defines the count of the possible configurations for $pa(X_i)$, r_i is the number of states for X_i, and m_{ijk} corresponds to the sample count of the kth possible value of X_i given the jth possible configuration of $pa(X_i)$. Besides, $m_{ij*} = \sum_{k=1}^{r_i} m_{ijk}$, m is the total sample count.

BIC makes a tradeoff between the accuracy and complexity of the model, thus it searches for the best DBN structure that not only has high degree of fitting to the data, but also avoids overfitting owing to a penalty in (2). Moreover, the above formula for BIC can be rewritten as follows:

$$score_{BIC} = \sum_{i=1}^{n} score_{BIC}(i). \tag{3}$$

where

$$score_{BIC}(i) = \sum_{i=1}^{q_i} \sum_{k=1}^{r_i} m_{ijk} \log \frac{m_{ijk}}{m_{ij*}} - \frac{q_i(r_i - 1)}{2} \log m. \tag{4}$$

In (4), $score_{BIC}(i)$ is defined as the family BIC score of X_i.

The score of the DBN structure is usually recalculated when the new individuals have been generated in the previous algorithms. However, in the evolutionary process, the difference between parent population and offspring is not big, and it is likely that the parents of only one or two nodes have changed. Hence, there is the reiteration calculation problem using the previous methods, resulting in high time cost. From this perspective, we contribute to the adaptation of the traditional scoring approach by means of family inheritance to greatly simplify the computation of scoring for DBN structures.

3   The Proposed Approach

Due to the drawbacks of previous evolutionary methods presented above, DBN structure learning has usually been limited to few nodes, and has taken high time cost. Therefore, corresponding to the evolutionary process, we give our unsupervised evolutionary algorithm in this section.

3.1   The Representation

At the beginning of the evolution, we have to design an encoding scheme to represent an individual, which is also called chromosome. Allow for the inefficiency of the previous coding scheme mentioned in Sect. 2.3, this paper provides a novel way based on the fact that a BN can at least correspond to one sequence of nodes [24]. In this new scheme, a chromosome consists of three sections: the order of nodes at discrete time t, the linkages across time slots, and the linkages at discrete time t based on the order of nodes given in the first section of the chromosome. The detailed design is given as follows:

$$\begin{aligned} (i_1, i_2, \ldots, i_n, a_{11}, a_{12}, \ldots, a_{1n}, a_{21}, a_{22}, \ldots, a_{2n}, a_{n1}, \ldots, a_{nn}, \\ e_{12}, e_{13}, e_{23}, e_{14}, e_{24}, e_{34}, \ldots, e_{1n}e_{2n}, \ldots, e_{(n-1)n}) \end{aligned} \tag{5}$$

where there are n nodes within the same time slot for a DBN, i_k denotes the kth node in the order, a_{jk} describes whether there is an arc directed from $X_j[t-1]$ to $X_k[t]$, when it is true, $a_{jk} = 1$; otherwise $a_{jk} = 0$, and e_{jk} represents if there is

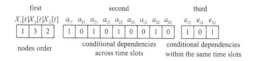

Fig. 3. Representation of an individual corresponding to a DBN transition structure depicted in Fig. 1.

an edge directed from jth node to kth node based on the order in the first section of the chromosome, when it is true, $e_{jk} = 1$; otherwise $e_{jk} = 0$. Notice that in the same time slot, we only allow the preceding nodes to direct to the posterior nodes according to the sequence that is given by the first part of the individual so as to guarantee the acyclicity of the network. Figure 3 shows the encoding of an individual corresponding to a DBN transition structure depicted in Fig. 1.

We can see that $X_2[t]$ only has one parent $X_3[t]$ at discrete time t, and at the same time, $X_3[t]$ has $X_1[t]$ as its parent, so in the first section of the chromosome, $X_1[t]$ should be placed in the first position, while $X_2[t]$ must come last. Besides, the adjacent matrix of these three nodes at discrete time t can be defined as follows:

$$intranet = (e_{ij})_{3\times3} = \begin{bmatrix} 0 & 0 & 1 \\ 0 & 0 & 0 \\ 0 & 1 & 0 \end{bmatrix}$$

Hence, corresponding to the order given by the first section obtained above, the third section which describes the conditional dependencies of nodes at discrete time t should be written as 101. The second section of the chromosome can be easily presented by the adjacent matrix that defines the connections across the time slots given by the following expression:

$$internet = (a_{ij})_{3\times3} = \begin{bmatrix} 1 & 0 & 0 \\ 0 & 1 & 1 \\ 1 & 0 & 0 \end{bmatrix}$$

It is evident that each individual generated with the above scheme has one and only one directed acyclic graph to correspond to. Additionally, this design is not only timesaving for no need to test the acyclicity, but also effective in reducing the encoding length, since using the traditional scheme, when there are n nodes within the same time slot, the length of the third section of the chromosome is $l(n) = n^2$, while by means of the proposed approach, it should be $l'(n) = n(n+1) \div 2 \leq l(n) = n^2$. Therefore, we claim that the new coding scheme has obvious superiority with the growth of nodes.

3.2   The Initial Population

Before the crossover and mutation operators, an initial population needs to be defined, which is the base of the evolution. And if this initial population gets closer to the optimal solution, we will spend less time finding out the true network. However, since the scale of a DBN structure becomes extremely large with the increase of the number of nodes and the introduction of temporal relations, it is troublesome for us to select a proper initial population in the exponentially growing search space. Therefore, to filter useless network structures with the computer in this section, we employ mutual information which is an excellent means of the interpretation of the dependence between nodes to establish initial population.

Given two random variables X and Y, their mutual information can be defined as follows:

$$I(X;Y) = \sum_{i=1}^{r} \sum_{j=1}^{s} p(a_i, b_j) \log \frac{p(a_i, b_j)}{p(a_i)p(b_j)} \qquad (6)$$

where r and s are the number of possible values of X and Y, respectively. a_i is the ith possible value of X, b_j is the jth possible value of Y. It can be seen that if X is completely unrelated to Y, $I(X;Y) = 0$, and the higher value of mutual information indicates the stronger relationship between variables. Consequently, we take advantage of this property to find out one edge for each node, which corresponds to the highest mutual information. And the next step is to search the solution space to locate the network structures containing these edges given by the previous step, and save the number of these networks as n_info. Then, there will be three kinds of situations: when n_info equals pop_size that denotes the size of the population, the individuals transformed from the networks obtained in the second step will directly serve as the initial population. While if $n\_info >$ pop_size, we need to sort these n_info individuals by BIC scores and choose the first pop_size ones as the initial population. The final situation is $n\_info <$ pop_size, and then the initial population should be formed of two parts: these n_info individuals and $pop\_size - n\_info$ arbitrarily selected individuals.

3.3   The Genetic Operators

After obtaining the initial population based on the new encoding scheme designed in Sect. 3.1, we provide the corresponding genetic operators in the following section, including the principles of crossover, mutation and selection.

Crossover. In the crossover operator, only one randomly selected position is defined on each individual. Provided that this position appears on the second or third section of the chromosome, the latter part from this locus is directly exchanged. While in case it presents to the first section which is the sequence of nodes, we need to first exchange the following two sections of the chromosome, and then implement the crossover for the code of the order. In order to explain the principle of the crossover for the first section of the chromosome, we take two sequences containing 4 nodes as an example. There are two arrays: $S_1 =$ $(2, 3, 1, 4)$, $S_2 = (3, 1, 2, 4)$. When the crossover site is 3, we should first translate S_1 and S_2 into $S_1' = (2, 3, 2, 4, 1, 4)$ and $S_2' = (3, 1, 1, 4, 2, 4)$, respectively. Then, each node of S_1' and S_2' will be scanned from left to right. Supposing that one node has emerged in the left-side substring, it has to be eliminated. Therefore, two new sequences has finally been formed: $S_1' = (2, 3, 4, 1)$ and $S_2' = (3, 1, 4, 2)$.

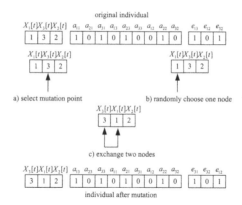

Fig. 4. Mutation operator. When the mutation position lies on the section of the nodes order, there are three stages. Stage 1: select one nutation point. Stage 2: randomly choose a node. Stage 3: exchange two nodes.

Mutation. In this case we give an example in Fig. 4 to clarify how the mutation operator works. There are two kinds of situations. One is that the mutation position locates in the latter two sections of the individual, we have to transfer the figure in this position from 0 to 1 or from 1 to 0. Another circumstance is that selected site lies on the section of the nodes order, and then it should be exchanged with another randomly chosen node for the same chromosome. Hence, we have three options in the mutation operator: insertion or deletion of an arc, or exchange of two nodes. Notice that if there is an illegal network generated after the insertion or deletion, i.e., producing loop circuit within the same time slot, the mutation is not allowed to be executed.

Selection. The selection operator is a strategy used in the population updating process to determine the outstanding individuals from the previous generation as the new offspring. Usually, we make decisions according to the rank of the individual fitness values. In this paper, roulette algorithm is employed to choose excellent individuals, that is, the selected probability of each chromosome is proportional to its fitness value. However, since this choice is random, it has potential for the loss of better individuals. Hence, we introduce the elite mechanism to guarantee the best individual from the previous population has directly been selected to the next generation.

The Fitness Function Based on Family Inheritance. Finally, in order to evaluate the given chromosome, we compute the fitness value of each individual, which is obtained by mapping the BIC score to the closed interval $[0, 1]$, and the fitness value of ith individual is given by the following expression:

$$fit(i) = \frac{score_i - \min_{1 \le j \le n}\left(score_j\right)}{\max_{1 \le j \le n}\left(score_j\right) - \min_{1 \le j \le n}\left(score_j\right)} \tag{7}$$

where $score_i$ represents the BIC score of ith individual, n is the size of the population. As we mentioned in Sect. 2.4, it is inefficient to use previous genetic algorithm for the recalculation of BIC score for each individual when the new population has been generated. As a result, we adopt two matrixes called *old_parents* and *old_famS* to respectively store the parents and family BIC score of each node for the previous population. In particular, *old_parents* $= \{pa_{ij}\}_{n \times m}$, where n is the size of the population, m denotes the number of nodes in the network, and pa_{ij} defines the parents set of jth node for ith individual. Note that if jth node has no parent in ith individual, we record it as $pa_{ij} = \{0\}$. *old_famS* $= (score_{ij})_{n \times m}$, where $score_{ij}$ represents the family BIC score of jth node for ith individual, thereby computing the score of ith individual by summing the ith row of *old_famS*.

In case of obtaining the family score of ith node, we should first take a comparison between ith node of the offspring and ith node of the parent generation, confirming whether its parents set stays the same during the period of evolution. If it is true, the family score of ith node is able to inherit from the previous population. While supposing that there is no same parents set for ith node, we have to scan sample set again to calculate the score. And the pseudocode for the above updating algorithm of *old_parents* and *old_famS* is described in Algorithm 1.

```
Algorithm 1. pa_famS_update
Input: the parents set of the previous population(old_parents);
       the family score of each node for the previous
       population(old_famS); the new population(population);
       the test data(data); the number of states for each
       node(node_size)
Output: new parents set(new_pa), new family score of each node
        for new population(new_famS)
1.  Get the size of the population:pop_size
2.  Get new parents of each node for new population
3.  for i = 1:pop_size
4.    for j = 1:n nodes
5.      Make the sign of the family inheritance:tag_inherit=false
6.      for k = 1:pop_size
7.        if new_pa{i,j}==old_parents{k,j}
8.          new_famS(i,j)=old_famS(k,j)
9.          tag_inherit=true
10.         break
11.       end if
12.     end for
13.     if tag_inherit==false
14.         Scan the data set to calculate new_famS(i,j)
15.     end if
16.   end for
17. end for
```

4   Experimental Results

We have implemented several experiments based on a set of simulated data to validate the effectiveness of the proposed algorithm in this section, including the accuracy and the efficiency. And all experiments were carried out on a PC with AMD Athlon 3.10 Ghz, 32 bits architecture, 4 GB RAM memory and under Windows 7, we used the Matlab software release 7.14.

In the following simulations, we consider a three-node DBN transition model as shown in Fig. 1, which is our target structure. Each node is discrete and observable. $X_1[t]$, $X_2[t]$ and $X_3[t]$ have 2, 3, and 2 states, respectively. Our test data set used as observations consists of 500 sample sequences of $X[t]$ drawn from the target structure, and the parameters of the network is given as follows:

$$\theta_1[t] = \begin{bmatrix} 0.3\ 0.7 \\ 0.9\ 0.1 \\ 0.8\ 0.2 \\ 0.4\ 0.6 \end{bmatrix} \quad \theta_2[t] = \begin{bmatrix} 0.2\ 0.1\ 0.7 \\ 0.3\ 0.4\ 0.3 \\ 0.3\ 0.3\ 0.4 \\ 0.2\ 0.5\ 0.3 \\ 0.5\ 0.1\ 0.4 \\ 0.6\ 0.1\ 0.3 \end{bmatrix} \quad \theta_3[t] = \begin{bmatrix} 0.2\ 0.8 \\ 0.3\ 0.7 \\ 0.3\ 0.7 \\ 0.2\ 0.8 \\ 0.5\ 0.5 \\ 0.6\ 0.4 \end{bmatrix}$$

In all experiments the crossover and mutation rates are set to $cross\_rate = 0.65$ and $mutate\_rate = 0.001$, respectively. $pop\_size = 30$, $generation\_size = 200$, and the sign of elite mechanism: $elitism = true$. In order to measure the effectiveness of the proposed algorithm, in Sects. 4.2 and 4.3, we have made a comparison of four methods, including unsupervised evolutionary algorithm proposed in this paper (UEA), an evolutionary method using randomly selected initial population (RIEA), an approach without utilizing family inheritance to compute the fitness (NIEA), and K2 algorithm using BIC score-based approach presented in Bayesian toolbox (K2A). These four methods are run ten times. And in all experiments we have achieved unsupervised learning adaptation, which is a key advantage of UEA. Therefore, in order to ensure fair comparison, the order of nodes as input in K2A is given stochastically without the priori information.

4.1   The Search Space

Due to the rapid growth of the dimension for DBN structures with the increasing number of nodes, the size of the search space becomes completely huge, and it limits the extension of DBN when the number of nodes is high. Therefore, one of the contributions for this paper is to reduce the size of the search space with the help of dependency analysis approach by excluding networks which do not contain the connections that have the maximum mutual information. And on this basis, we can select the initial population from a smaller space at the beginning of the unsupervised evolutionary algorithm. In order to measure the efficiency of the proposed method in compressing search space, we use the normalized rate given by:

$$r = \frac{n\_total\_struct - n\_info\_struct}{n\_total\_struct} \tag{8}$$

Table 3. Number of structures using mutual information

n	n_total_struct	n_info_struct	r
2	33	12	63.636 %
3	10591	960	90.836 %

where n_total_struct denotes the total number of all possible structures for a three-node DBN, and n_info_struct represents the number of structures containing the connections that have the maximum mutual information. It is shown in Table 3 that when the number of nodes is 2 or 3, the corresponding total number of networks is 33 and 10591, respectively. But after using mutual information to filter the networks, the size of the space has a sharp drop to 12 and 960, respectively.

We can see from the fourth column of Table 3 that the size of the search space has significantly shrunk when the number of nodes increases, and this operation is able to accelerate the speed of locating the target, i.e., find the optimal solution more quickly for the limitation to a small range.

4.2 The Accuracy

We first examine the accuracy of four algorithms mentioned above by means of the Structural Hamming Distance (SHD), which is a principled measurement on how well the estimated network matches the target structure, and this value is defined as follows:

$$s(i) = A(i) + D(i) + T(i) \tag{9}$$

where $s(i)$ is the SHD between the estimated optimal structure and target network in ith time, so $1 \leq i \leq 10$; $A(i)$, $D(i)$, and $T(i)$ denotes the number of mistakenly added arcs, deleted arcs and reversed arcs, respectively, compared with target structure in ith time. From this perspective, the smaller value of SHD implies that the evaluated structure is more similar with true network and indicates the better learning ability. The resultant divergences of SHD in ten-time experiments for 4 methods have been listed in Table 4.

It is seen from Table 4 that there are two algorithms which achieve the lowest divergences from target structure, including UEA and NIEA. Both of these two methods have taken advantage of mutual information when selecting the initial population. While when we try to complete the initialization with stochastic individuals, large estimation error is always incurred, i.e., the phenomenon of adding redundant arcs happens every time. Because there is no guarantee that random synthetic networks reflect real-structure data, it is usually difficult to converge to correct structure at the end of the evolutionary process. Furthermore, we can see that there are always errors with target structure when we employ K2A method. This is because this algorithm requires the initial order of nodes as input, which is strongly dependent on prior knowledge, and in our simulations,

Table 4. Resultant divergences of SHD in ten-time experiments for 4 methods

UEA				RIEA				NIEA				K2A			
$A(i)$	$D(i)$	$T(i)$	$s(i)$	$A(i)$	$D(i)$	$T(i)$	$s(i)$	$A(i)$	$D(i)$	$T(i)$	$s(i)$	$A(i)$	$D(i)$	$T(i)$	$s(i)$
0	0	0	0	2	1	1	4	0	0	0	0	0	0	2	2
0	0	0	0	2	2	0	6	0	0	0	0	0	0	1	1
0	0	0	0	2	1	1	4	0	0	0	0	0	1	1	2
0	0	0	0	2	1	0	3	0	0	0	0	0	0	2	2
0	0	0	0	1	1	2	4	0	0	0	0	0	1	1	2
0	0	0	0	1	1	2	4	0	0	0	0	0	1	1	2
0	0	0	0	2	2	0	4	0	0	0	0	0	1	1	2
0	0	0	0	3	3	0	6	0	0	0	0	0	1	1	2
0	0	0	0	1	2	1	4	0	0	0	0	0	1	1	2
0	0	0	0	2	3	0	5	0	0	0	0	0	0	1	1

we run ten times with random order that is probably not correct, then it is easy to converge to a wrong mode with deletion or reverse of some arcs.

4.3   The Efficiency

To validate the efficiency of unsupervised evolutionary algorithm proposed in this paper, we have carried out an experiment under the conditions described at the beginning of Sect. 4 to measure the running time (in seconds). Four different methods are compared in Table 5.

Table 5. Running time of ten-time experiments for 4 methods

UEA	RIEA	NIEA	K2A
5.1894	3.2423	11.5464	0.1644
5.1507	3.1699	11.4676	0.1702
5.1511	3.2564	11.6401	0.1689
4.9878	3.1201	11.4841	0.1706
5.2007	3.0384	11.9823	0.1688
4.9331	2.9512	11.5268	0.1820
4.9261	2.9232	11.2611	0.1738
4.8871	2.9306	11.2592	0.1735
4.9020	2.8843	11.4429	0.1757
4.8894	2.8582	11.1807	0.1900

According to the resultant data listed Table 5, we can compute the average running time of these four methods. As a result, RIEA using randomly selected initial population obtains results in the second least average running time because there is no filter process in this method and it employs the family inheritance to evaluate new individuals for saving time. While NIEA without utilizing family inheritance to compute the fitness spends the most time getting results, since it has to recalculate BIC score of each individual for new generation, and it adds a filter process using mutual information. Furthermore, we can see that the computation time of the proposed algorithm UEA is much lower than that of NIEA. This is because family inheritance takes the decomposable property of BIC into consideration so as to avoid the recalculation in the evolutionary process.

In combination with the accuracy experiments in Sect. 4.2, it also can be seen that though two approaches, including RIEA without utilizing mutual information and K2A using K2 algorithm based on BIC, take less running time than UEA, it is usually divergent or easy to converge to a wrong structure which has been validated in the previous section, since in RIEA initial population is arbitrarily selected from a large solution space without a filter process, and K2A method performs poorly when given wrong initial node ordering, moreover the time costs will be bound to increase if it needs the correct order as input [24].

Given that a DBN model contains not just three variables, we carry out several experiments based on UEA proposed in this paper for multiple nodes, including four, six and ten nodes. However, before the filter process to complete the initialization, the computer runs out of memory frequently when we calculate the solution space under the current computing environment. We think that this is because original search space grows so large with the increasing number of nodes that it consumes too much memory, resulting in a program crash. For example, the total number of possible transition structures for a four-node DBN is 32382465. As future work we plan to explore alternative initialization in the evolutionary process to adapt to multiple nodes.

5   Conclusion and Perspectives

In this paper an unsupervised evolutionary algorithm to tackle with the structural learning problems for DBN transition network is proposed. According to the analysis of previous studies, we found that many researchers only focus on one particular type of DBN, such as DNBC and Gaussian DBN, due to the disadvantage of exponential growth of search space for DBN structures with the increasing number of nodes. Therefore, to deal with this problem we make use of mutual information to design initial population in a small search space. Moreover, a novel encoding scheme and its corresponding adapted operators are employed in the proposed algorithm since there is no need to test the acyclicity of new individuals formed in the evolutionary process. Additionally, the family inheritance is introduced to update the BIC score of each individual for a new generation in order to avoid the recalculation. We have validated the effectiveness of the proposed method compared with other three approaches without using these above techniques. The experimental results demonstrate the superiority of our method which has excellent performance in accuracy and efficiency. Furthermore, it is worth mentioning that the proposed algorithm in this paper is able to apply to many types of DBN for its fewer assumptions than other methods, and the key advantage of the proposed method is its unsupervised character, i.e., it does not rely heavily on priori information but be based on data.

Future work is aimed at adapting our method to learn time-varying DBN models which have a DBN as base structure for the sake of modeling nonstationary sequences. Additional experiments are planned to extend the proposed method to multiple nodes and analyze the robustness of the proposed algorithm when some part of the data set is missing or wrong as well as to analyze this approach in small sample size problem.

Acknowledgments. This paper was supported by the International S&T Cooperation Projects of China (2015DFR10510), the National Natural Science Foundation of China (61162010; 61440048; 61562018), and the Natural Science Foundation of Hainan Province, China (614227).

References

1. Heckerman, D.: Bayesian networks for data mining. Data Min. Knowl. Disc. **1**, 79–119 (1997)
2. Zhang, L., Zhang, J., Sun, Y.: The construction and application of Bayesian network in data mining. In: 6th IEEE International Conference on Information Management, Innovation Management and Industrial Engineering, pp. 501–503. IEEE Press, New York (2013)
3. Wachsmuth, S., Brandt-Pook, H., Socher, G., Kummert, F., Sagerer, G.: Multilevel integration of vision and speech understanding using Bayesian networks. In: Christensen, H.I. (ed.) ICVS 1999. LNCS, vol. 1542, pp. 231–254. Springer, Heidelberg (1998)
4. Chickering, D.M.: Learning Bayesian networks is NP-complete. In: Fisher, D., Lenz, H. (eds.) Learning from Data. LNS, vol. 112, pp. 121–130. Springer, Heidelberg (1999)
5. Xia, J., Richard, E.N., Michael, B., Shyam, V.: Learning genetic epistasis using Bayesian network scoring criteria. BMC Bioinform. **12**, 1–12 (2011)
6. Liu, Z., Malone, B., Yuan, C.: Empirical evaluation of scoring functions for Bayesian network model selection. BMC Bioinform. **S15**, 14 (2012)
7. Wang, S., Xu, G., Du, R.: Restricted Bayesian classification networks. Sc. China Inf. Sci. **56**, 1–15 (2013)
8. Song, W., Yu, J.X., Cheng, H., Liu, H., He, J., Du, X.: Bayesian network structure learning from attribute uncertain data. In: Gao, H., Lim, L., Wang, W., Li, C., Chen, L. (eds.) WAIM 2012. LNCS, vol. 7418, pp. 314–321. Springer, Heidelberg (2012)
9. Tsamardinos, I., Brown, L.E., Aliferis, C.F.: The max-min hill-climbing Bayesian network structure learning algorithm. Mach. Learn. **65**, 31–78 (2006)
10. Wong, M.L., Leung, K.S.: An efficient data mining method for learning Bayesian networks using an evolutionary algorithm-based hybrid approach. IEEE Trans. Evol. Comput. **8**, 378–404 (2004)

11. Srinivasa, K.G., Seema, S., Jaiswal, M.: Modelling of time series microarray data using dynamic Bayesian network. Retrovirology **84**, 489–492 (2009)
12. Shibata, K., Nakano, H., Miyauchi, A.: A learning method for dynamic Bayesian network structures using a multi-objective particle swarm optimizer. Artif. Life Robot. **16**, 329–332 (2011)
13. Shin, J., Lee, T., Kim, J., Lee, H.: Stochastic model of production and inventory control using dynamic Bayesian network. Artif. Life Robot. **13**, 148–154 (2008)
14. Palacios-Alonso, M.A., Brizuela, C.A., Sucar, L.E.: Evolutionary learning of dynamic Naive Bayesian classifiers. J. Autom. Reason. **45**, 21–37 (2010)
15. Wu, X., Wen, X., Li, J., Yao, L.: A new dynamic Bayesian network approach for determining effective connectivity from fMRI data. Neural Comput. Appl. **24**, 91–97 (2014)
16. Wei, Z., Xu, H., Li, W., Gui, X., Wu, X.: Improved Bayesian network structure learning with node ordering via K2 algorithm. In: Huang, D.-S., Jo, K.-H., Wang, L. (eds.) ICIC 2014. LNCS, vol. 8589, pp. 44–55. Springer, Heidelberg (2014)
17. Wang H., Yu K., Yao H.: Learning dynamic Bayesian networks using evolutionary MCMC. In: 2nd IEEE International Conference on Computational Intelligence and Security, pp. 45–50. IEEE Press, Guangzhou (2006)
18. Salama, K.M., Freitas, A.A.: Learning Bayesian network classifiers using ant colony optimization. Swarm Intell. **7**, 229–254 (2013)
19. Li, J., Chen, J.: A hybrid optimization algorithm for Bayesian network structure learning based on database. J. Comput. **9**, 2787–2791 (2014)
20. Bac, F.Q., Perov, V.L.: New evolutionary genetic algorithms for NP-complete combinatorial optimization problems. Biol. Cybern. **69**, 229–234 (1993)
21. Lee, J., Chung, W., Kim, E., Kim, S.: A new genetic approach for structure learning of Bayesian networks: matrix genetic algorithm. Int. J. Control Autom. Syst. **8**, 398–407 (2010)
22. Ross, B.J., Zuviria, E.: Evolving dynamic Bayesian networks with multi-objective genetic algorithms. Appl. Intell. **26**, 13–23 (2007)
23. Robinson, R.W.: Counting unlabeled acyclic digraphs. In: Little, C.H.C. (ed.) Combinatorial Mathematics V. LNM, vol. 622, pp. 28–43. Springer, Heidelberg (1977)
24. Chen, X.W.: Improving Bayesian network structure learning with mutual information-based node ordering in the K2 algorithm. IEEE Trans. Knowl. Data Eng. **20**, 628–640 (2007)

A Fast Clique Maintenance Algorithm
for Optimal Triangulation of Bayesian Networks

Chao Li[✉] and Maomi Ueno

University of Electro-Communications,
1-5-1 Chofugaoka, Chofu, Tokyo 182-8585, Japan
{ricyou,ueno}@ai.is.uec.ac.jp

Abstract. The junction tree algorithm is currently the most popular algorithm for exact inference on Bayesian networks. To improve the time and space complexity of the junction tree algorithm, we must find an optimal total table size triangulations. For this purpose, Ottosen and Vomlel proposed a depth-first search (DFS) algorithm for optimal triangulation. They also introduced several techniques for improvement of the DFS algorithm, including dynamic clique maintenance and coalescing map pruning. However, their dynamic clique maintenance might compute some duplicate cliques. In this paper, we propose a new dynamic clique maintenance that only computes the cliques that contain a new edge. The new approach explores less search space and runs faster than the Ottosen and Vomlel method does. Some simulation experiments show that the new dynamic clique maintenance improved the running time of the optimal triangulation algorithm.

Keywords: Optimal triangulation · Junction tree algorithm · Dynamic clique maintenance

1 Introduction

Bayesian networks are graphical models that encode probabilistic relations among variables [1]. A Bayesian network is a directed acyclic graph in which vertices represent random variables, and the arcs represent conditional dependencies. Two vertices that are not connected by an arc represent the two variables that are conditionally independent of each other. Each variable is associated with a conditional probability table conditioning on its parents. Bayesian networks provide a neat and compact representation of joint probability distributions.

Probabilistic inference is an extremely common task that is conducted on Bayesian networks. However, probabilistic inference using Bayesian network is known to be NP-hard [2]. The network size limitation of the inference algorithm obstructs the more widespread application of Bayesian Networks. Many studies have been undertaken to improve inference algorithms in the past two decades. The most influential exact inference algorithm is the junction tree propagation algorithm [3–5]. In this algorithm, a Bayesian network is first converted into a

© Springer International Publishing Switzerland 2015
J. Suzuki and M. Ueno (Eds.): AMBN 2015, LNAI 9505, pp. 152–167, 2015.
DOI: 10.1007/978-3-319-28379-1_11

special data structure called junction tree; then belief is propagated on that tree. A junction tree can be formed if and only if the moral graph of the Bayesian network is triangulated. If the graph is not triangulated, then it is necessary to add extra edges to it until it becomes so. This process is called triangulation. A Bayesian network allows several different triangulations. The triangulation is expected to affect the structure of the junction tree and the performance of subsequent belief propagation. In this paper, we especially focus on the optimal triangulation of Bayesian networks. Unfortunately, finding an optimal triangulation is NP-hard [6]. However, this defect is not crucially important because the triangulation can often be done off-line and can be saved for inference algorithms.

Previous investigations of triangulation problems have been conducted by researchers from various fields for different purposes. Their triangulation algorithms are designed to optimize various criteria. The commonly used criteria are the fill-in, the treewidth, and the total table size. Of all these criteria, the total table size criterion yields the most exact bounds of the memory and time requirements of probabilistic inference. Thus, for inference on a Bayesian network, a triangulation is optimal if the triangulation has the minimum total table size. Finding an optimal triangulation is important because the junction algorithm provides the best performance from optimal triangulation. Moreover, optimal triangulation is required for an embedded system that is often with real-time computing constraints and with limited memory usage. We solve the optimal triangulation problem by searching the space of all possible triangulations. This search is conducted by enumerating all possible elimination orders to find the order that has the minimum total table size.

To obtain an optimal triangulation for total table size criterion, Ottosen and Vomlel investigated depth-first search and best-first search algorithms [7]. They claimed that the depth-first search uses less memory than the best-first search does. Moreover, they demonstrated that the two methods have almost equal run time in computational experiments. The best-first search with theoretically better order does not necessarily run faster than the depth-first search in practice. Although the depth-first search expands more search nodes than the best-first search does, the best-first search has heavy overhead costs for maintaining a priority queue (In this paper, the term "node" is used exclusively for a point in the search space for the optimal triangulation algorithm. The term "vertex" is used exclusively for a point in the graph being triangulated.). For optimal triangulation algorithms, it is necessary to make the overhead as low as possible when reducing the search space. To reduce the overhead cost, Ottosen and Vomlel introduced dynamic clique maintenance. In the optimal triangulation algorithm, it is necessary to compute total table size of each search node, which is a lower bound of the node. Therefore, we must also ascertain the set of cliques of each node. It is necessary to maintain a set of cliques in a dynamic graph. The dynamic graph means that the edges can be removed and added but the set of vertices is invariant. To compute the cliques of the updated graph, a simple approach is to run the Bron–Kerbosch (BK) algorithm [8] on the graph. However, the BK algorithm suffers from heavy computational costs. For a graph with n vertices, the worst-case running time of the BK algorithm is $\mathcal{O}(3^{n/3})$ [9]. To resolve this problem, Ottosen and Vomlel proposed a dynamic clique maintenance [7], which runs the BK algorithm on a smaller subgraph where all the new cliques can be found and all the existing cliques are removed. The dynamic clique maintenance reduced the overhead of each node and made the optimal triangulation algorithm faster. However, the dynamic clique maintenance proposed by Ottosen and Vomlel might compute some duplicate cliques. The method presents shortcomings in computational costs as the number of duplicate cliques becomes large. In the elimination process for triangulating a graph, it is well know that a new fill-in edge cannot connect to the vertex that has been eliminated. Based on this observation, Li and Ueno [10] proposed an improved dynamic clique maintenance. The Li and Ueno method reduced the search range of the BK algorithm by removing eliminated vertices from the graph that the Ottosen and Vomlel method explores. However, the method still computes many duplicated cliques. A new clique in the updated graph must include a new edge. However, these methods might compute some cliques that do not contain a new edge. Those cliques are computed both in the original graph and in the updated graph.

In this paper, we propose a new dynamic clique maintenance algorithm for optimal triangulation of a Bayesian network. When some new edges are inserted in a graph for triangulation purpose, we must update the set of cliques. A new clique in the updated graph must contain a new edge. The idea of our method is to avoid recomputing the cliques that do not contain a new edge. We only explore an even smaller subgraph than the graph that the Ottosen and Vomlel method explores. The subgraph only contains the new edges and their neighboring vertices. We run the BK algorithm on the subgraph where all the new cliques can be found. The new algorithm explores less search space and runs faster than the Ottosen and Vomlel method does. The computational cost of dynamic clique maintenance is inherent in the calculation of the lower bound at each node. Thereby, the improvement of the dynamic clique maintenance algorithm can decrease the overhead of each node. Some simulation experiments show that the new dynamic clique maintenance improved the running time of the optimal triangulation algorithm.

The remainder of this paper is organized as follows. Section 2 presents the triangulation problem and describes the formulation of the search space of the optimal triangulation algorithm. Section 3 reviews the depth-first search algorithm presented in [7]. In Sect. 4, we propose a new dynamic clique maintenance algorithm. Section 5 provides some experiments that are useful to evaluate the proposed method. Section 6 concludes the paper.

2   Triangulation Problem

We first introduce some notations and definitions for description of triangulation problem. Then we formulate the search space of the optimal triangulation algorithm.

2.1   Notation and Definitions

Let $G = (V, E)$ be an undirected graph with vertex set V and edge set E. For a set of vertices $W \subseteq V$, $G[W] = (W, \{(v, w) \in E | v, w \in W\})$ is the subgraph of G induced by W. For a set of edges F, $\mathcal{V}(F)$ denotes the set of vertices $\{v, w | (v, w) \in F\}$.

Two vertices v and w are said to be adjacent if $(v, w) \in E$. The neighbors $\mathcal{N}(v, G)$ of a vertex v is the set $W \subseteq V$ such that each $u \in W$ is adjacent to v. The family $\mathcal{FA}(W, G)$ of a set of vertices W is defined as the set $(\cup_{w \in W} \mathcal{N}(w, G)) \cup W$. Let $e = (v, w) \in E$ be an edge, we define the neighbors $\mathcal{N}(e, G)$ of an edge e as the set of vertices $U \subseteq V$ such that each $u \in U$ is adjacent to v and w. The family $\mathcal{FA}(F, G)$ of a set of edges F is defined as the set $(\cup_{f \in F} \mathcal{N}(f, G)) \cup \mathcal{V}(F)$. Note the family $\mathcal{FA}(F, G)$ of a set of edges F is a subset of the family $\mathcal{FA}(\mathcal{V}(F), G)$ of a set of vertices $\mathcal{V}(F)$. For example, see the left graph of Fig. 1, we have $F = \{(a,c),(b,c)\}$, and $\mathcal{V}(F) = \{a,b,c\}$, then $\mathcal{FA}(F, G) = \{a,b,c\}$ is a subset of $\mathcal{FA}(\mathcal{V}(F), G) = V$.

A graph G is complete if all pairs of vertices (u,v) $(u \neq v)$ are adjacent in G. A set of vertices $W \subseteq V$ is complete in G if $G[W]$ is a complete graph. If W is a complete set and no complete set U exists such that $W \subset U$, then W is a *clique*. (Remark: Any complete set is called a clique in some literatures. In that case, what we have defined as a clique is called a maximal clique.) The set of all cliques of graph G is denoted as $\mathcal{C}(G)$. For a set of vertices $W \subseteq V$, $\mathcal{C}(W, G)$ denotes the set of cliques that intersects W. Let $G' = (V, E \cup F)$ $(F \cap E = \emptyset)$ be the right graph obtained by adding a set of new edges F to $G = (V, E)$, $\mathcal{RC}(G,G') = \mathcal{C}(G) \backslash \mathcal{C}(G')$ denotes the set of removed cliques, and $\mathcal{NC}(G,G') = \mathcal{C}(G) \backslash \mathcal{C}(G)$ denotes the set of new cliques. For example, in Fig. 1, let G be the graph on the left, and G' be the graph obtained by adding edge (c,d) to G, then in this example, we can compute $\mathcal{C}(G) = \{\{a,b,c\},\{b,d\},\{d,e\},\{c,e\}\}$ and $\mathcal{C}(G') = \{\{a,b,c\},\{b,c,d\},\{c,d,e\}\}$. Therefore, we have $\mathcal{RC}(G,G') = \{\{b,d\}, \{d,e\},\{c,e\}\}$ and $\mathcal{NC}(G,G') = \{\{b,c,d\},\{c,d,e\}\}$.

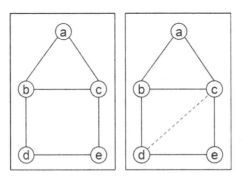

Fig. 1. Left: Initial graph $G = (V, E)$. Right: Updated graph G' obtained by adding one edge (c,d) to G.

The table size of a clique C is defined as $ts(C) = \prod_{(v \in C)} |sp(v)|$, where $sp(v)$ denotes the state space of the variable corresponding to v in the Bayesian network. The total table size (TTS) of a graph G is defined as $tts(G) = \sum_{C \in \mathcal{C}(G)} ts(C)$.

A undirected graph G is said to be triangulated if every cycle of length greater than 3 has a chord, that is an edge connecting two nonconsecutive vertices in the cycle. A triangulation of G is defined as a set of edges T such that $T \cap E = \emptyset$ and graph $H = (V, E \cup T)$ is triangulated. For example, in Fig. 1, the graph on the left is not triangulated because a chord-less cycle {b,c,e,d} exists. The graph on the right is triangulated because the edge (c, d) is added, which is a triangulation for the graph on the left.

Elimination of a vertex $v \in V$ from graph $G = (V, E)$ is the process of adding necessary edges F to make the set $\mathcal{N}(v, G)$ complete, then removing v and all the incident edges from G. The edges F that are added during the elimination process are called fill-in edges. If $F = \emptyset$, then v is called a simplicial vertex of G. An elimination order for graph G is a total ordering π of the vertices of G, where $\pi(i)$ denotes the i-th vertex in the ordering. Let τ be the partial elimination order, which is a sequence of vertices. Let $\mathcal{V}(\tau)$ denotes the set of vertices presented in τ. Let T be all the fill-in edges that result from eliminating vertices from graph G according to order π. We will then use G^π to denote the graph that results from adding these fill-in edges T to $G = (V, E)$ and write $G^\pi = (V, E \cup T)$. Given any elimination order π, if all vertices are eliminated sequentially from G according to π, then the union of all the fill-in edges is a triangulation of G and G^π is a triangulated graph.

We present one example for elimination of vertices from a moral graph of Asia [4] Bayesian network in Fig. 2. Consider an elimination order starting with the sequence $\langle D, S \rangle$. Because eliminating vertex D does not add fill-in edge, D is a simplicial vertex. This process induces two associated graphs (filled-in graph and remaining graph). Let $\tau = \langle X \rangle$ denote the partial elimination order, We also refer to the filled-in graph G^τ as partially triangulated graph, which is shown in Fig. 2(a). The remaining graph $G^\tau[V/\mathcal{V}(\tau)]$ ($\mathcal{V}(\tau) = \{D\}$) is shown in Fig. 2(b). Then we eliminate vertex S. Eliminating vertex S adds a fill-in edge (L,B). This process also induces two associated graphs. Let partial order τ' be the vertex sequence $\langle D, S \rangle$, $F = \{(L,B)\}$ be all fill-in edges when we eliminated along τ'. The corresponding partially triangulated graph $G^{\tau'} = (V, E \cup F)$ is shown in Fig. 2(c). The corresponding remaining graph $G^{\tau'}[V/\mathcal{V}(\tau')]$ ($\mathcal{V}(\tau') = \{D,S\}$) is shown in Fig. 2(d). If we continue to eliminate vertices until no vertex was left. The final partially triangulated graph (also called filled-in graph) is a correct triangulated graph such that there is no chordless cycle on it. Therefore, triangulation using vertex elimination is simple, but the determination of a good elimination order is the most important step. In this paper, we try to find the order π that induces a triangulated graph with minimum total table size.

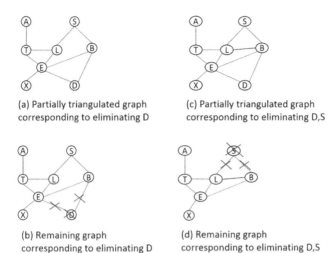

(a) Partially triangulated graph
corresponding to eliminating D

(c) Partially triangulated graph
corresponding to eliminating D,S

(b) Remaining graph
corresponding to eliminating D

(d) Remaining graph
corresponding to eliminating D,S

Fig. 2. Example of eliminating vertices from the moral graph of Asia network.

2.2 Search Space of the Optimal Triangulation Algorithm

To find the optimal triangulation, we can conduct a search in the space of all possible elimination orders of the moralized graph of Bayesian network [7]. For this purpose, we generate a search graph that includes all elimination orders for a Bayesian network. Figure 3 depicts the search graph for a network with five vertices. The search graph is a tree with root node corresponding to the start search node and leaf nodes corresponding to all distinct elimination orders. In this search tree, each node is labeled using a partial elimination order τ. We also associate the intermediate partially triangulated graph with each node for reasons of computational convenience in the optimal triangulation algorithm. Each leaf node is labeled using a complete order and is associated with a triangulated graph. For a node labeled τ, the successor node can be generated by the elimination of a vertex from remaining graph $G^\tau[V/\mathcal{V}(\tau)]$. Given the search tree, we can explore all possible elimination orders to find the order that has minimum total table size.

3   The Optimal Triangulation Algorithm

This section presents a review of the depth-first search algorithm for optimal triangulation presented by Ottosen and Vomlel [7].

3.1   The Depth-First Search Algorithm for Optimal Triangulation

The naive depth-first branch and bound algorithm for optimal triangulation operates as follows. First, we initialize the upper bound (UB) on total table size

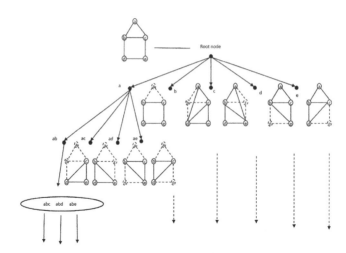

Fig. 3. Search tree of the optimal triangulation algorithm for a graph of five vertices.

(TTS) with the triangulation using minimum fill-in heuristic, which greedily selects the next vertex to eliminate if the vertex leads to add the minimum fill-in edges. Next, it traverses all search tree nodes in a depth-first manner. For each tree node, we calculate the TTS of the partially triangulated corresponding to the node. The TTS is a lower bound of a search node because by adding an edge to graph G, the TTS of G cannot decrease [7]. If we find a node of which TTS is greater than the TTS of UB, then we prune all the successors from the node. On the other hand, if we find a leaf node (labeled using a complete ordering) that is better than UB, we update the UB by replacing the leaf node to UB. The search continues until all nodes have been explored. It is noteworthy that the algorithm performs a search in the space of all elimination orders.

We intend to use the TTS upper bound for pruning nodes in depth-first search triangulations. Therefore, we must compute the TTS of each node in the search tree. The TTS is easy to compute if we know the cliques of the partially triangulated graph corresponding to the node. Therefore, we must also associate the set of cliques with each node. In the Ottosen and Vomlel algorithm [7], for computing the TTS lower bound, each node t is represented as a tuple $(\tau,\mathrm{H},\mathcal{C},\mathrm{tts},\mathrm{R})$.

- $t.\tau$: an ordered list of vertices representing the partial elimination order.
- $t.\mathrm{H} = (\mathrm{V},\ \mathrm{E} \cup \mathrm{F})$: partially triangulated graph obtained by adding all fill-in edges accumulated along the τ to the original moral graph.
- $t.\mathcal{C}$: A set of cliques for H, $\mathcal{C}(\mathrm{H})$.
- $t.\mathrm{tts}$: Total table size of graph H, which is a lower bound for node t.
- $t.\mathrm{R}$: The remaining graph, $\mathrm{R} = \mathrm{H}[\mathrm{V} \backslash \mathcal{V}(\tau)]$, where $\mathcal{V}(\tau)$ denotes the set of vertices that lie in τ.

To compute $t.\mathrm{tts}$, we must calculate all the cliques $t.\mathcal{C}$ first. For this purpose, we can use a standard clique enumeration algorithm such as the well-known

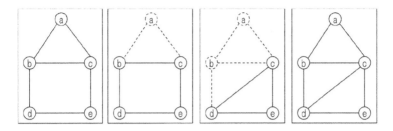

Fig. 4. Example of the vertex elimination and partially triangulated graphs induced by an elimination order that starts with sequence {a,b}. Left: Initial graph. Middle left: Partially triangulated graphs correspond to elimination partial order {a}. Middle right: Partially triangulated graphs correspond to the elimination partial order {a,b}. Right: Final triangulated graph.

Bron–Kerbosch algorithm (BK algorithm) [8]. Below, we present an example to explain the lower bound and related computations.

Example 1. Fig. 4 depicts the vertex elimination process according to the left-most path in Fig. 3. The path corresponds to sequential elimination of vertices a and then b. The *root node* r corresponds to the graph on the left in Fig. 4(initial graph), where no vertex has been eliminated. We can compute the cliques of the root node's graph $r.\mathcal{C} = \{\{a,b,c\},\{b,d\},\{d,e\},\{c,e\}\}$ using the BK algorithm. In this case, the TTS (assuming all binary variables) is $3 \cdot 2^2 + 2^3 = 20$, which is a lower estimate of TTS of optimal triangulation.

The successor node t of r (induced by elimination of vertex a) corresponds to the graph on the middle-left in Fig. 4. The partially triangulated graph t.H is the same as the initial one. Therefore, we can derive t.tts = 20.

We expand the successor node t' of t (corresponding to the elimination of vertex b). The induced partially triangulated graph t'.H corresponds to the middle-right graph in Fig. 4, which includes the fill-in edge (c, d). This process continues until the graph is triangulated. The resulting triangulated graph corresponds to the right graph in Fig. 4, which is the same as t'.H. Finally, the cliques of the triangulated graph are $t'.\mathcal{C} = \{\{a,b,c\},\{b,c,d\},\{c,d,e\}\}$. Their t'.tts is $3 \cdot 2^3 = 24$. In this example, we can see that the TTS of a node is never higher than the TTS of its successor nodes. This key property makes sure the correctness of applying branch and bound technique in the optimal triangulation algorithm.

However, the BK algorithm suffers from heavy computational cost. For a graph with n vertices, the worst-case running time of the BK algorithm is $\mathcal{O}(3^{n/3})$. Indeed, the BK algorithm engenders many redundant computations. To tackle this problem, Ottosen and Vomlel [7] proposed a more efficient algorithm for computation of the set of cliques $\mathcal{C}$ in a dynamic graph. We will explain the dynamic clique maintenance algorithm in Sect. 3.2.

We explained the search tree of depth-first search and how to compute a lower bound for each node. The depth-first search algorithm presented by Ottosen and

Algorithm 1. Depth-first search with coalescing and upper-bound pruning.

```
 1: function TRIANGULATIONBYDFS(G)
 2:     Let s = (G,𝒞(G),tts(G),V)
 3:     EliminateSimplicial(s)                        ▷ Simplicial vertex rule
 4:     if |𝒱(s.R)| = 0 then
 5:         return s
 6:     Let best = MinFill(s)                          ▷ Best path
 7:     Let map = ∅                                    ▷ Coalescing map
 8:     ExpandNode(s,best,map)        ▷ Start recursive call return best
 9: procedure EXPANDNODE(n,&best,&map)
10:     for all v ∈ 𝒱(n.R) do
11:         Let m = Copy(n)
12:         EliminateVertex(m, v)            ▷ Update graph, cliques and TTS
13:         EliminateSimplicial(m)                ▷ Simplicial vertex rule
14:         if |𝒱(m.R)| = 0 then
15:             if m.tts<best.tts then
16:                 Set best = m
17:         else
18:             if m.tts≥best.tts then
19:                 continue                           ▷ Branch and bound
20:             if map[m.R].tts≤m.tts then
21:                 continue
22:             Set map[m.R] = m
23:             ExpandNode(m,best,map)
```

Vomlel can be implemented in $O(|V|)$ space and $O(|V|!)$ time. A pseudo code of the Ottosen and Vomlel algorithm is shown in Algorithm 1. The Eliminate Vertex(m,v) procedure eliminates vertex v from the remaining graph of node m. To prune unnecessary search nodes further, Ottosen and Vomlel also introduced the following pruning rules: (1) Graph reduction techniques called the simplicial vertex rule [11,12], and (2) pruning based on a coalescing map. The procedure EliminateSimplicial(m,v) sequentially removes all simplicial vertices from the remaining graph of node m. Coalescing map uses $\mathcal{O}(n^2)$ memory space to prune unnecessary search nodes; see [13] for details. Although it is well known that the depth-first search runs in $O(|V|!)$ time, the algorithm combined with these techniques described above merely hits the upper bound. Ottosen and Vomlel [7] claimed that their algorithm runs in $\mathcal{O}(2^{|V|})$ time in practice.

3.2 Previous Works on Dynamic Clique Maintenance

Ottosen and Vomlel [7] observed that recomputing all cliques of a graph using the BK algorithm is unnecessary. Then they proposed the following dynamic clique maintenance algorithm. The main idea behind the algorithm is the following. Instead of searching for all cliques in the whole graph, as the BK algorithm does, their algorithm runs a clique enumeration algorithm simply on a smaller subgraph on which all the new cliques can be found and all the existing cliques are

Algorithm 2. Dynamic clique maintenance proposed by Ottosen and Vomlel.

```
 1: procedure CLIQUEUPDATE(G, C(G), F)
 2:     Let G' = (V, E ∪ F)
 3:     C(G') = C(G)
 4:     Let U = V(F)
 5:     for each clique C ∈ C(G) do                          ▷ Remove old cliques
 6:         if C ∩ U ≠ ∅ then
 7:             Set C(G') = C(G')\{C}
 8:     let C^{new} = BKalgorithm(G'[FA(U, G')])
 9:     for each clique C ∈ C^{new} do                       ▷ Add new cliques
10:         if C ∩ U ≠ ∅ then
11:             Set C(G') = C(G') ∪ {C}
```

removed. This dynamic clique maintenance is presented in Algorithm 2, where G is the initial graph, $C(G)$ is the set of cliques of G, F signifies the fill-in edges, and G' is derived by adding F to G. BKalgorithm(G) returns a set of cliques of the graph G. The Ottosen and Vomlel algorithm is derived based on the following theorem:

Theorem 1 ([7]). *Let* $G = (V, E)$ *be an undirected graph, and let* $G' = (V, E \cup F)$ *be the graph result from adding a set of new edges F to G. Let* $U = V(F)$, *and let* $C(G') = C(G)$. *We remove the cliques of* $C(U, G)$ *from* $C(G')$ *and add cliques of* $C(U, FA(U, G')])$ *to* $C(G')$,*then* $C(G')$ *is the set of all cliques of G'.*

(a) Partially triangulated graph
corresponding to eliminating D

(b) Partially triangulated graph
corresponding to eliminating D,S

Fig. 5. A sequence of graphs corresponding to eliminating of vertices D and S. (L,B) is the fill-in edge.

Next, we provide an example to trace Algorithm 2.

Example 2. Consider the left graph G in Fig. 5. $C(G)$ is the set of cliques of G, {{A,T}, {T,L,E}, {E,X}, {S,L}, {S,B}, {B,D,E}}. We add fill-in edges F = {(L, B)} to graph G, resulting in new graph G'(corresponding to the right graph in Fig. 5). The set U = {L,B} and we let $C(G') = C(G)$.

First, we iterate through the cliques in $C(G')$ to remove the cliques that intersect with U, which is the set of cliques {{T,L,E}, {S,L}, {S,B}, {B,D,E}}.

Next, we run the BK algorithm on a subgraph G'[$\mathcal{FA}$(U, G')]. Thereby, we obtain C^{new} = {{T,L,E}, {S,L,B}, {L,B,E}, {B,D,E}}.

Finally, we add to $\mathcal{C}$(G') all the cliques found in the subgraph G'[$\mathcal{FA}$(U, G')] that intersect with U. Now the $\mathcal{C}$(G') = {{A,T}, {E,X},{T,L,E}, {S,L,B}, {L,B,E}, {B,D,E}}, which is the cliques of new graph G'.

Algorithm 3. Dynamic clique maintenance proposed by Li and Ueno (2012).

1: **procedure** CLIQUEUPDATE1(G, $\mathcal{C}$(G), F, W)
2:     Let G' = (V, E∪F)
3:     $\mathcal{C}$(G') = $\mathcal{C}$(G)
4:     Let U = $\mathcal{V}$(F)
5:     **for** each clique C ∈ $\mathcal{C}$(G) **do**   ▷ Remove old cliques
6:         **if** C∩U ≠ ∅ **then**
7:             **if** C∩W = ∅ **then**
8:                 Set $\mathcal{C}$(G') = $\mathcal{C}$(G')\{C}
9:     let C^{new} = BKalgorithm(G'[$\mathcal{FA}$(U, G')\W])
10:    **for** each clique C ∈ C^{new} **do**   ▷ Add new cliques
11:        **if** C∩U ≠ ∅ **then**
12:            Set $\mathcal{C}$(G') = $\mathcal{C}$(G')∪{C}

The example shows that the algorithm sometimes removes and adds the same cliques again. Although the Ottosen and Vomlel method reduces the search range of the BK algorithm from the whole graph to a small subgraph G'[$\mathcal{FA}$(U, G')], the method might present shortcomings in performance when the number of duplicated cliques becomes large. In this example, we observed that vertex D has been eliminated. It is well known that a new fill-in edge cannot connect to the vertex that has been eliminated. The neighbors of D are invariant in G and G'. Therefore, any clique containing D in the initial graph should remain a clique in the updated graph. Generally, no clique containing one of the eliminated vertices should be calculated again. Based on this observation, Li and Ueno [10] proposed an improved dynamic clique maintenance. The improved dynamic clique maintenance is shown in Algorithm 3, where G, $\mathcal{C}$(G), F are defined in the same manner as presented in Algorithm 2, and W is the set of vertices that have been eliminated before. The improved dynamic clique maintenance runs BK algorithm on the graph G'[$\mathcal{FA}$(U, G')\W], which is a subgraph of G'[$\mathcal{FA}$(U, G')] which the Ottosen and Vomlel method explores. As the BK algorithm is the most time consuming part in the dynamic clique maintenance procedure. For a graph with n vertices, the worst-case running time of the BK algorithm is $\mathcal{O}(3^{n/3})$. Therefore, this reduction of the search range is important to improve the performance of the dynamic clique maintenance. In the Li and Ueno method, when we remove an old clique C, one more conditional check is necessary to ascertain whether clique C is disjoint W. This check is usually not a problem because the complexity of comparison of cliques is constant if we store a clique using BitSet Object in JAVA programming language.

4 Proposed Dynamic Clique Maintenance

In the depth-first search triangulation algorithm, it is necessary to compute the TTS lower bound of each search node. Therefore, the computational cost of dynamic clique maintenance is inherent in calculation of the lower bound at each node. To lower the overhead cost in the triangulation algorithm, we must compute the cliques of each graph efficiently.

Algorithm 4. Proposed dynamic clique maintenance.

1: **procedure** CLIQUEUPDATE2(G, $\mathcal{C}$(G), F)
2:     Let G' = (V, E∪F)
3:     $\mathcal{C}$(G') = $\mathcal{C}$(G)
4:     Let W = $\mathcal{FA}$(F, G')
5:     **for** each clique C ∈ $\mathcal{C}$(G) **do**   ▷ Remove old cliques
6:         **if** C ⊆ W **then**
7:             Set $\mathcal{C}$(G') = $\mathcal{C}$(G')\{C}
8:     $\mathcal{C}$(G') = $\mathcal{C}$(G') ∪ BKalgorithm(G'[W])   ▷ Add new cliques

---

In Sect. 3.2, we have demonstrated by example that the Ottosen and Vomlel approach might compute some duplicated clique. To resolve this problem, we propose a new dynamic clique maintenance algorithm. The main idea of our method is to avoid recomputing the cliques that do not contain a new edge. When some new edges are inserted to a graph, a new clique must contain a new edge. We find that all new cliques and removed cliques are included in the vertex set $\mathcal{FA}$(F, G'), where F is the set of new edges. Therefore, we can only run the BK algorithm on the subgraph G[$\mathcal{FA}$(F, G')]. The proposed dynamic clique maintenance algorithm is shown in Algorithm 4, where U, G, F, $\mathcal{C}$(G) are defined in the same manner as presented in Algorithm 2, and W = G[$\mathcal{FA}$(F, G')] denotes the family of a set of edges F. The new algorithm is based on the following Theorem:

Theorem 2. *Let G = (V, E) be an undirected graph, and let G' = (V, E∪F) be the graph resulting from adding a set of new edges F to G. Let U = $\mathcal{V}$(F), and let $\mathcal{C}$(G') = $\mathcal{C}$(G). We remove all the cliques included in $\mathcal{FA}$(F, G') from $\mathcal{C}$(G') and add cliques of $\mathcal{C}$(G'[$\mathcal{FA}$(F, G')]) to $\mathcal{C}$(G'), then $\mathcal{C}$(G') is the set of all cliques of G'.*

Proof. If a clique C is a new clique in G', C ∈ $\mathcal{NC}$(G,G'), then C must contain at least one new edge f ∈ F; otherwise C is not a new clique. Because C contains a new edge, any vertex v ∈ C must be a neighbor of one of the new edges. For any vertex v ∈ C, v must in the set $\mathcal{FA}$(F, G'). Therefore, C ⊆ $\mathcal{FA}$(F, G'). That is to say, all the new cliques can be found on the subgraph G[$\mathcal{FA}$(F, G')].

    If a clique C is a removed clique, C ∈ $\mathcal{RC}$(G,G'), then there exists a new clique K such that C ⊆ K. Because the only way we remove a clique is by replacing the old clique by a new clique K such that C ⊆ K. From the result presented above, a new clique $K \subseteq \mathcal{FA}(F, G')$. Therefore, each removed clique C is included in $\mathcal{FA}(F, G')$.

We remove all the old cliques by removing all the cliques included in $\mathcal{FA}(F, G')$ from $\mathcal{C}(G')$, and then add all the new cliques which can be found on the subgraph $G[\mathcal{FA}(F, G')]$ to $\mathcal{C}(G')$. Then, $\mathcal{C}(G')$ is the set of all cliques of G'.

The following example explains the algorithm:

Example 3. Consider the graph G and updated graph G' in Fig. 5. $\mathcal{C}(G)$ is the set of cliques of G, {{A,T}, {T,L,E}, {E,X}, {S,L}, {S,B}, {B,D,E}}. We let $\mathcal{C}(G') = \mathcal{C}(G)$. We first compute a family of edge set F, $W = \mathcal{FA}(F, G')$, $W = \{S,E,L,B\}$. Next, we remove from $\mathcal{C}(G')$ all the cliques that are included in W, which is the set of cliques {{S,L}, {S,B}}.

Then, we run the BK algorithm on a subgraph G'[W]. We obtain $C^{new} = \{\{S,L,B\}, \{L,B,E\}\}$. In the Ottosen and Vomlel method, we run the BK algorithm on G'$[\mathcal{FA}(U, G')]$, where $\mathcal{FA}(U, G') = \{S,T,E,D,L,B\}$. However, in our new method, we run the BK algorithm on G'[W], where $W = \{S,E,L,B\}$. It can be easily proved that vertex set $W = \mathcal{FA}(F, G')$ is always a subset of $\mathcal{FA}(\mathcal{V}(F), G')$ that is used in Ottosen and Vomlel algorithm. Our method makes the BK algorithm explore less search space for updating cliques than the Ottosen and Vomlel method does. The BK algorithm is the most time-consuming part for the dynamic clique maintenance. Therefore, this reduction of the search range is important to improve the performance of the dynamic clique maintenance.

Finally, we simply add all new cliques C^{new} to $\mathcal{C}(G')$. In the Ottosen and Vomlel approach, it is necessary to check each clique in G'$[\mathcal{FA}(U, G')]$ to ascertain whether it intersects with U, or not. However, we relax this conditional check in our algorithm. In this example, we only remove cliques $\mathcal{RC}(G,G')$ and add cliques $\mathcal{NC}(G,G')$. On the other hand, the Ottosen and Vomlel method removes some duplicated cliques and adds those again.

The new dynamic clique maintenance algorithm performs the BK algorithm on $W = G[\mathcal{FA}(F, G')]$, which is a subgraph of G'$[\mathcal{FA}(U, G')]$ on which the Ottosen and Vomlel method does. The clique enumeration algorithm (BK algorithm) entails exponential costs with the number of vertices. In the dynamic clique maintenance algorithm, the running of BK algorithm is the most time consuming part. Therefore, the reduction of search range is effective to reduce calculation costs of dynamic clique maintenance. In the Ottosen and Vomlel approach, a newly found clique has to be check whether it intersects U, or not. Our method relaxes this conditional check and simply adds the all new cliques found in W. To conclude, the new algorithm explores less search space and runs faster than the Ottosen and Vomlel method does. We demonstrate the performance superiority of the new algorithm by simulation experiments in Sect. 6.

5   Experiments

We conducted computational experiments to evaluate the performance of our proposed dynamic clique maintenance. We also compared our algorithm with the

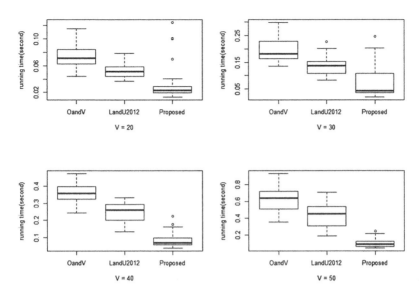

Fig. 6. Comparison of the running times of the Ottosen and Vomlel method (OandV), the Li and Ueno method (LandU2012) and the proposed method for dynamic clique maintenance.

Ottosen and Vomlel method (OandV)[7], and the Li and Ueno (LandU2012)[10]. All algorithms described in this paper were implemented in the Java language in the same manner. All experiments were conducted on a 3.0 GHz processor (Xeon-5675; Intel Corp.) with 12 GB of RAM.

5.1 Dynamic Clique Maintenance

This section presents a comparison of several dynamic clique maintenance methods. For this purpose, we generated 40 random Bayesian networks each for 20, 30, 40, and 50 vertices with various density using BNGenerator[1]. For each graph in the dataset, we triangulated the graph 1,000 times by sequentially eliminating all vertices (with different random elimination order on each run). The set of cliques of the graph is updated after each vertex is eliminated. We chose this experimental scenario because it shows the expected speedup of our proposed method for the triangulation problem. Figure 6 shows the total running time of 1,000 times triangulation for all random Bayesian networks in the dataset. It is clear that the proposed dynamic clique maintenance is faster than OandV and LandU2012.

5.2 Optimal Triangulation

This section describes the experimentally obtained results for the optimal triangulation algorithm for total table size criteria. To examine the effectiveness of our

---

[1] http://sites.poli.usp.br/pmr/ltd/Software/BNGenerator/.

Table 1. Comparison of depth-first search triangulation algorithms among various dynamic clique maintenance methods.

Bayesian networks		DFS	DFS-1	DFS-2
Name	V	Time (s)	Time (s)	Time (s)
Insurance	27	1.656	1.134	0.776
Water	32	10.415	6.247	4.767
Mildew	35	13.285	8.193	5.246
Alarm	37	0.013	0.010	0.006
Hailfinder	56	8.869	7.493	4.337
WIN95PTS	76	60.082	44.122	27.146

proposed dynamic clique maintenance in the optimal triangulation algorithms, we implemented the following algorithm:

DFS: depth-first search (DFS) optimal triangulation algorithm obtained by introducing OandV dynamic clique maintenance.

DFS-1: improved DFS obtained by introducing LandU2012 dynamic clique maintenance.

DFS-2: improved DFS obtained by introducing the proposed dynamic clique maintenance.

We used six well-known graphs in the Bayesian network repository[2]. The running times of optimal triangulation algorithms are presented in Table 1. Results show that our proposed dynamic clique maintenance dramatically improves the running time of triangulation. This result suggests that the proposed method can extend the available network size of Bayesian network inference.

6   Conclusion

In this paper, we proposed a fast clique maintenance algorithm for optimal triangulation of Bayesian Networks. The performance of the proposed algorithm was compared with the state-of-the-art, the Ottosen and Vomlel method, and the Li and Ueno method. Theoretically analysis and experiments reveal that the new method is superior to the previous proposed method.

Given graph G, new edges F, and eliminated vertex set W, consider the problem of updating cliques of new graph. The Ottosen and Vomlel method runs BK algorithm on $G'[\mathcal{FA}(\mathcal{V}(F), G')]$. The Li and Ueno method runs BK algorithm on $G'[\mathcal{FA}(\mathcal{V}(F), G')\backslash W]$. The proposed method runs BK algorithm on $G'[\mathcal{FA}(F, G')]$. The proposed method is faster than the Li and Ueno method. The Li and Ueno method is faster than the Ottosen and Vomlel method. The main reason for the results is that the BK algorithm suffers from heavy computational cost and the proposed method reduces search space for BK algorithm because $G'[\mathcal{FA}(\mathcal{V}(F), G')] \supseteq G'[\mathcal{FA}(\mathcal{V}(F), G')\backslash W] \supseteq G'[\mathcal{FA}(F, G')]$.

---

[2] http://compbio.cs.huji.ac.il/Repository/networks.html.

The Li and Ueno method (2012) assumes eliminating processes in which some edges are added in a step-by-step manner. Application of the method in other areas such as protein interaction network is expected to create a problem. However, the proposed method does not assume this eliminating process. The proposed dynamic clique maintenance is more generally applicable.

References

1. Pearl, J.: Probabilistic Reasoning in Intelligent Systems: Networks of Plausible Inference. Morgan Kaufmann Publishers Inc., San Francisco (1988)
2. Cooper, G.F.: The computational complexity of probabilistic inference using Bayesian belief networks. Artif. Intell. **42**, 393–405 (1990)
3. Lauritzen, S.L., Spiegelhalter, D.J.: Local computations with probabilities on graphical structures and their application to expert systems. J. Roy. Stat. Soc. Ser. B (Methodol.) **50**, 157–224 (1988)
4. Jensen, F., Lauritzen, S., Olesen, K.: Bayesian updating in causal probabilistic networks by local computations. Computat. Stat. Q. **4**, 269–282 (1990)
5. Shenoy, P.P., Shafer, G.: Axioms for probability and belief-function propagation. In: Uncertainty in Artificial Intelligence. Elsevier, Amsterdam (1990)
6. Wen, W.: Optimal decomposition of belief networks. In: Proceedings of the Sixth Conference Annual Conference on Uncertainty in Artificial Intelligence (UAI 1990), pp. 245–256. Elsevier Science, New York (1990)
7. Ottosen, T., Vomlel, J.: All roads lead to rome: new search methods for the optimal triangulation problem. Int. J. Approx. Reason. **53**, 1350–1366 (2012)
8. Cazals, F., Karande, C.: A note on the problem of reporting maximal cliques. Theor. Comput. Sci. **407**, 564–568 (2008)
9. Tomita, E., Tanaka, A., Takahashi, H.: The worst-case time complexity for generating all maximal cliques. In: Chwa, K.-Y., Munro, J.I. (eds.) COCOON 2004. LNCS, vol. 3106, pp. 161–170. Springer, Heidelberg (2004)
10. Chao, L., Maomi, U.: A depth-first search algorithm for optimal triangulation of Bayesian network. In: Proceedings of the Sixth European Workshop on Probabilistic Graphical Models (2012)
11. Bodlaender, H.L., Koster, A.M.C.A., van den Eijkhof, F.: Preprocessing rules for triangulation of probabilistic networks. Comput. Intell. **21**, 286–305 (2005)
12. van den Eijkhof, F., Bodlaender, H.L., Koster, A.M.C.A.: Safe reduction rules for weighted treewidth. Algorithmica **47**, 139–158 (2007)
13. Darwiche, A.: Modeling and Reasoning with Bayesian Networks. Cambridge University Press, Cambridge (2009)

Factorization of ZDDs for Representing Bayesian Networks Based on d-Separations

Shan Gao$^{(\boxtimes)}$ and Shin-ichi Minato

Graduate School of Information Science and Technology,
Hokkaido University, Sapporo, Japan
gaoshan@ist.hokudai.ac.jp
http://art.ist.hokudai.ac.jp

Abstract. Multi-Linear Functions (MLFs) is a well known way of probability calculation based on Bayesian Networks (BNs). For a given BN, we can calculate the probability in a linear time to the size of MLF. However, the size of MLF grows exponentially with the size of BN, so the computation requires exponential time and space. Minato et al. have shown an efficient method of calculating the probability by using Zero-Suppressed BDDs (ZDDs). This method is more effective than the conventional approach of Darwiche et al. which encodes BNs into Conjunctive Normal Forms (CNFs) and then translates CNFs into factored MLFs. In this article, we present an improvement of Minato's method by factoring ZDDs of MLFs into more factored form utilizing weak divison operation based on d-separation structure of BNs.

Keywords: Bayesian Network · Multi-Linear Functions · ZDD · d-separation

1 Introduction

Basyesian networks (BNs) [1] are directed acyclic graphs and used for representing uncertain knowledge across a number of fields. Recently, compiling BNs becomes a hot topic within probabilistic modeling and processing. Algorithms of compiling BNs enable propagation of probability calculation and inference in networks with a reasonable number of variables. However, the Bayesian propagation computations, even for an small example, are very complex and cannot be calculate manually. How to compile a BN into a condensed form has been attracting much attention. One of the approaches is known as Multi-Linear Functions (MLFs) [2], which represent a BN as a polynomial and probabilistic queries are answered by evaluating the polynomial. However, the MLFs itself is exponential in size, it can not be represented explicitly. Minato et al. [3] have shown an efficient method of compiling BNs into factored forms of MLFs based on Zero-suppressed Binary Decision Diagrams [4], which is a graph-based representation first used for VLSI logic design applications. In that method, they produce a set of ZDDs each of which represents MLF of each BN nodes and shows this method

© Springer International Publishing Switzerland 2015
J. Suzuki and M. Ueno (Eds.): AMBN 2015, LNAI 9505, pp. 168–183, 2015.
DOI: 10.1007/978-3-319-28379-1_12

is more effective than conventional approach [2] in some cases. However, since ZDDs are still too large for the realistic BNs. There is a need for more compact representation of MLFs.

In this paper, we present an improvement of Minato's methods of compiling MLFs into more factored forms based on ZDDs to accelerate calculation and inference in BNs since Minato et al. [3] have shown that these operations can be executed in a time almost linear with the ZDD size. We first introduce the weak division algorithm [5] the most successful and prevalent technique of logic synthesis and optimization and show that if we treat MLFs as logic polynomials, we can use this algorithm to factor MLFs into compressed forms. Then we explain that for this algorithm, finding a good *divisor* to factor MLFs is the key to success. Finally, we illustrate that the structure of d-separation used to check some conditional independence in BNs is also effective to help us find a good divisor to execute this factoring. We show the details of our method and experiments results in Sects. 4 and 5.

2 Preliminaries

2.1 Bayesian Networks and MLFs

BayesianNetwork (BN) is a directed acyclic graph which defines a joint distribution over a set of random variables [1]. BNs are used for representing uncertain knowledge across a number of fields. For a given BN and observed data, we calculate the probability distribution of the entire network by substituting the observed data into a portion of the BN.

Each BN node has a network variable X whose domain is a discrete set of values. Each BN node also has a *Conditional Probability Table* (CPT) to describe the conditional probabilities of the value of X given the values of its parent BN nodes. We can use the CPT to represent the probability distribution of the random variable and to predict the likelihood of uncertain events. Figure 1 shows a small example of a BN with its CPTs. In this BN, it has four nodes resulting to 24 different variable instantiations. If want to know the probability of $D = d_1$ given $A = a_1$ (we call $A = a_1$ as *evidence*). First we need to calculate the probability of all cases of B and C. Then using the result to multiply the addition of items in CPT(D) that satisfies $D = d_1$. Although we can use CPT to answer queries, it is usually prohibitive if a BN is huge enough since the size of CPT grows exponentially with the number of variables.

Multi-Linear Functions (MLFs) are well known way of probability calculation based on BNs [2]. An MLF consists of two types of variables, an *indicator variable* λ_x for each value $X = x$, and a *parameter variable* $\theta_{x|\mathbf{u}}$ for each CPT parameter $Prb(x|\mathbf{u})$. The MLF contains a term for each instantiation of the BN variables, and the term is the product of all indicators and parameters that are consistent with the instantiation. For the example in Fig. 1, the MLF has the following

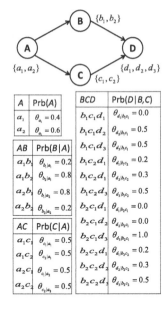

Fig. 1. An example of BN

form:

$$\lambda_{a_1}\lambda_{b_1}\lambda_{c_1}\lambda_{d_1}\theta_{a_1}\theta_{b_1|a_1}\theta_{c_1|a_1}\theta_{d_1|b_1c_1}$$
$$+ \lambda_{a_1}\lambda_{b_1}\lambda_{c_1}\lambda_{d_2}\theta_{a_1}\theta_{b_1|a_1}\theta_{c_1|a_1}\theta_{d_2|b_1c_1}$$
$$+ \lambda_{a_1}\lambda_{b_1}\lambda_{c_1}\lambda_{d_3}\theta_{a_1}\theta_{b_1|a_1}\theta_{c_1|a_1}\theta_{d_3|b_1c_1}$$
$$\cdots$$
$$+ \lambda_{a_2}\lambda_{b_2}\lambda_{c_2}\lambda_{d_3}\theta_{a_2}\theta_{b_2|a_2}\theta_{c_2|a_2}\theta_{d_3|b_2c_2}$$

Once we have generated the MLF for a given BN, the probability of instantiation e can be calculated by setting indicators that contradict e to 0 and other indicators to 1. Namely, we can calculate the probability in time linear in the size of MLF. Apparently, the size of MLF grows exponentially with the size of BN, the calculation is quite time consuming. However, if we factor the MLF into a compact arithmetic expression, it is possible to speed up the probability calculation. One way to do this factorization using Zero-suppressed BDDs has been proposed by Minato at [3].

2.2   Zero-Suppressed BDDs

A Binay Decision Diagram (BDD) is a directed graph representation of a Boolean function, as shown in Fig. 2(a). BDDs have two terminal nodes, which we call *0-terminal node* and *1-terminal node*, and many decision nodes with two edges, called *0-edge* and *1-edge*. A BDD is derived by reducing a binary tree graph as shown in Fig. 2(b). The reduction is based on the two rules shown in Fig. 3.

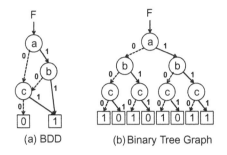

(a) BDD   (b) Binary Tree Graph

Fig. 2. BDD and Binary Decision Tree.

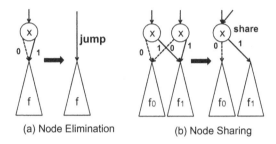

(a) Node Elimination   (b) Node Sharing

Fig. 3. Reduction Rules on BDDs.

BDDs were originally developed for handling Boolean function data, however, they can also be used for implicit representation of *combinatorial itemset*. A combinatorial item set consists of elements each of which is a combination of a number of items. There are 2^n possible combinations of n items, so we have 2^{2^n} possible combinatorial itemset. For example, for a domain of five items a, b, c, d, and e, some combinatorial item sets are:

$$\{ab, e\}, \{abc, cde, bd, acde, e\}, \{1, cd\}, \emptyset.$$

Here "1" denotes a combination of no items, and $\emptyset$ means the empty set.

A combinatorial itemset can be mapped into Boolean space of n input variables. For the truth table of the Boolean function $F = (ab\bar{c}) \vee (\bar{b}c)$, it also represents the combinatorial itemset $S = \{ab, ac, c\}$, which is the set of input combinations for which F is 1. Using BDDs for the corresponding Boolean functions, we can implicitly represent and manipulate combinatorial itemset.

Zero-suppressed BDDs (ZDDs) [4] are variant of BDDs for efficient manipulations of combinatorial itemset. An example of a ZDD is shown in Fig. 4 on the left. ZDDs are based on the following special reduction rules.

– Delete all nodes whose 1-edge directly points to a 0-terminal node, and jump through to the 0-edge's destination, as shown in Fig. 4 on the right.
– Share equivalent nodes as in ordinary BDDs.

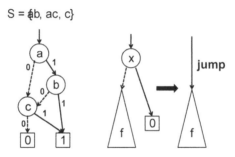

$$S = \{ab, ac, c\}$$

Fig. 4. An example of a ZDD and ZDD reduction rules

According to the reduction rules, nodes of irrelevant items (never chosen in any combination) are automatically deleted. So ZDD can be much more efficient in dealing with combinatorial itemset than BDD. Furthermore, The basic set operations such as intersection and union can be carried out efficiently in ZDDs (the details of algorithms in Minato et al. [3] are omitted). The calculation can be performed in the time approximately proportional to the compressed ZDD size but not the number of terms of the combinatorial itemset.

2.3   ZDD-Based MLF Representation

An MLF is a polynomial in the indicator and parameter variables. It can be regarded as a combinatorial itemset. Since each term is simply a combination of variables, it can be represented compactly by a ZDD. For example, the MLF at node B in Fig. 1 can be written as follows:

$$MLF_B = \lambda_{a_1} \lambda_{b_1} \theta_{a_1} \theta_{b_1|a_1} + \lambda_{a_1} \lambda_{b_2} \theta_{a_1} \theta_{b_2|a_1}$$
$$+ \lambda_{a_2} \lambda_{b_1} \theta_{a_2} \theta_{b_1|a_2} + \lambda_{a_2} \lambda_{b_2} \theta_{a_2} \theta_{b_2|a_2}.$$

Here, we rename the parameter variables with actual number of probability in the CPT of Fig. 1 so that equal parameters share the same variable.

$$MLF_B = \lambda_{a_1} \lambda_{b_1} \theta_{a(0.4)} \theta_{b(0.2)} + \lambda_{a_1} \lambda_{b_2} \theta_{a(0.4)} \theta_{b(0.8)}$$
$$+ \lambda_{a_2} \lambda_{b_1} \theta_{a(0.6)} \theta_{b(0.8)} + \lambda_{a_2} \lambda_{b_2} \theta_{a(0.6)} \theta_{b(0.2)}.$$

The ZDD for MLF_B is shown in Fig. 5. In this example, there are four paths from the root node to the 1-terminal node, each of which corresponds to a term of the MLF. It is an implicit factored representation of the MLF. At the same time, it accelerates probability calculation by using the structure of ZDD of sharing nodes because the calculation is performed in the time proportional to the ZDD size [3].

A ZDD for the entire network in Fig. 1 are shown in Fig. 6 and the experiment results of their approach are shown in Table 1. In this table, the first three columns show the network specifications such as BN name, the number of BN nodes, literals and items. The last four columns presents the ZDD size, literals, items and time consumption when compiling MLFs to ZDDs. Since the size of ZDD changes a lot along with different orders of ZDD node, we use the numbers of literals and items of MLFs, which do not change along with different orders of ZDD nodes, to show upper limits of the calculate and inference consumption as we know that the if we use no technique to condense MLFs size, the consumption is in time linear in the size of literals and items. As the results shown in Minato et al. [3], their method is more effective than conventional approach in some cases.

Although using shared ZDDs we can condense the size of ZDDs to some extent, in some cases the size of ZDDs are still too large. Therefore we consider to factor ZDDs into more condensed size to represent BNs.

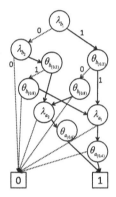

Fig. 5. An example of a ZDD for the MLF_B.

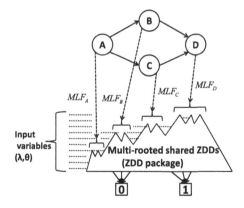

Fig. 6. ZDD construction procedure for BN

Table 1. Original MLFs and ZDDs before factorization.

Dataset	BN specifications			Before factorization			
	BN nodes	Indicators	Parameters	ZDD size	Total terms	Total literals	Time of generating ZDD from LF
alarmN36	37	105	187	4,551	>100 billion	>500 billion	0.647 s
hailfinder43	56	223	835	73,700	>2 billion	>210 million	45.902 s
insuranceN5	27	–	–	6,182	628,992	>10 million	11.768 s
insuranceN14	27	–	–	56,490	>70 million	>2 billion	11.603 s
insuranceN19	27	–	–	17,893	>500 million	>2 billion	11.613 s
MildewN14	35	616	6,709	80,248	>2 billion	>2 billion	946.35 s
MildewN20	35	616	6,709	107,828	>2 billion	>2 billion	947.73 s

3   ZDD Factorization for MLF Representation

3.1   Basic Method of ZDD Factorization

Weak divison algorithm [5] is the most successful and prevalent technique of logic synthesis and optimization. For optimizing a two-level logic (a form of the Boolean expressions with the AND-OR two level structure), we first generate multi-level logics from it and then apply weak divison algorithm to factor the two-level logics. When we determine a good intermediate logic, we make a new variable to present it and regard it as a *divisor*. Then we reduce the other existing logics by factoring them with the *divisor*. Eventually, we construct a multi-level logic that consists of a number of small two-level logics.

The weak division algorithm is executed to computing the common part of quotients for respective items in the *divisor*. For example, suppose the two expressions are

$$f = abd + abe + abg + cd + ce + ch, \text{ and } p = ab + c.$$

If we write f as:
$$f = ab(d + e + g) + c(d + e + h).$$

We factor f by *divisor* p and the quotient (f/p) can then be computed as

$$(f/p) = (f/(ab)) \cap (f/c) = (d + e + g) \cap (d + e + h) = d + e.$$

The remainder $(f\%p)$ is computed using the quotient:

$$(f\%p) = f - p(f/p) = abg + ch.$$

Using the quotient and the remainder, we can rewrite f as follows:

$$f = p(p/f) + (f\%p) = pd + pe + abg + ch.$$

In this example, $f = abd + abe + abg + cd + ce + ch$ with 15 literals is reduced to 12 literals ($f = pd + pe + abg + ch$ has 9 literals and $p = ab + c$ has 3 literals).

procedure (P / Q) {
if (Q = 1) return P ;
if (P = 0 or P = 1) return 0 ;
if (P = Q) return 1 ;
R ← cache (" P / Q ") ; if (R exists) return R ;
v ← Q.top ; / * the highest variable in Q */
(P_0, P_1) ← factors of P by v ;
(Q_0, Q_1) ← factors of Q by v ; /* ($Q_1 \neq 0$) */
R ← P_1 / Q_1 ;
if (R ≠ 0) if ($Q_0 \neq 0$) R ← R ∩ P_0 / Q_0 ;
cache (" P / Q ") ← R ;
return R ;
}

Fig. 7. Fast weak division algorithm [5].

Then if we have *divisor* like $d + e$, we can continue factorization as above to condense the size of f furthermore.

Minato [5] has proposed a fast weak division algorithm to refine this algorithm as described in Fig. 7. They implicitly represent logics using ZDDs and manipulate them using ZDD operations. The fast weak division algorithm is computed in a time almost proportional to the number of nodes in ZDDs, which are usually much smaller than the number of literals in logics, and is much faster than conventional methods [5]. Thus if we have a proper *divisor*, we can quickly condense the size of a given polynomial by using this fast weak division algorithm. So we consider to use this approach to factor a given MLF and the main problem now is how to find a proper *divisor* for a MLF.

3.2   Problem in Factoring MLFs

For an MLF of a given BN, if we consider it as a polynomail, we can extract repeatedly appeared variables by using the fast weak division algorithm to condense its size. Here we use MLF_B in Fig. 1 as an example.

$$MLF_B = \lambda_{a_1} \lambda_{b_1} \theta_{a_1} \theta_{b_1|a_1} + \lambda_{a_1} \lambda_{b_2} \theta_{a_1} \theta_{b_2|a_1}$$
$$+ \lambda_{a_2} \lambda_{b_1} \theta_{a_2} \theta_{b_1|a_2} + \lambda_{a_2} \lambda_{b_2} \theta_{a_2} \theta_{b_2|a_2}.$$

If we use $\lambda_{b_1}\theta_{b_1|a_1} + \lambda_{b_2}\theta_{b_2|a_1}$ as a *divisor* P, according to the algorithm, MLF_B can be factored as follows:

$$MLF_B/(\lambda_{b_1}\theta_{b_1|a_1} + \lambda_{b_2}\theta_{b_2|a_1})$$
$$= (MLF_B/(\lambda_{b_1}\theta_{b_1|a_1}) \cap (MLF_B/\lambda_{b_2}\theta_{b_2|a_1})$$
$$= (\lambda_{a_1}\theta_{a_1}) \cap (\lambda_{a_1}\theta_{a_1})$$
$$= \lambda_{a_1}\theta_{a_1}$$

Finally, MLF_B can be rewritten as

$$MLF_B = \lambda_{a_1}\theta_{a_1} * P + \lambda_{a_2}\lambda_{b_1}\theta_{a_2}\theta_{b_1|a_2} + \lambda_{a_2}\lambda_{b_2}\theta_{a_2}\theta_{b_2|a_2}$$

However, if we factor MLF_B using factor $P = \lambda_{a_1}\theta_{a_1} + \lambda_{a_2}\theta_{a_2}$, the quotient will be the empty set so MLF_B can not be rewritten by *divisor* P. Therefore, the quality of the results of this algorithm greatly depends on the choice of *divisors*.

3.3   Divisor Extraction Based on BN Nodes

The MLF of a node in a given BN is based on its parents nodes. For the node B in Fig. 1, MLF_B contains information about node A. Here we refer to this information with parameters a_1 and a_2. Also, if the number of parameters of node A and B are given, we can forecast the size of MLF_B and the frequency of characters λ and θ. Therefore, we consider factoring an MLF of a node with the MLF of its parents directly. But, this fails when we implement MLF_B/MLF_A. We give the details next.

$$MLF_B/MLF_A$$
$$= MLF_B/(MLF_{a_1} + MLF_{a_2})$$
$$= (MLF_B/(MLF_{a_1}) \cap (MLF_B/(MLF_{a_2})$$
$$= \{(\lambda_{a_1}\lambda_{b_1}\theta_{a_1}\theta_{b_1|a_1} + \lambda_{a_1}\lambda_{b_2}\theta_{a_1}\theta_{b_2|a_1})/\lambda_{a_1}\theta_{a_1}\} \cap$$
$$\quad \{(\lambda_{a_1}\lambda_{b_1}\theta_{a_1}\theta_{b_1|a_1} + \lambda_{a_1}\lambda_{b_2}\theta_{a_1}\theta_{b_2|a_1})/\lambda_{a_2}\theta_{a_2}\}$$
$$= (\lambda_{b_1}\theta_{b_1|a_1} + \lambda_{b_2}\theta_{b_2|a_1}) \cap (\lambda_{b_1}\theta_{b_1|a_2} + \lambda_{b_2}\theta_{b_2|a_2})$$
$$= \emptyset.$$

We refer to the division of MLF_B/MLF_A as blotting out information about node A. Why we get the empty set is that though we try to blot out a_1 by MLF_B/MLF_{a_1}, a_1 is still left in $\theta_{b_1|a_1}$ and $\theta_{b_2|a_1}$. The same applies to a_2. When we intersect the quotients, which are obtained by factoring MLF_B with MLF_{a_1} and MLF_{a_2}, a_1 and a_2 are contrary, hence we obtain the empty set. But, if we omit the intersection, which means we perform the factorization as MLF_B/MLF_{a_1}, MLF_B/MLF_{a_2}, MLF_B can be rewritten as

$$MLF_B = MLF_{a_1}(\lambda_{b_1}\theta_{b_1|a_1} + \lambda_{b_2}\theta_{b_2|a_1}) + MLF_{a_2}(\lambda_{b_1}\theta_{b_1|a_2} + \lambda_{b_2}\theta_{b_2|a_2})$$

However, even this works only in the case of a node which has only one parent node like node B. If it has more than one parent node, for example, node

C in Fig. 1, the representations of MLF_{b_1}, MLF_{b_2} are not capable of factoring MLF_C.

$$
\begin{aligned}
& MLF_D/MLF_{b_1} \\
&= (\lambda_{a_1}\lambda_{b_1}\lambda_{c_1}\lambda_{d_1}\theta_{a_1}\theta_{b_1|a_1}\theta_{c_1|a_1}\theta_{d_1|b_1c_1} + \\
& \quad \lambda_{a_1}\lambda_{b_1}\lambda_{c_1}\lambda_{d_2}\theta_{a_1}\theta_{b_1|a_1}\theta_{c_1|a_1}\theta_{d_2|b_1c_1} + \\
& \quad \lambda_{a_1}\lambda_{b_1}\lambda_{c_1}\lambda_{d_3}\theta_{a_1}\theta_{b_1|a_1}\theta_{c_1|a_1}\theta_{d_3|b_1c_1} + \\
& \quad \cdots \\
& \quad \lambda_{a_2}\lambda_{b_2}\lambda_{c_2}\lambda_{d_3}\theta_{a_2}\theta_{b_2|a_2}\theta_{c_2|a_2}\theta_{d_3|b_2c_2}) \\
& \quad /(\lambda_{a_1}\lambda_{b_1}\theta_{a_1}\theta_{b_1|a_1} + \lambda_{a_2}\lambda_{b_1}\theta_{a_2}\theta_{b_1|a_2}) \\
&= (\lambda_{c_1}\lambda_{d_1}\theta_{c_1|a_1}\theta_{d_1|b_1c_1} + \ldots + \lambda_{c_2}\lambda_{d_3}\theta_{c_2|a_1}\theta_{d_3|b_1c_2}) \cap \\
& \quad (\lambda_{c_1}\lambda_{d_1}\theta_{c_1|a_2}\theta_{d_1|b_1c_1} + \ldots + \lambda_{c_2}\lambda_{d_3}\theta_{c_2|a_2}\theta_{d_3|b_1c_2}) \\
&= \emptyset
\end{aligned}
$$

The reason we get the empty set is since MLF_B is based on node A, when we try to blot out the information about MLF_{b_1} by MLF_C/MLF_{b_1}, we are also blotting out information about a_1 and a_2 contained in MLF_D. The blotting out is inadequate because for a_1 and a_2 are also contained in MLF_C and they contradict to each other when we intersect the quotients. Thus, this motivates us to find a node set that can separate node A and node D as independent nodes so that after we factoring MLF_D, the information about node A can be cleared up thoroughly.

In this paper, we propose an idea of factoring MLFs using the combinations of variables of d-separation node sets so solve the problems mentioned above.

4 Divisor Extraction Based on d-Separations

4.1 d-Separations

The structure of d-separation is used to check conditional independence between variables in Bayesian Networks. It can be presented as three graph patterns [6] in Fig. 8. The d-separation has an important property that if we substitute the observed values to the d-separation nodes, nodes in both sides cut by the d-separation become independent so the calculation of probability inference is simplified. The approach we propose is based on d-separation of serial pattern.

For a given MLF to be factored which is represented by ZDD, first we try to find suitable d-separation node set and multiply their MLFs of these nodes. Then we consider the result of multiplication as a divisor to factor the MLF using fast weak algorithm. We first try from the most simple d-separation which consists of only one node (one-node d-separation). Since the one d-separations not always exist in BNs, we use d-separations which consist of two or three nodes (multi-node d-separations). However, the multi-node d-separations are found manually in this paper.

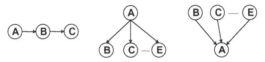

(a) serial pattern (b) diverging pattern(c) converging pattern

Fig. 8. An example of d-separation.

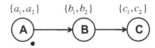

Fig. 9. An example of simple-node d-separation.

4.2   Divisor Selection by One-Node d-Separations

For an MLF to be factored by one-node d-separation, we use the MLF of every variable of this node as its divisor. For the example in Fig. 9, the MLF of node C is:

$$MLF_C = \lambda_{a_1}\lambda_{b_1}\lambda_{c_1}\theta_{a_1}\theta_{b_1|a_1}\theta_{c_1|b_1} + \lambda_{a_1}\lambda_{b_1}\lambda_{c_2}\theta_{a_1}\theta_{b_1|a_1}\theta_{c_2|b_1}$$
$$+ \lambda_{a_1}\lambda_{b_2}\lambda_{c_1}\theta_{a_1}\theta_{b_2|a_1}\theta_{c_1|b_2} + \lambda_{a_1}\lambda_{b_2}\lambda_{c_2}\theta_{a_1}\theta_{b_2|a_1}\theta_{c_2|b_2}$$
$$+ \lambda_{a_2}\lambda_{b_1}\lambda_{c_1}\theta_{a_2}\theta_{b_1|a_2}\theta_{c_1|b_1} + \lambda_{a_2}\lambda_{b_1}\lambda_{c_2}\theta_{a_2}\theta_{b_1|a_2}\theta_{c_2|b_1}$$
$$+ \lambda_{a_2}\lambda_{b_2}\lambda_{c_1}\theta_{a_2}\theta_{b_2|a_2}\theta_{c_1|b_2} + \lambda_{a_2}\lambda_{b_2}\lambda_{c_2}\theta_{a_2}\theta_{b_2|a_2}\theta_{c_2|b_2}$$

The MLFs of variables of node B are:

$$MLF_{b_1} = \lambda_{a_1}\lambda_{b_1}\theta_{a_1}\theta_{b_1|a_1} + \lambda_{a_2}\lambda_{b_1}\theta_{a_2}\theta_{b_1|a_2}$$
$$MLF_{b_2} = \lambda_{a_1}\lambda_{b_2}\theta_{a_1}\theta_{b_2|a_1} + \lambda_{a_2}\lambda_{b_2}\theta_{a_2}\theta_{b_2|a_2}$$

We can factor MLF_C using MLF_{b_2} and MLF_{b_1} because B is the single node that separates node A and C.

$$MLF_C/MLF_{b_1} = (\lambda_{a_1}\lambda_{b_1}\lambda_{c_1}\theta_{a_1}\theta_{b_1|a_1}\theta_{c_1|b_1} + \ldots \lambda_{a_2}\lambda_{b_2}\lambda_{c_2}\theta_{b_2}\theta_{b_2|a_2}\theta_{c_2|b_2})$$
$$/(\lambda_{a_1}\lambda_{b_1}\theta_{a_1}\theta_{b_1|a_1} + \lambda_{a_2}\lambda_{b_1}\theta_{a_2}\theta_{b_1|a_2})$$
$$= (\lambda_{c_1}\theta_{c_1|b_1} + \lambda_{c_2}\theta_{c_2|b_1}) \cap (\lambda_{c_1}\theta_{c_1|b_1} + \lambda_{c_2}\theta_{c_2|b_1})$$
$$= \lambda_{c_1}\theta_{c_1|b_1} + \lambda_{c_2}\theta_{c_2|b_1}$$

$$MLF_C/MLF_{b_2} = (\lambda_{a_1}\lambda_{b_1}\lambda_{c_1}\theta_{a_1}\theta_{b_1|a_1}\theta_{c_1|b_1} + \ldots \lambda_{a_2}\lambda_{b_2}\lambda_{c_2}\theta_{b_2}\theta_{b_2|a_2}\theta_{c_2|b_2})$$
$$/(\lambda_{a_1}\lambda_{b_2}\theta_{a_1}\theta_{b_2|a_1} + \lambda_{a_2}\lambda_{b_2}\theta_{a_2}\theta_{b_2|a_2})$$
$$= (\lambda_{c_1}\theta_{c_1|b_2} + \lambda_{c_2}\theta_{c_2|b_2}) \cap (\lambda_{c_1}\theta_{c_1|b_2} + \lambda_{c_2}\theta_{c_2|b_2})$$
$$= \lambda_{c_1}\theta_{c_1|b_2} + \lambda_{c_2}\theta_{c_2|b_2}$$

Finally, we rewrite MLF_C as follows:

$$MLF_C = MLF_{b_1}(\lambda_{c_1}\theta_{c_1|b_1} + \lambda_{c_2}\theta_{c_2|b_1}) + MLF_{b_2}(\lambda_{c_1}\theta_{c_1|b_2} + \lambda_{c_2}\theta_{c_2|b_2})$$

4.3   Divisor Selection by Multi-Node d-Separations

We use the BN of Fig. 1 to show how to perform the factorization based on multi-node d-separation. For nodes A and D, nodes B and C are the d-separation node set that separates them as independent nodes. Since both node B and C have two values, there are four combinations of their information $b_1 c_1$, $b_1 c_2$, $b_2 c_1$ and $b_2 c_2$. According to [3], we multiply their MLFs as follows. There are two terms in each of these MLFs, so the number of terms after multiplication should be $2 * 2 = 4$. But since the parameters λ are eliminated if they contradict each other, only two of the four terms are left. Following shows the details of the multiplication.

$$MLF_{b_1} MLF_{c_1}$$
$$= (\lambda_{b_1} \lambda_{a_1} \theta_{a_1} \theta_{b_1|a_1} + \lambda_{b_1} \lambda_{a_2} \theta_{a_2} \theta_{b_1|a_2})$$
$$(\lambda_{c_1} \lambda_{a_1} \theta_{a_1} \theta_{c_1|a_1} + \lambda_{c_1} \lambda_{a_2} \theta_{a_2} \theta_{c_1|a_2})$$
$$= \lambda_{b_1} \lambda_{c_1} \lambda_{a_1} \theta_{a_1} \theta_{b_1|a_1} \theta_{c_1|a_1} + \lambda_{b_1} \lambda_{c_1} \lambda_{a_2} \theta_{a_2} \theta_{b_1|a_2} \theta_{c_1|a_2}.$$

$$MLF_{b_1} MLF_{c_2}$$
$$= (\lambda_{b_1} \lambda_{a_1} \theta_{a_1} \theta_{b_1|a_1} + \lambda_{b_1} \lambda_{a_2} \theta_{a_2} \theta_{b_1|a_2})$$
$$(\lambda_{c_2} \lambda_{a_1} \theta_{a_1} \theta_{c_2|a_1} + \lambda_{c_2} \lambda_{a_2} \theta_{a_2} \theta_{c_2|a_2})$$
$$= \lambda_{b_1} \lambda_{c_2} \lambda_{a_1} \theta_{a_1} \theta_{b_1|a_1} \theta_{c_2|a_1} + \lambda_{b_1} \lambda_{c_2} \lambda_{a_2} \theta_{a_2} \theta_{b_1|a_2} \theta_{c_2|a_2}.$$

$$MLF_{b_2} MLF_{c_1}$$
$$= (\lambda_{b_2} \lambda_{a_1} \theta_{a_1} \theta_{b_2|a_1} + \lambda_{b_2} \lambda_{a_2} \theta_{a_2} \theta_{b_2|a_2})$$
$$(\lambda_{c_1} \lambda_{a_1} \theta_{a_1} \theta_{c_1|a_1} + \lambda_{c_1} \lambda_{a_2} \theta_{a_2} \theta_{c_1|a_2})$$
$$= \lambda_{b_2} \lambda_{c_1} \lambda_{a_1} \theta_{a_1} \theta_{b_2|a_1} \theta_{c_1|a_1} + \lambda_{b_2} \lambda_{c_1} \lambda_{a_2} \theta_{a_2} \theta_{b_2|a_2} \theta_{c_1|a_2}.$$

$$MLF_{b_2} MLF_{c_2}$$
$$= (\lambda_{b_2} \lambda_{a_1} \theta_{a_1} \theta_{b_2|a_1} + \lambda_{b_2} \lambda_{a_2} \theta_{a_2} \theta_{b_2|a_2})$$
$$(\lambda_{c_2} \lambda_{a_1} \theta_{a_1} \theta_{c_2|a_1} + \lambda_{c_2} \lambda_{a_2} \theta_{a_2} \theta_{c_2|a_2})$$
$$= \lambda_{b_2} \lambda_{c_2} \lambda_{a_1} \theta_{a_1} \theta_{b_2|a_1} \theta_{c_2|a_1} + \lambda_{b_2} \lambda_{c_2} \lambda_{a_2} \theta_{a_2} \theta_{b_2|a_2} \theta_{c_2|a_2}.$$

After these multiplication, we factor the MLF_D with the four combinations respectively. We give an example of $MLF_D/MLF_{b_1} MLF_{c_1}$ in details.

$$MLF_D/MLF_{b_1} MLF_{c_1}$$
$$= MLF_D/(\lambda_{b_1} \lambda_{c_1} \lambda_{a_1} \theta_{a_1} \theta_{b_1|a_1} \theta_{c_1|a_1} +$$
$$\lambda_{b_1} \lambda_{c_1} \lambda_{a_2} \theta_{a_1} \theta_{b_1|a_2} \theta_{c_1|a_2})$$
$$= MLF_D/(\lambda_{b_1} \lambda_{c_1} \lambda_{a_1} \theta_{a_1} \theta_{b_1|a_1} \theta_{c_1|a_1}) \cap$$
$$MLF_D/(\lambda_{b_1} \lambda_{c_1} \lambda_{a_2} \theta_{a_1} \theta_{b_1|a_2} \theta_{c_1|a_2})$$
$$= \lambda_{d_1} \theta_{d_1|b_1 c_1} + \lambda_{d_2} \theta_{d_2|b_1 c_1} + \lambda_{d_3} \theta_{d_3|b_1 c_1}.$$

Finally, we can rewrite MLF_D as follows.

$$
\begin{aligned}
MLF_D = \; & MLF_{b_1} MLF_{c_1} (\lambda_{d_1}\theta_{d_1|b_1c_1} + \lambda_{d_2}\theta_{d_2|b_1c_1} + \lambda_{d_3}\theta_{d_3|b_1c_1}) + \\
& MLF_{b_1} MLF_{c_2} (\lambda_{d_1}\theta_{d_1|b_1c_2} + \lambda_{d_2}\theta_{d_2|b_1c_2} + \lambda_{d_3}\theta_{d_3|b_1c_2}\cdot) + \\
& MLF_{b_2} MLF_{c_1} (\lambda_{d_1}\theta_{d_1|b_2c_1} + \lambda_{d_2}\theta_{d_2|b_2c_1} + \lambda_{d_3}\theta_{d_3|b_2c_1}\cdot) + \\
& MLF_{b_2} MLF_{c_2} (\lambda_{d_1}\theta_{d_1|b_2c_2} + \lambda_{d_2}\theta_{d_2|b_2c_2} + \lambda_{d_3}\theta_{d_3|b_2c_1}\cdot).
\end{aligned}
$$

5   Experiments and Results

To show how our method works to condense the size of ZDDs representing MLFs [3], we implement our experiments on the platform of a Intel Core Quad CPU Q9550@2.83 GHz * 4 PC with Ubuntu 12.04 LTS and 3.8 GiB of main memory. We manipulate up to 40,000,000 nodes of ZBDDs. We use data set of BN Benchmark [8] *alarm*, *hailfinder*, *insurance* and *Mildew* to implement our experiment.

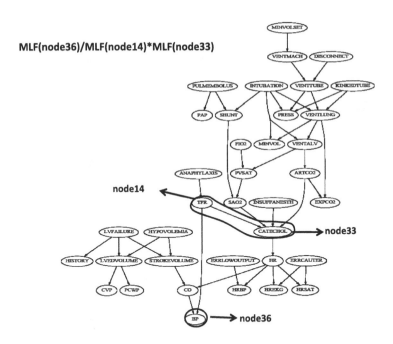

Fig. 10. the example of $alarmN36(n14n33)$

In our experiment, we compile nodes to ZDDs which we concern to avoid unnecessary calculation of redundant terms. However, to compare with the method in [3] which compile a whole BN, we try to choose the nodes which have the biggest size of ZDD in *alarm* and *hailfinder*. For *insurance* and *Mildew*. considering time consumption, we do not use the nodes with biggest

ZDD size but choose the nodes which seem to have suitable d-separation node sets. To compare with our method, we use a simple criterion to extract divisors from MLFs, that is abstracting variables that appear more than twice in the MLFs. The experiment results are shown in Table 2. The first column shows the number of node we choose as our data, for example, $alarmN36$ means we factor the MLFs of node 36 in $alarm$. The other columns show the ZDD specifications after factoring and time consumption with the simple divisor extracting algorithm to show the upper limits of consumption of calculation and inference based on MLFs without any compression techniques.

For the nodes we use in Table 2, we use algorithm to find one-node d-separation and manually find multi-node d-separation and use our method to factor the MLFs. The experiment results are shown in Table 3. The first column in Table 3 lists the number of node and d-separation set we choose in BN. For instance, $alarmN36(n14n33)$ means factoring the MLF of node 36 with d-separation set MLFs of node 14 and node 33 (Fig. 10). According to Table 3, we could always achieve quite smaller ZDDs and condense MLFs quite efficiently using our method comparing to Tables 1 and 2. However, we get a bigger ZDDs due to the factorization in 'AlarmN36'. This is because the number of newly introduces variables to represent $divisors$ are more than the reduction of ZDD nodes. Since the size of ZDD is depended on the structure of BN itself and the probability of every instance, we can not precisely tell how much we can condense MLFs. We hope to find the d-separation node sets with minimum number of combinations of variables to reduce the newly introduces variables. This is the reason why we just use one or two nodes d-separations but not more. For $alarmN36$ and $hailfinderN43$, we get different results with different d-separation sets. That is to say, if we can find suitable d-separation, we will further condense ZDD size which is important because it determines the time and space requirements for online inference which is linear in this size [2].

6 Related Work

Darwiche et al. [9] have shown a different efficient approach to factor MLFs of BN based on an arithmetic circuit called Conjunctive Normal Forms (CNFs). In their method, they compile MLFs into CNFs and show an efficient way to evaluate and differentiate CNFs in time and space which is linear in their size. They also prove their effectivity in calculation and inference in BNs because CNFs subsume the famous structure $jointree$, one of the most influential methods for inference in BNs. Since the efficiency of algorithms using $jointree$ algorithm for probabilistic inference in BNs can be improved by a careful exploitation of the d-separation properties, in our method, we introduce the idea of using one pattern of d-separation into ZDDs. We want to try to make full use of the three patterns of d-separation to explore the relationship between factorization of MLFs using ZDDs and $jointrees$. We hope this will bring a significant improvement to ZDD-based methods in compiling BNs as well as Darwiche's method.

Table 2. Experimental results without d-separation.

Dataset and node ID	Factorization without d-separation			
	ZDD size	Total terms	Total literals	Time of factorization
alarmN36	6,784	3,500	13,512	1969.77 s
hailfinderN43	Overflow	–	–	–
insuranceN5	5,256	3,026	10,483	2.169 s
insuranceN14	Overflow	–	–	–
MildewN14	Overflow	–	–	–
MildewN20	Overflow	–	–	–

Table 3. Experimental results with d-separation.

Dataset and node ID and d-separation node set	After factorization based on d-separation			
	ZDD size	Total terms	Total literals	Time of factorization
alarmN36(n14n33)	5,178	2662	9,796	51.425 s
alarmN36(n20n32)	4,247	2,423	8,776	4.825 s
alarmN36(n14n33/n20n32)	4,133	2,342	8,422	3.427 s
hailfinderN43(n14n20)	24,353	>2 G	>2 G	391.648 s
hailfinderN43(n14n20/n4n12)	22,833	14,512	50,113	303.641 s
insuranceN5(n4n8n9)	1,985	1,359	3,924	0.24 s
insuranceN14(n2n3n9)	34,222	22,088	72,882	2427.79 s
MildewN14(n11n12)	14,727	11,785	33,355	619.391 s
MildewN20(n17n18)	22,356	16,784	51,242	754.577 s
MildewN20(n17n24)	Overflow	–	–	–

7 Conclusion and Future Work

We represented an improvement of compiling MLFs using ZDDs by combining weak division algorithm with d-separation. In our method, we use the d-separation structure in BN to quickly find a good divisor to factor MLFs into compact representations and we get much more compact MLFs than the method in Minato et al. [3].

In our method, we first generate ZDDs for the whole given BN, and then factor the ZDDs using d-separation nodes. The process of factoring costs too much time and sometime the ZDD for a BN is too large to factor, such as $MildewN20(n17n24)$ in Table 3, though we could find proper d-separation nodes. For future work, we want to improve our method by developing simple and fast heuristic algorithms to find d-separation sets as divisors instead of finding d-separation sets manually. For example, using the *jointree* algorithm, or straightly use the formula recursively constructed in Minato et al. [3] by controlling its trade-off. Also we want to consider not to generate the ZDD for the whole network but just generate the ZDD for the newfound divisor, we may avoid the time consumption and get quite compact ZDDs. For example, $alarmN36(n14n33/n20n32)$ shows that if we first factor node 36 with set of node 14 and 33, then factor the result with set of node 20 and 32, we can get a smaller ZDD using this two-level *d*-separations. It gives us a hint that in the process of generating the ZDD from the MLF of *alarm*, we can first generate a ZDD for *divisor* of node 20 and 32, then using this ZDD to generate a ZDD of divisor node 14 and 33. Finally, we use the two ZDDs to generate the ZDD of the whole network so that we avoid the time consumption of factoring and also reduce the ZDD.

References

1. Pearl, J.: Probabilistic Reasoning in Intelligent Systems: Networks of Plausible Inference. Morgan Kaufmann, San Mateo (2014)
2. Chavira, M., Darwiche, A.: Compiling Bayesian networks with local structure. IJCAI **5**, 1306–1312 (2005)
3. Minato, S, Satoh, K, Sato, T.: Compiling Bayesian networks by symbolic probability calculation based on zero-suppressed BDDs. In: IJCAI 2007, pp. 2550–2555 (2007)
4. Minato S. Zero-suppressed BDDs for set manipulation in combinatorial problems. In: 30th Conference on Design Automation, pp. 272–277. IEEE (1993)
5. Minato, S.: Zero-suppressed BDDs and their applications. Int. J. Softw. Tools Technol. Transf. **3**(2), 156–170 (2001). Springer
6. Word Press Entries (RSS) and Comments (RSS), *d*-separation. http://www.bayesnets.com/d-separation.html
7. Tarjan, R.: Depth-first search and linear graph algorithms. SIAM J. Comput. **1**(2), 146–160 (1972)
8. Marco, S.: Bayesian network Repository. http://www.bnlearn.com/bnrepository/
9. Darwiche, A.: A differential approach to inference in Bayesian networks. J. ACM (JACM) **50**(3), 280–305 (2003)

Missing Data from a Causal Perspective

Karthika Mohan[(✉)] and Judea Pearl

Department of Computer Science, University of California,
Los Angeles, CA 90095, USA
karthika@cs.ucla.edu

Abstract. This paper applies graph based causal inference procedures for recovering information from missing data. We establish conditions that permit and prohibit recoverability. In the event of theoretical impediments to recoverability, we develop graph based procedures using auxiliary variables and external data to overcome such impediments. We demonstrate the perils of model-blind recovery procedures both in determining whether or not a query is recoverable and in choosing an estimation procedure when recoverability holds.

Keywords: Graphical models · Causal inference · Missing data

1 Introduction

The missing data (or incomplete data) problem, characterized by the absence of values for one or more variables in a dataset is a major impediment to both theoretical and empirical research and leaves no branch of experimental science untouched. The vast amount of literature on missing data problems in such diverse fields as computer science, geology, archeology, biology, statistics and epidemiology attests to both its extent and pervasiveness [8,12,15,32]. Simply ignoring the problem by deleting all tuples with missing values will, in most cases, significantly distort the outcome of a study, regardless of the size of the dataset [1,6].

Existing methods of dealing with missing data such as Expectation Maximization Algorithm and Multiple Imputation are based on the theoretical work of Rubin [27] and Little and Rubin [28] who formulated conditions under which the damage of missingness would be minimized. However, theoretical guarantees are provided only for a subset of problems falling into the Missing At Random (MAR) category thereby leaving the vast space of MNAR problems relatively unexplored.

In this paper we view missingness from a causal perspective and take the following steps to answer questions pertaining to consistent estimation of queries of interest. Given an incomplete dataset our first step is to postulate a model based on causal assumptions of the underlying data generation process. Our second step is to determine whether the data rejects the postulated model by identifiable testable implications of that model. Our third and final step, which

© Springer International Publishing Switzerland 2015
J. Suzuki and M. Ueno (Eds.): AMBN 2015, LNAI 9505, pp. 184–195, 2015.
DOI: 10.1007/978-3-319-28379-1_13

is also the primary focus of this paper, is to determine from the postulated model if any method exists that produces consistent estimates of the queries of interest? A negative answer confirms the presence of a theoretical impediment to estimation. In other words, a bias is inevitable.

2   Missingness Graphs

Missingness graphs as discussed below was first defined in [17] and we adopt the same notations. Let $G(\mathbb{V}, E)$ be the causal DAG where $\mathbb{V} = V \cup U \cup V^* \cup \mathbb{R}$. V is the set of observable nodes. Nodes in the graph correspond to variables in the data set. U is the set of unobserved nodes (also called latent variables). E is the set of edges in the DAG. We use bi-directed edges as a shorthand notation to denote the existence of a U variable as common parent of two variables in $V \cup \mathbb{R}$. V is partitioned into V_o and V_m such that $V_o \subseteq V$ is the set of variables that are observed in all records in the population and $V_m \subseteq V$ is the set of variables that are missing in at least one record. Variable X is termed as *fully observed* if $X \in V_o$, *partially observed* if $X \in V_m$ and *substantive* if $X \in V_o \cup V_m$. Associated with every partially observed variable $V_i \in V_m$ are two other variables R_{v_i} and V_i^*, where V_i^* is a proxy variable that is actually observed, and R_{v_i} represents the status of the causal mechanism responsible for the missingness of V_i^*; formally,

$$v_i^* = f(r_{v_i}, v_i) = \begin{cases} v_i & \text{if } r_{v_i} = 0 \\ m & \text{if } r_{v_i} = 1 \end{cases} \tag{1}$$

V^* is the set of all proxy variables and $\mathbb{R}$ is the set of all causal mechanisms that are responsible for missingness. R variables may not be parents of variables in $V \cup U$. We call this graphical representation **Missingness Graph** (or m-graph). An example of an m-graph is given in Fig. 1. We use the following shorthand. For any variable X, let X' be a shorthand for $X = 0$. For any set $W \subseteq V_m \cup V_o \cup R$, let W_r, W_o and W_m be the shorthand for $W \cap R$, $W \cap V_o$ and $W \cap V_m$ respectively. Let R_w be a shorthand for $R_{V_m \cap W}$ i.e. R_w is the set containing missingness mechanisms of all partially observed variables in W. Note that R_w and W_r are not the same. $G_{\underline{X}}$ and $G_{\overline{X}}$ represent graphs formed by removing from G all edges leaving and entering X, respectively.

A *manifest distribution* $P(V_o, V^*, R)$ is the distribution that governs the available dataset. An *underlying distribution* $P(V_o, V_m, R)$ is said to be compatible with a given manifest distribution $P(V_o, V^*, R)$ if the latter can be obtained from the former using Eq. 1. Manifest distribution P_m is compatible with a given underlying distribution P_u if $\forall X, X \subseteq V_m$ and $Y = V_m \setminus X$, the following equality holds true.

$$P_m(R_x', R_y, X^*, Y^*, V_o) = P_u(R_x', R_y, X, V_o)$$

where R_x' denotes $R_x = 0$ and R_y denotes $R_y = 1$.

3   Recoverability

Given a manifest distribution $P(V^*, V_o, R)$ and an m-graph G that depicts the missingness process, query Q is recoverable if we can compute a consistent estimate of Q as if no data were missing. Formally,

Definition 1 (Recoverability). *Given a m-graph G, and a target relation Q defined on the variables in V, Q is said to be recoverable in G if there exists an algorithm that produces a consistent estimate of Q for every dataset D such that $P(D)$ is (1) compatible with G and (2) strictly positive i.e. $P(V_o, V^*, \mathbb{R}) > 0$.*

For an introduction to the notion of recoverability see, [17, 20].

3.1   Recovering from MCAR and MAR Data

Examine the m-graph in Fig. 1, X is the treatment and Y is the outcome. Let us assume that some patients who underwent treatment are not likely to report the outcome, and hence the arrow $X \rightarrow R_y$. Under these circumstances, can we recover $P(X, Y)$?

From the manifest distribution, we can compute $P(X, Y^*, R_y)$. From the m-graph G, we see that Y^* is a collider and X is a fork. Hence by d-separation, $Y \perp\!\!\!\perp R_y | X$. Thus

$$P(X, Y) = P(Y|X)P(X)$$
$$= P(Y|X, R_y = 0)P(X) \text{ (using } Y \perp\!\!\!\perp R_y | X)$$
$$= P(Y^*|X, R_y = 0)P(X) \text{ (using Eq. 1)}$$

Since both factors in the estimand are estimable from the manifest distribution, $P(X, Y)$ is recoverable.

The scenario discussed above is a typical instance of Missing At Random (MAR). When data are Missing At Random (MAR), we have $\mathbb{R} \perp\!\!\!\perp V_m | V_o$. Therefore $P(V) = P(V_m|V_o)P(V_o) = P(V_m|V_o, R = 0)P(V_o)$. In other words, the joint distribution $P(V)$ is recoverable given MAR data. Estimation methods applicable to MAR are applicable to MCAR as well because by the weak union axiom of graphoids, Missing Completely at Random (MCAR: $(V_m, V_o) \perp\!\!\!\perp R$) implies Missing At Random (MAR: $V_m \perp\!\!\!\perp R | V_o$). Therefore, it implicitly follows that queries (such as joint distribution and (identifiable) causal effects) that are recoverable given MAR datasets are recoverable given MCAR datasets as well.

Fig. 1. An m-graph depicting MAR category missingness

4   Recoverability Procedures for MNAR Data

Data that are neither MAR nor MCAR fall into the Missing Not At Random (MNAR) category. In this section we will detail with examples three distinct recovery procedures.

4.1   Sequential Factorization

Consider an observational study that measured the variables X, Y, W and Z where we wish to estimate the effect of treatment (X) on outcome (Y). The interactions between the variables and the underlying missingness process are depicted in Fig. 2. We notice that all variables are corrupted by missing values. The least bothersome missingness is that of Y which is caused by a random process such as an accidental deletion of cases while the most troubling missingness is that of W which is caused by its own underlying value- a typical example is the case of very rich and very poor people being reluctant to reveal their income.

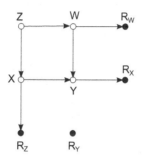

Fig. 2. MNAR model in which $P(Y|do(x))$ is recoverable by sequential factorization

Recovering Causal Effect of X on Y: By backdoor criterion [19], we have two admissible sets, $\{Z\}$ and $\{W\}$ which yield the following estimands, respectively:

$$P(y|do(x)) = \sum_z P(y|xz)P(z)$$
$$= \sum_w P(y|xw)P(w)$$

We choose the first estimand over the second because the latter contains $P(W)$ which we know to be non-recoverable [17].[1] Therefore, to recover the causal effect we have to recover both $P(y|xz)$ and $P(z)$.

---

[1] The presence of a non-recoverable factor in a summand does not always imply the non-recoverability of the summand. See Example-3 in [18].

Recovering P(z) In order to d-separate Z from R_z, one needs to condition of X and to d-separate X from R_x one needs to condition on Y. Therefore, we can write:

$$P(z) = \sum_{x,y} P(z,x,y)$$

$$= \sum_{x,y} P(z|x,y)P(x|y)P(y) \qquad (2)$$

$$= \sum_{x,y} P(z|x,y,R_x=0,R_y=0,R_z=0)P(x|y,R_x=0,R_y=0)P(y|R_y=0)$$

(Using $Z \perp\!\!\!\perp (R_z, R_x, R_y)|(X,Y)$, $X \perp\!\!\!\perp (R_x, R_y)|Y$ and $Y \perp\!\!\!\perp R_y$, respectively)

$$= \sum_{x,y} P(z^*|x^*,y^*,R_x=0,R_y=0,R_z=0)P(x^*|y^*,R_x=0,R_y=0)P(y^*|R_y=0)$$

In the process of recovering $P(z)$ we have in fact recovered $P(x,y,z)$. Therefore it follows that $P(y|x,z)$ is recoverable. Finally, the causal effect may be recovered as:

$$P(y|do(x)) = \sum_z \frac{P(z^*|x^*,y^*,R_x=0,R_y=0,R_z=0)P(x^*|y^*,R_x=0,R_y=0)P(y^*|R_y=0)}{\sum_y P(z^*|x^*,y^*,R_x=0,R_y=0,R_z=0)P(x^*|y^*,R_x=0,R_y=0)P(y^*|R_y=0)}$$

$$\times$$

$$\sum_{x,y} P(z^*|x^*,y^*,R_x=0,R_y=0,R_z=0)P(x^*|y^*,R_x=0,R_y=0)P(y^*|R_y=0)$$

Recovery Procedure: Given an m-graph with no edges between R variables, a sufficient condition for recoverability of query Q is that it be decomposable into sub-queries of the form $P(Y|X)$ such that $Y \perp\!\!\!\perp (R_x, R_y)|X$. This recovery procedure called as seuential factorization (generalized in Theorem 1 below) is sensitive to the ordering of variables in the factorization, which in turn is dictated by the graph. For instance, in Eq. 2 had we factorized $P(x,y,z)$ as $P(y|x,z)P(x|z)P(z)$, we would not have had the permission to insert the R terms in any factor.

Recovering in the presence of edges between R variables: A quick inspection reveals that the factorization in Eq. 2 guarantees recoverability even when an edge $R_x \rightarrow R_z$ is added. However, addition of the (reversed) edge $R_z \rightarrow R_x$ would require conditioning on R_z and Y to d-separate X from R_x. The procedure for recovering the marginal distribution $P(Z)$ is presented below:

$$P(z) = \sum_{x,y,r_z} P(z,x,y,r_z)$$

$$= \sum_{x,y,r_z} P(z|x,y,r_z)P(x|y,r_z)P(y|r_z)P(r_z) \qquad (3)$$

$$= \sum_{x,y} P(z|x,y,R_x=0,R_y=0,r_z=0) \sum_{r_z} P(x|y,R_x=0,R_y=0,r_z)P(y|R_y=0,r_z)P(r_z)$$

(Using $Z \perp\!\!\!\perp (R_z, R_x, R_y)|(X,Y)$, $X \perp\!\!\!\perp (R_x, R_y)|(Y,R_z)$ and $Y \perp\!\!\!\perp R_y|R_z$, respectively)

The following definition and theorem in [18] formalizes the preceding recovery procedure.

Definition 2 (General Ordered factorization). *Given a graph G and a set O of ordered $V \cup R$ variables $Y_1 < Y_2 < \ldots < Y_k$, a general ordered factorization relative to G, denoted by $f(O)$, is a product of conditional probabilities*

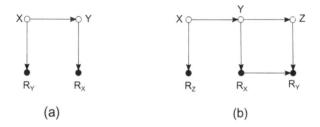

Fig. 3. MNAR Model in which $P(Y, X)$ and $P(X, Y, Z)$ are recoverable

$f(O) = \prod_i P(Y_i|X_i)$ where $X_i \subseteq \{Y_{i+1}, \ldots, Y_n\}$ is a minimal set such that $Y_i \perp\!\!\!\perp (\{Y_{i+1}, \ldots, Y_n\} \backslash X_i)|X_i$ holds in G.

Theorem 1 (Sequential Factorization). *A sufficient condition for recoverability of a relation Q defined over substantive variables is that Q be decomposable into a general ordered factorization, or a sum of such factorizations, such that every factor $Q_i = P(Y_i|X_i)$ satisfies, (1) $Y_i \perp\!\!\!\perp (R_{y_i}, R_{x_i})|X_i \backslash \{R_{y_i}, R_{x_i}\}$, if $Y_i \in (V_o \cup V_m)$ and (2) $R_z \perp\!\!\!\perp R_{X_i}|X_i$ if $Y_i = R_z$ for any $Z \in V_m$, $Z \notin X_i$ and $X_r \cap R_{X_m} = \emptyset$.*

4.2 R-Factorization

Consider the model in Fig. 3(a) in which missingness in X is caused by Y and vice-versa. This type of missingness model is called **entangled** because in order to d-separate any variable from its missingness mechanism one needs to condition on the other. Factorizing $P(x, y)$ as $P(x|y)P(y)$ or $P(y|x)P(x)$ does not satisfy sequential factorization criterion since neither $X \perp\!\!\!\perp (R_x, R_y)|Y$ nor $Y \perp\!\!\!\perp (R_x, R_y)|X$ holds in the graph. This deadlock can however be disentangled by the following method:

$$P(X, Y) = P(X, Y)\frac{P(R_x = 0, R_y = 0|X, Y)}{P(R_x = 0, R_y = 0|X, Y)}$$

$$= \frac{P(R_x = 0, R_y = 0)P(X, Y|R_x = 0, R_y = 0)}{P(R_x = 0, R_y = 0|X, Y)}$$

$$= \frac{P(R_x = 0, R_y = 0)P(X, Y|R_x = 0, R_y = 0)}{P(R_x = 0|Y, R_y = 0)P(R_y = 0|X, R_x = 0)}$$

$$\text{(using } R_x \perp\!\!\!\perp (R_y, X)|Y \text{ and } R_y \perp\!\!\!\perp (R_x, Y)|X)$$

$$= \frac{P(R_x = 0, R_y = 0)P(X^*, Y^*|R_x = 0, R_y = 0)}{P(R_x = 0|Y^*, R_y = 0)P(R_y = 0|X^*, R_x = 0)}$$

The following theorem generalizes this recovery procedure:

Theorem 2 (R-factorization). *Given a m-graph G with no edges between the R variables and no latent variables as parents of R variables, a necessary and sufficient condition for recovering the joint distribution $P(V)$ is that no variable X be a parent of its missingness mechanism R_X. Moreover, when recoverable, $P(V)$ is given by*

$$P(v) = \frac{P(R = 0, v)}{\prod_i P(R_i = 0 | pa_{r_i}^o, pa_{r_i}^m, R_{Pa_{r_i}^m} = 0)}, \tag{4}$$

where $Pa_{r_i}^o \subseteq V_o$ and $Pa_{r_i}^m \subseteq V_m$ are the parents of R_i.

Interestingly, given a model in which R variables are connected by an edge sometimes we have to use a combination of sequential and R factorization. Examine the model in Fig. 3(b). The query of interest is the joint distribution $P(x, y, z)$ and the recovery procedure inspired by Theorem 2 follows:

$$P(x, y, z) = \frac{P(x, y, z, r_x = 0, r_y = 0, r_z = 0)}{P(r_x = 0 | y) P(r_z = 0 | x, r_x = 0) P(r_y = 0 | z, r_x = 0, r_z = 0)}$$

In order to recover $P(r_x = 0 | y)$ we rely on sequential factorization as shown below:

$$P(y, r_x) = \sum_{x,z} P(x, y, z, r_x)$$

$$= \sum_{x,z} \frac{P(x, y, , z, r_x, r_z = 0, r_y = 0)}{P(r_z = 0 | x, r_x = 0) P(r_y = 0 | z, r_x, r_z = 0)}$$

$$= \sum_{x,z} \frac{P(x | y, z, r_x = 0, r_z = 0, r_y = 0) P(y, z, r_x, r_z = 0, r_y = 0)}{P(r_z = 0 | x, r_x = 0) P(r_y = 0 | z, r_x, r_z = 0)}$$

$$\text{(using } X \perp\!\!\!\perp R_x | (Y, Z, R_y, R_z) \text{ i.e. sequential factorization)}$$

Recoverability of $P(y, r_x)$ implies that $P(r_x = 0 | y)$ is recoverable. Hence joint distribution $P(x, y, z)$ is recoverable given Fig. 3(b).

4.3   Interventional Factorization

Consider the model in Fig. 4. Let the query of interest be $P(w, x, y, z)$. We will first factorize $P(w, x, y, z)$ in a manner similar to R factorization:

$$P(w, x, y, z) = \frac{P(w, x, y, z, r_x = 0, r_y = 0)}{P(r_x = 0 | r_y = 0, y, z) P(r_y = 0 | x, z)}$$

The recovery of the joint distribution depends on the recovery of $P(r_y = 0 | x, z)$. We notice that

$$P(R_y | do(Z = z), X) = P(R_y | Z = z, X) \text{(using rule-2 of do-calculus)}$$

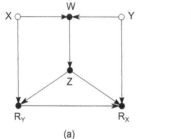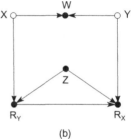

Fig. 4. (a) MNAR model in which joint distribution is recoverable, (b) mutilated model corresponding to (a) obtained by intervening on Z

The interventional distribution can be computed as given below:

$$P(x^*, y^*, w, r_x, r_y | do(z)) = \frac{P(x^*, y^*, w, r_x, r_y, z)}{P(z|w)}$$

$$P(r_y, x^*, r_x | do(z)) = \sum_{w, y^*} \frac{P(x^*, y^*, w, r_x, r_y | do(z))}{P(z|w)} \tag{5}$$

In order to recover $P(r_y = 0|x, z)$, we will recover $P(x, r_y | do(z))$ and express it in terms of proxy variables.

$$P(x, r_y | do(z)) = P(x|r_y, r_x = 0, do(z)) P(r_y | do(z))$$
$$= P(x^*|r_y, r_x = 0, do(z)) P(r_y | do(z)) \tag{6}$$

Each factor in Eq. 6 can be computed from the intervential distribution derived in Eq. 5.

A general algorithm incorporating all these three recovery procedures in a slightly more relaxed setting is discussed in [26].

5   Recourses to Non-recoverability

Joint distribution is not recoverable given the m-graphs in Fig. 5 [18]. In this section we will show how auxiliary variables and external data can be utilized to aid recoverability.

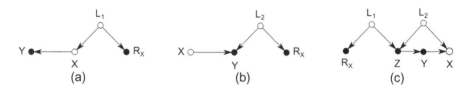

Fig. 5. MNAR models in which the joint distribution is not recoverable. Variables denoted by L serve as candidates for auxiliary variables.

Auxiliary variables are variables that are anciliary to the substantive research questions but are potential correlates of missingness mechanisms or partially observed variables [6]. However as noted in [29], not all variables satisfying this criterion may be used as auxiliary variables.

Selection Criteria For Auxiliary Variables: Firstly an auxiliary variable should not be a collider or a descendant of a collider on the path from a partially observed variable to its missingness mechanism. For example in Fig. 5(b) neither Y nor its descendants may serve as auxiliary variables while recovering $P(X)$. Secondly, in the presence of an inducing path between X and R_x as shown in Fig. 5(c), the ideal auxiliary variables are latent variables L_1 or L_2. Conditioning on either of these will d-separate X from R_x and facilitate the recovery of $P(X)$.

Recovery Aided By External Data: It is often the case that incorporating data from external sources can aid recovery. For example, consider a manifest distribution in which age is a partially observed variable. Distribution of age for a given population may be easily available from an external agency such as the census bureau. The question we ask is how can this data be combined with the existing missing data set to recover a query of interest.

Consider the Fig. 5(a), suppose the query of interest is $P(X,Y)$. $P(Y|X)$ is recoverable by sequential factorization. If from an external source we obtain $P(X)$, then $P(y,x)$ may be recovered as $P(y|x^*, r_x = 0)P(x)$. In Fig. 5(b) however, $P(Y)$ and $P(X)$ are recoverable. If we can obtain either $P(y|x)$ or $P(x|y)$ from an external source, then $P(x,y)$ can be recovered.

6   Perils of Model Blind Recovery Procedures

Model-blind algorithms are algorithms that attempt to handle missing-data problems on the basis of the data alone, without making any assumptions about the structure of the missingness process. We unveil a fundamental limitation of model-blind algorithms by presenting two statistically indistinguishable models such that a given query is recoverable in one and non-recoverable in the other.

The two graphs in Fig. 6(a) and (b) cannot be distinguished by any statistical means, since Fig. 6(a) has no testable implications [16] and Fig. 6(b) is a complete graph. However in Fig. 6(a) $P(X,Y) = P(X^*|Y, R_x)P(Y)$ is recoverable while in Fig. 6(b) $P(X,Y)$ is not recoverable (by Theorem 2 in [17]).

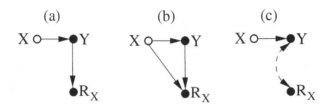

Fig. 6. Statistically indistinguishable graphs. (a) $P(X,Y)$ is recoverable (b) $P(X,Y)$ is not recoverable (c) $P(X)$ is recoverable

An even stronger limitation is demonstrated below. We show that no model-blind algorithm exists even in those cases where recoverability is feasible. We exemplify our claim below by constructing two statistically indistinguishable models, G_1 and G_2, dictating different estimation procedure E_1 and E_2 respectively; yet Q is not recoverable in G_1 by S_2 or in G_2 by S_1.

Consider the graphs in Fig. 6(a) and (c); they are statistically indistinguishable since neither has testable implications. Let the target relation of interest be $Q = P(X)$. In Fig. 6(a), Q may be estimated as $P(X) = \sum_y P(X|Y, R_x = 0)P(Y)$ since $X \perp\!\!\!\perp R_x | Y$ and in Fig. 6(b), Q can be derived as $P(X) = P(X|R_x = 0)$ since $X \perp\!\!\!\perp R_x$.

7   Related Work

Deletion based methods such as listwise deletion that are easy to understand as well as implement, guarantee consistent estimates only for certain categories of missingness such as MCAR [24]. Maximum Likelihood method is known to yield consistent estimates under MAR assumption; expectation maximization algorithm and gradient based algorithms are widely used for searching for ML estimates under incomplete data [4,5,10,11]. Most work in machine learning assumes MAR and proceeds with ML or Bayesian inference. However, there are exceptions such as recent work on collaborative filtering and recommender systems which develop probabilistic models that explicitly incorporate missing data mechanism [13–15].

Other methods for handling missing data can be classified into two: (a) Inverse Probability Weighted Methods and (b) Imputation based methods [23]. Inverse Probability Weighing methods analyze and assign weights to complete records based on estimated probabilities of completeness [22,32]. Imputation based methods substitute a reasonable guess in the place of a missing value [1] and Multiple Imputation [12] is an imputation method that is less sensitive to a bad starting point.

Missing data is a special case of coarsened data and data are said to be coarsened at random (CAR) if the coarsening mechanism is only a function of the observed data [9]. [21] introduced a methodology for parameter estimation from data structures for which full data has a non-zero probability of being fully observed and their methodology was later extended to deal with censored data in which complete data on subjects are never observed [31].

The use of graphical models for handling missing data is a relatively new development. [3] used graphical models for analyzing missing information in the form of missing cases (due to sample selection bias). Attrition is a common occurrence in longitudinal studies and arises when subjects drop out of the study [7,25,30] analysed the problem of attrition using causal graphs. [27,28] cautioned the practitioner that contrary to popular belief (as stated in [2,6]), not all auxiliary variables reduce bias. Both [7,28] associate missingness with a single variable and interactions among several missingness mechanisms are unexplored.

[17] employed a formal representation called Missingness Graphs to depict the missingness process, defined the notion of recoverability and derived conditions under which queries would be recoverable when datasets are categorized as Missing Not At Random (MNAR). Tests to detect misspecifications in the m-graph are discussed in [16].

8   Conclusions

This chapter presents the missing data problem from a causal perspective and provided procedures for estimating queries of interest for datasets falling into the MNAR (Missing Not At Random) Category. We demonstrated how auxiliary variables and data from external sources can be used to circumvent theoretical impediments to recoverability. Finally we showed that model-blind recovery techniques such as Multiple Imputation are prone to error and are insufficient to guarantee consistent estimates.

References

1. Allison, P.D.: Missing Data Series: Quantitative Applications in the Social Sciences (2002)
2. Collins, L.M., Schafer, J.L., Kam, C.-M.: A comparison of inclusive and restrictive strategies in modern missing data procedures. Psychol. Methods **6**(4), 330 (2001)
3. Daniel, R.M., Kenward, M.G., Cousens, S.N., De Stavola, B.L.: Using causal diagrams to guide analysis in missing data problems. Stat. Methods Med. Res. **21**(3), 243–256 (2012)
4. Darwiche, A.: Modeling and Reasoning with Bayesian Networks. Cambridge University Press, New York (2009)
5. Dempster, A.P., Laird, N.M., Rubin, D.B.: Maximum likelihood from incomplete data via the em algorithm. J. Roy. Stat. Soc. B. (Methodol.) **39**(1), 1–38 (1977)
6. Enders, C.K.: Applied Missing Data Analysis. Guilford Publications, New York (2010)
7. Garcia, F.M.: Definition and diagnosis of problematic attrition in randomized controlled experiments. Working paper, April 2013. http://ssrn.com/abstract=2267120
8. Graham, J.W.: Missing Data: Analysis and Design. Statistics for Social and Behavioral Sciences. Springer, New York (2012)
9. Heitjan, D.F., Rubin, D.B.: Ignorability and coarse data. Ann. Stat. **19**(4), 2244–2253 (1991)
10. Koller, D., Friedman, N.: Probabilistic Graphical Models: Principles and Techniques. Cambridge University Press, New York (2009)
11. Lauritzen, S.L.: The EM algorithm for graphical association models with missing data. Comput. Stat. Data Anal. **19**(2), 191–201 (1995)
12. Little, R.J.A., Rubin, D.B.: Statistical Analysis with Missing Data. Wiley, New York (2002)
13. Marlin, B.M., Zemel, R.S.: Collaborative prediction and ranking with non-random missing data. In: Proceedings of the Third ACM Conference on Recommender Systems, pp. 5–12. ACM (2009)

14. Marlin, B.M., Zemel, R.S., Roweis, S., Slaney, M.: Collaborative filtering and the missing at random assumption. In: UAI (2007)
15. Marlin, B.M., Zemel, R.S., Roweis, S.T., Slaney, M.: Recommender systems: missing data and statistical model estimation. In: IJCAI (2011)
16. Mohan, K., Pearl, J.: On the testability of models with missing data. In: Proceedings of AISTAT (2014)
17. Mohan, K., Pearl, J., Tian, J.: Graphical models for inference with missing data. Adv. Neural Inf. Process. Syst. **26**, 1277–1285 (2013)
18. Mohan, K., Pearl J.: Graphical models for recovering probabilistic and causal queries from missing data. In: Ghahramani, Z., Welling, M., Cortes, C., Lawrence, N.D., Weinberger, K.Q. (eds.) Advances in Neural Information Processing Systems 27, pp. 1520–1528 (2014)
19. Pearl, J.: Causality: Models, Reasoning and Inference. Cambridge University Press, New York (2009)
20. Pearl, J., Mohan, K.: Recoverability and testability of missing data: Introduction and summary of results. Technical report R-417, UCLA (2013). http://ftp.cs.ucla.edu/pub/stat_ser/r417.pdf
21. Robins, J.M., Rotnitzky, A.: Recovery of information and adjustment for dependent censoring using surrogate markers. In: Jewell, N.P., Dietz, K., Farewell, V.T. (eds.) AIDS Epidemiology, pp. 297–331. Springer, New York (1992)
22. Robins, J.M., Rotnitzky, A., Zhao, L.P.: Estimation of regression coefficients when some regressors are not always observed. J. Am. Stat. Assoc. **89**(427), 846–866 (1994)
23. Rothman, K.J., Greenland, S., Lash, T.L.: Modern Epidemiology. Lippincott Williams & Wilkins, Philadelphia (2008)
24. Rubin, D.B.: Inference and missing data. Biometrika **63**, 581–592 (1976)
25. Shadish, W.R.: Revisiting field experimentation: field notes for the future. Psychol. Methods **7**(1), 3 (2002)
26. Shpitser, I., Mohan, K., Pearl, J.: Missing data as a causal and probabilistic problem. In: Proceedings of the Thirty-First Conference on Uncertainty in Artificial Intelligence (2015)
27. Thoemmes, F., Mohan, K.: Graphical representation of missing data problems. Struct. Equ. Model. Multi. J. **37**(1), 1–13 (2015)
28. Thoemmes, F., Rose, N.: Selection of auxiliary variables in missing data problems: Not all auxiliary variables are created equal. Technical report R-002, Cornell University (2013)
29. Thoemmes, F., Mohan, K.: Graphical representation of missing data problems. Struct. Equ. Model. Multi. J. **22**(4), 1–13 (2015)
30. Twisk, J., de Vente, W.: Attrition in longitudinal studies: how to deal with missing data. J. clin. epidemiol. **55**(4), 329–337 (2002)
31. Van Der Laan, M.J., Robins, J.M.: Locally efficient estimation with current status data and time-dependent covariates. J. Am. Stat. Assoc. **93**(442), 693–701 (1998)
32. Van der Laan, M.J., Robins, J.M.: Unified Methods for Censored Longitudinal Data and Causality. Springer, New York (2003)

Learning Maximal Ancestral Graphs with Robustness for Faithfulness Violations

Takashi Isozaki[1]([✉]) and Manabu Kuroki[2]

[1] Sony Computer Science Laboratories, Inc.,
3-14-13 Higashigotanda, Shinagawa, Tokyo 141-0022, Japan
isozaki@csl.sony.co.jp
[2] Department of Data Science, The Institute of Statistical Mathematics,
10-3 Midori-cho, Tachikawa, Tokyo 190-8562, Japan
mkuroki@ism.ac.jp

Abstract. Discovering causal models hidden in the background of observational data has been a difficult issue. It is often necessary to deal with latent common causes and selection bias for constructing causal models in real data. Ancestral graph models are effective and useful for representing causal models with latent variables. The causal faithfulness condition, which is usually assumed for determining the models, is statistically known to often be weakly violated for finite data. One of the authors developed a constraint-based causal learning algorithm that is robust against the violations while assuming no latent variables. In this study, we applied and extended the thoughts of the algorithm to the inference of ancestral graphs. The practical validity and effectiveness of the algorithm are also confirmed by using some standard datasets in comparison with FCI and RFCI algorithms.

Keywords: Causal discovery · Latent variables · Maximal ancestral graphs · Causal faithfulness condition · Robust inference

1 Introduction

For many complex systems, appropriate interventions are expected or needed in order to improve their current situations. The following examples are often regarded as such systems: manufacturing processes, business marketing, healthcare, and global economics. The cause-effect relationships around objective variables must be known to perform effective interventions. However, in such systems, controlled experiments for clarifying cause-effect relationships often cannot be performed because of various reasons such as the complexity of systems and moral issues. Therefore, we have few choices but to infer causal relations from observational data. However, observing all related variables is probably impossible in practice, so some representations and techniques for dealing with unobserved variables are necessary. Common causal latent variables (i.e., latent confounders) and selection bias are particularly important for distinguishing direct and indirect correlations that arise from causations in real data. Directed

© Springer International Publishing Switzerland 2015
J. Suzuki and M. Ueno (Eds.): AMBN 2015, LNAI 9505, pp. 196–208, 2015.
DOI: 10.1007/978-3-319-28379-1_14

Acyclic Graphs (DAGs) [9], which are assumed to represent hidden true causal relations in the background of observational data, are extended as ancestral graph representations [13], which can provide information about latent common causes and selection bias.

In the inference of ancestral graphs, the constraint-based (CB) approach [14] using conditional independence tests is often used because the approach is based on the relationships between causality and probability, first found by Reichenbach via a research of the direction of time [12]. However, the causal faithfulness condition [14], an assumption necessary for perfect correspondence of a graphical representation to conditional independence relations, is statistically pointed out to be weakly violated for finite data [11,18]. Ramsey et al. proposed a conservative method of orientations and showed the effectiveness of decreasing false positive errors [11].

Isozaki considered that statistical errors due to violations could be decreased by reducing the number of unnecessary conditional independence tests performed, and he proposed an algorithm that avoids doing such tests without a loss of theoretical correctness. The effectiveness of his algorithm was experimentally shown for both Bayesian networks and causal structural equation models by using some standard datasets in comparison with representative algorithms [6]. In this study, we extend the algorithm to one that is applicable to the inference of ancestral graph models.

This paper is organized as follows. In the next section, we provide a background of causal discovery with and without latent common causes and selection bias. In Sect. 3, the Minimal Blocker Condition introduced by Isozaki [6] is described and extended for the inference of ancestral graphs. An extended algorithm is also provided. In Sect. 4, practical effectiveness is shown by experimental comparison with well known FCI [14] and recently proposed RFCI [5] algorithms with some standard datasets.

2   Background

When we infer causal relationships hidden in the background of observational data that we use for the purpose of understanding and utilizing such relationships, we may need an infinite number of unobserved variables. At a minimum, inferring causal models while assuming no latent common causes and selection bias is probably difficult unless we carefully choose variables to sufficiently construct causal models. However, it is not advantageous to explicitly represent so many latent variables. Thus, representing causal models that explicitly use only observed variables and indicating the existence of unobserved variables necessary for consideration is preferable [13]. In correlational patterns that emerge from causal relations, for variables X, Y, and Z, direct causation as $X \to Y$ and indirect causation between X and Z as $X \to Y \to Z$ are both represented by the usual DAG models with probability distributions. The systems that DAGs can represent only with observed variables are called causal sufficiency systems [14]. Meanwhile, latent common causal variables and selection bias generate pseudo correlations, which cannot be represented by the usual DAG model only with observed variables. Such systems are called causally insufficient. Ancestral graph models [13] are suitable for representing the causal models in insufficient systems, and we thereby focus on these models. Ancestral graph models are a natural extension of DAG models.

We define graph theoretical notions for representing causal models with ancestral graphs [13,17]. A graph consists of a set of vertices and edges. A mixed graph $\mathscr{G}$ can have at most one of three kinds of edges: directed ($\rightarrow$), bi-directed ($\leftrightarrow$), and undirected ($-$), between two vertices. For two vertices X and Y in $\mathscr{G}$, if $X \rightarrow Y$, we call X a parent of Y and Y a child of X; if $X \leftrightarrow Y$, we call X and Y a spouse of each other; if $X - Y$, we call X and Y a neighbor of each other; if there is any kind of edge, we call X and Y adjacent. For a sequence of adjacent vertices $\langle X_1, X_2, \cdots, X_{n+1} \rangle$ in $\mathscr{G}$, a path is the sequence, that is, X_i and X_{i+1} are adjacent for $i = 1, 2, \cdots, n$, and a directed path in $\mathscr{G}$ is a path such that X_i is a parent of X_{i+1} for $i = 1, 2, \cdots, n$. If there is a directed path from X to Y, X is called an ancestor of Y, and Y is called a descendant of X. Let the set of ancestors of X in $\mathscr{G}$ be denoted by $\mathrm{An}_{\mathscr{G}}(X)$. For three variable sets $\langle X, Z, Y \rangle$, when two directed edges exist from both X and Y to Z, Z is called a collider, otherwise a non-collider, and when X and Y are adjacent, the triple is called a shielded triple, otherwise called an unshielded triple and also a v-structure. When X is an ancestor of Y, and a directed edge $Y \rightarrow X$ exists in $\mathscr{G}$, it is called a directed cycle in $\mathscr{G}$. When X is an ancestor of Y, and a bi-directed edge $Y \leftrightarrow X$ exists in $\mathscr{G}$, it is called an almost directed cycle. A mixed graph is ancestral if the following conditions hold: (i) there are no directed cycles, (ii) there are no almost directed cycles, and (iii) for any pairs such that there is an undirected edge between them, both variables have no parents or spouses.

Next, we should define the cause-effect relationships that are inferable from observational data. The causal Markov condition (CMC) plays a role. We use vertices and variables interchangeably hereafter. CMC is defined as follows: each variable in an ancestral graph is probabilistically independent of non-descendants given its parent sets [14]. The validity of CMC is due to the following: CMC can be regarded as an extension of Reichenbach's principle found in the domain of physics [12], which first connects (conditional) independences in probability distributions with causal relationships. In this respect, inferable causation from observational data should have the form of DAG models with CMC, and then we focus on the models. D-separation criterion describes in more detail the conditional independences on DAGs. We thus assume that hidden true causal models have DAG-forms, but some variables are not observed. M-separation in ancestral graph models is an extension of d-separation in DAG models [13]. The definition is as follows:

Definition 1. *In an ancestral graph $\mathscr{G}$, a path p between two vertices X and Y is called blocked if the following conditions hold about a vertex set $\mathbf{Z}$ ($X, Y \notin \mathbf{Z}$) that has the possibility of being the empty set in $\mathscr{G}$: (i) each non-collider is in $\mathbf{Z}$ on p, and (ii) each collider and its descendants are not in $\mathbf{Z}$ on p. If all the paths between X and Y are blocked, X and Y are m-separated by $\mathbf{Z}$.*

We call the set which m-separates a pair of vertices m-separators. As seen in the definition of m-separations, colliders (or v-structures) have critical information from the perspective of learning causal models because we are able to identify causal directions if we are able to detect colliders (or v-structures).

An ancestral graph is called maximal if each non-adjacent pair is m-separated by a vertex set. DAGs are maximal in this respect, and we focus on the maximal ancestral graphs (MAGs). The causal faithfulness condition (CFC) is defined as follows: all conditional independence relationships are involved by the CMC in a MAG $\mathcal{G}$. A Markov equivalence class in MAGs is defined for a graphical group which has the same set of m-separators as in DAGs with d-separations [13]. There are cases in which the end-mark of an edge is undecidable due to Markov equivalence. We denote the end-mark as a circle ($\circ$). The Markov equivalence class graphs for MAGs are called partial ancestral graphs (PAGs), which are output forms of learning algorithms such as FCI, RFCI, and our algorithm. We denote a conditional independence relationship between two vertices X and Y given a conditioning set S as $X \perp\!\!\!\perp Y \mid S$.

3   Robust Algorithm for Violations of Faithfulness

We usually use conditional independence (CI) tests for determining the CMC or m-separation relations from data. As usual in recognizing CI relationships, we follow the way that an edge between two variables is removed if the corresponding test fails to reject the null hypothesis of CI relations and not removed if the test rejects the null. Due to the statistical limitations of detecting small dependencies between variables in finite sizes of samples, errors concerning missing edges that should not be removed often happen. These kinds of errors were pointed out [11], and we call them statistical violations of the CFC.

3.1   Minimal Blocker Conditions and Outline of CS* Algorithm

If we can detect CI tests that are unnecessary to perform and avoid them, an increase in inference accuracy for finding causal models is expected because the influence of violations of the CFC is expected to decrease. Isozaki proposed a new algorithm along with that thought. This algorithm combines two stages for detecting unnecessary CI tests: adjacency and (partial) orientation identification stages-the Combining Stage (CS) algorithm [6]. CS algorithm uses the Minimal Blocker Condition (MBC) defined by Isozaki [6], and then we extend it for MAGs. In Definition 1, if Z blocks a path between two vertices X and Y, Z is called a blocker of the path. Then, among the set of blockers of a path, the minimal size set of blockers is called the minimal blockers.

Definition 2. *In a MAG, a vertex set S satisfies the Minimal Blocker Condition for a pair of vertices X and Y if there is a path between X and Y that contains a vertex in S and no colliders.*

The following two properties are valid in MAGs as in DAGs.

Property 1. *A collider on path u between two vertices X and Y in a MAG is not a blocker on the path for the pair.*

Proof 1. *The proof immediately follows from Definition 1.* □

Property 2. *If path u between two vertices X and Y in a MAG has colliders, any vertex Z on u is not the minimal blocker on u for the pair.*

Proof 2. *The proof immediately follows from the proof of Property 2 in [6] with replacing the definition of d-separation by that of m-separation.* □

The following designation is used in this paper. Given conditioning set S, m-th order CI tests mean that $|S| = m$. When we use the definitions of colliders and m-separations in a MAG, the following theorem is valid as in a DAG.

Theorem 1. *Suppose that an algorithm searches for a MAG consisting of a vertex set V, and the algorithm starts from a complete graph[1], performs CI tests in ascending order of conditioning sets S starting from $|S| = 0$ for any currently connecting pair $X, Y \in V$, and removes the edge between X and Y if it detects a set S such that $X \perp\!\!\!\perp Y \mid S$. In such an algorithm, if a conditioning set S has a condition $|S| > 0$, and S does not satisfy MBC, S is not a m-separator candidate at the order of CI tests in the algorithm.*

Proof 3. *In a CI test for two vertices $X, Y \in V$ given conditioning set $S \in V \setminus \{X, Y\}$, if each path between X and Y, which has vertex $Z \in S$, contains a v-structure, the common minimal blocker is the empty set due to Property 2. Thus, if $(X \perp\!\!\!\perp Y \mid S \setminus Z)$ is true, this relationship should be found in lower order CI tests rather than in the current order tests. So, we can skip the CI test in the algorithm. The v-structures are temporal, and thus those may be removed in the processing. However, if a vertex $Z \in S$ that is not on a path between X and Y in the true graph, Z cannot be a member of m-separators due to Definition 1, and thus we can still skip the tests in that case.* □

We can extend CS algorithm for learning causal DAGs to for MAGs by using Theorem 1 as the property of MBC for MAGs, and then we call the algorithm CS*. CS* algorithm combines two stages as CS does: adjacency identification and v-structure identification. CS* algorithm starts from a complete graph where each edge forms ○—○. First, 0th order CI tests are performed, and then v-structure identification is immediately done from the results of the 0th order CI tests. Then, the procedure for higher order in ascending order is repeated. The algorithm can avoid a part of CI tests using Theorem 1. The temporal v-structures may be removed by higher order CI tests, but those do not bring present problems as described in Theorem 1, and the algorithm keeps theoretical soundness as FCI algorithm does.

---

[1] A complete graph is a graph that has edges for any pairs.

3.2   Lower Reliable Directions

Isozaki pointed out that learning causal models using CI tests has another inference accuracy issue, which is related to contradictory directions in inference models [6]. This issue is due to the locality of CI tests. In a MAG $\mathcal{G}$ that consists of a vertex set $\boldsymbol{V}$, suppose that the following three relationships of independence/dependence among a set of four vertices $\{X, Y, Z, W\} \in \boldsymbol{V}$: (i) $(X \perp\!\!\!\perp W \mid \boldsymbol{S}_1) \wedge (Z \notin \boldsymbol{S}_1)$, (ii) $(Y \perp\!\!\!\perp W \mid \boldsymbol{S}_2) \wedge (Z \notin \boldsymbol{S}_2)$, and (iii) $(X \perp\!\!\!\perp Y \mid \boldsymbol{S}_3) \wedge (Z \in \boldsymbol{S}_3)$, where $\boldsymbol{S}_1, \boldsymbol{S}_2, \boldsymbol{S}_3 \subset \boldsymbol{V}$. When we perform CI tests in ascending order, we obtain the following relations in an inferred PAG: $X *\!\!\rightarrow Z$, $W *\!\!\rightarrow Z$, and $Y *\!\!\rightarrow Z$, where the asterisk denotes the wildcard. However, the three relations are prohibited in an inferred PAG, and it means that contradictory directional errors occur. When adjacency faithfulness is defined as: adjacency of a pair of vertices means that the pair is dependent conditional on any subset [11], the following property is valid:

Property 3. *When the adjacency faithfulness is valid, and one of the above three conditions (i), (ii), and (iii) does not hold, $W *\!\!-\!\!\rightarrow Z$ always holds.*

Proof 4. *It can easily be confirmed.* □

We define these kinds of errors as lower reliable directions as follows. When we assume the adjacency faithfulness and that a directional error occurs in the above three conditions, the edges between X and Z and Y and Z are lower reliable. We disorient these edges into Z to $*\!\!-\!\!\circ Z$. This procedure is expected to reduce successive errors of orientation due to the use of the orientation rules.

We show the CS* algorithm in 1^2, where ADJ_X denotes an adjacency set of vertex X, and the output graph is represented by a maximal informative PAG [14].

3.3   Related Work

FCI algorithm [14] is known for inference of MAG models, but it is not actually sufficiently accurate in inference; nor is PC, although FCI and PC are proven to have soundness assuming statistical correctness under the CFC. MBCS* algorithm has the same level of accuracy as FCI but with much better time efficiency [10]. Meanwhile, computation for inference is a hard task for large variable systems, so RFCI algorithm was proposed for reducing such time complexity with an approximation [5]. BCCD algorithm achieves better inference results due to using Bayesian statistics [4]. In a DAG learning context, the Recursively Autonomy Identification (RAI) algorithm decomposes a DAG into subgraphs for reducing the number of large order CI tests [16] and thus seems effective for large DAG models. A similar recursive algorithm was also proposed by Xie and Geng [15].

Ramsey et al. decomposed the CFC into the adjacency and orientation faithfulness, assumed the former condition holds, and proposed a conservative

---

[2] About Possible-D-Sep, refer Spirtes et al. [14].

Algorithm 1. CS* Algorithm

Input: V: a set of observed variables
Output: $\mathscr{G}$: a maximally informative PAG
1: form a complete graph $\mathscr{G}$ over V with edges $\circ\!\!-\!\!\circ$;
2: make empty sets for any pair of $X, Y \in V$ denoted by $Sepset(X, Y)$;
3: set $n = 0$;
4: **repeat**
5:   **for** each $X \in V$ **do**
6:     **for** each $Y \in \text{ADJ}_X$ **do**
7:       **for** each subset $S \subseteq \{\text{ADJ}_X \setminus Y\}$ such that $|S| = n$ **do**
8:         set $m = 0$
9:         **for** each $Z \in S$ **do**
10:            **if** a path containing Z between X and Y does not contain a v-structure **then**
11:              set $m = 1$
12:              break
13:            **end if**
14:          **end for**
15:          **if** $m = 0$ **then**
16:            continue
17:          **else if** $X \perp\!\!\!\perp Y \mid S$ **then**
18:            remove the edge $X *\!\!-\!\!* Y$ from $\mathscr{G}$
19:            add S to $Sepset(X, Y)$
20:          **end if**
21:        **end for**
22:      **end for**
23:    **end for**
24:    **for** each unshielded triple $\langle X, Z, Y \rangle$ in $\mathscr{G}$ **do**
25:      orient it as $X *\!\!\rightarrow Z \leftarrow\!\!* Y$ iff Z is not in $Sepset(X, Y)$
26:      add them to $Vstr(X, Z, Y)$
27:    **end for**
28:    set $n = n + 1$
29: **until** $|\text{ADJ}_X| \leq n$ for all $X \in V$
30: detect non-adjacencies due to Possible-D-Sep.
31: find all unreliable directions and disorient those edges
32: orient edges using the orientation rules
33: **return** $\mathscr{G}$

PC algorithm for violations of the latter condition [11]. The algorithm practically reduces false positive errors of orientations. Isozaki proposed the CS algorithm from a more aggressive perspective of improving accuracy even for edge-identification inference and showed that the aim was actually achieved for five standard datasets with a wide range of sample sizes [6].

4   Experimental Evaluation

CS* algorithm also has the soundness of accuracy for learning MAGs the same as FCI does. In this section, we demonstrate practical performance with finite

Table 1. Numbers of parameters in each DAG used in the experiments.

Database	Variables	Edges	Conditional probabilities
Alarm	37	46	509
Carpo	60	74	342
Hailfinder	56	66	2656
Insurance	27	52	1008
Water	32	66	10083

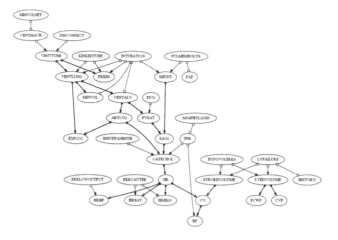

Fig. 1. PAG representation of Alarm network.

data of some standard datasets in comparison with well-known FCI and recently proposed RFCI algorithms. We used R package, pcalg, for performing FCI and RFCI [8]. All three algorithms used the orientation rules by Zhang [17]. The datasets we used here are sampled from well-known Alarm [2], Hailfinder [1], Insurance [3], Water [7], and Carpo datasets, which have known DAG-structures and discrete conditional probability distributions. Numbers of parameters of the DAGs (variables, edges, and conditional probabilities) are shown in Table 1. Then we sampled from the distributions with a sample size of 1000, 2000, 5000, and 10000. G^2 tests [14] with a significance level of 0.01 are used for all algorithms since the threshold value was best for all algorithms in pre-experiments. The maximum degree of conditioning sets in CI tests was set as 5. We evaluate inference accuracy by performing recovery of causal structures in averaged values with 10 runs for each dataset and sample size. The known structures are DAGs, so we transformed the DAGs to PAGs when evaluating accuracy because all of the algorithms provide inference results as PAGs. As examples, PAGs of Alarm and Insurance are shown in Figs. 1 and 2, where the edges with definite direction are denoted by bold lines for visual clarity. The structural Hamming Distance (SHD) was used for the evaluations [6]. SHD, which consists of extra/missing

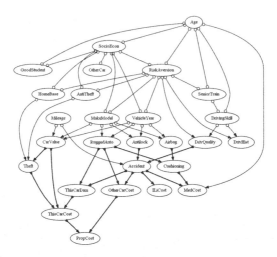

Fig. 2. PAG representation of Insurance network.

edge errors and directional errors, is usually used for DAG-evaluations. We used SHD for PAG-evaluations, where SHD is defined to have the same extra/missing errors and errors of marks of both sides of edges.

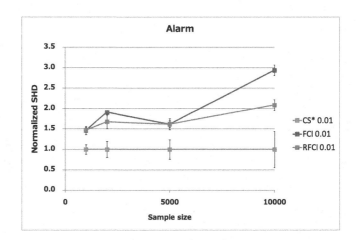

Fig. 3. Averaged and normalized SHD results for inferred PAGs from Alarm dataset by CS*, FCI, and RFCI.

The results of SHDs are shown in Figs. 3, 4, 5, 6, and 7, where the values are normalized by averaged values of CS* and the error bars denote the standard deviations. As smaller values than 1 indicate better results than CS*, CS* outperforms FCI/RFCI in terms of the averaged values for all sample sizes in Alarm, Carpo, Hailfinder, and Insurance, although the results of CS* for Water

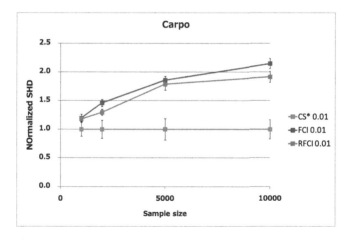

Fig. 4. Averaged and normalized SHD results for inferred PAGs from Carpo dataset by CS*, FCI, and RFCI.

performed slightly lower against FCI/RFCI. These results show that CS* algorithm is practically effective on average in addition to its theoretical soundness.

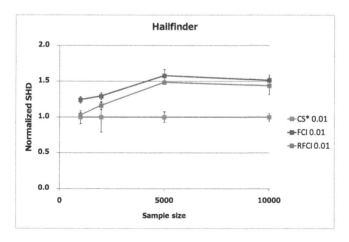

Fig. 5. Averaged and normalized SHD results for inferred PAGs from Hailfinder dataset by CS*, FCI, and RFCI.

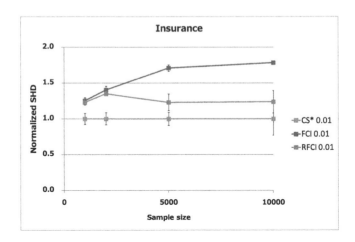

Fig. 6. Averaged and normalized SHD results for inferred PAGs from Insurance dataset by CS*, FCI, and RFCI.

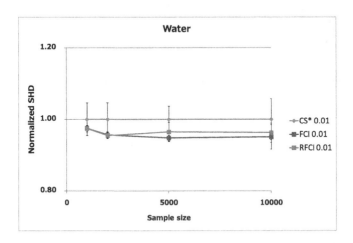

Fig. 7. Averaged and normalized SHD results for inferred PAGs from Water dataset by CS*, FCI, and RFCI.

5   Conclusion

In realistic observational data without careful design on obtaining the data, causal learning seems intractable due to latent variables that generate pseudo correlations, leading to a misunderstanding of the systems. Those variables are known to latent common causes and selection bias. Maximal ancestral graphical (MAG) representations are a natural extension of directed acyclic graphs (DAGs) and are suitable for such causal learning tasks, which can indicate the existence or absence of the latent variables. The causal faithfulness condition (CFC) is necessary for a perfect correspondence of conditional independence relations, which is assumed even in MAG learning, to the background causal graphical structures. However, the CFC is statistically recognized to often be weakly violated for finite samples. These situations naturally occur in MAG inference the same as in DAG. The CS algorithm was proposed for robust DAG inference against weak violations while keeping theoretical soundness.

In this study, we extend the CS to that which is applicable to MAG inference: the Minimal Blocker Condition and lower reliable directions are extended for latent variable systems that can be represented by MAGs, and we call the extended algorithm CS*. In experimental evaluations with five standard datasets in comparison with FCI/RFCI algorithms, CS* outperformed them in four out of the five datasets on a wide range of sample sizes, 1000–10000. The results shows the effectiveness of CS* algorithm for learning MAGs on average.

Acknowledgments. The authors thank Tomohide Haruguchi for assisting with the experiments. Isozaki thanks Hiroaki Kitano of Sony Computer Science Laboratories, Inc. for his support.

References

1. Abramson, B., Brown, J., Winkler, R.L.: Hailfinder: a Bayesian system for forecasting severe weather. Int. J. Forecast. **12**, 57–71 (1996)
2. Beinlich, I., Suermondt, H., Chavez, R., Cooper, G.: The ALARM monitoring system: a case study with two probabilistic inference techniques for belief networks. In: Proceedings of European Conference on Artificial Intelligence in Medicine (AIME 1989), pp. 247–256 (1989)
3. Binder, J., Koller, D., Russell, S., Kanazawa, K.: Adaptive probabilistic networks with hidden variables. Mach. Learn. **29**, 213–244 (1997)
4. Claassen, T., Heskes, T.: A Bayesian approach to constraint based causal inference. In: Proceedings of Conference on Uncertainty in Artificial Intelligence (UAI 2012), pp. 207–216 (2012)
5. Colombo, D., Maathuis, M.H., Kalisch, M., Richardson, T.S.: Learning high-dimensional directed acyclic graphs with latent and selection variables. Ann. Stat. **40**(1), 294–321 (2012)
6. Isozaki, T.: A robust causal discovery algorithm against faithfulness violation. Trans. Jpn Soc. Artif. Intell. **29**(1), 137–147 (2014)

7. Jensen, F.V., Kjærulff, U., Olesen, G., Pedersen, J.: An expert system for control of waste water treatment. Technical report, Judex Datasystemer A/S, Aalborg, Denmark (1989) (in Danish)
8. Kalisch, M., Mächler, M., Colombo, D., Maathuis, M.H., Bühlmann, P.: Causal inference using graphical models with the R package pcalg. J. Stat. Softw. **47**(11), 1–26 (2012)
9. Pearl, J.: Probabilistic Reasoning in Intelligent Systems. Morgan Kaufmann, San Mateo (1988)
10. Pellet, J., Elisseeff, A.: Finding latent causes in causal networks: an efficient approach based on markov blankets. In: Proceedings of Advances in Neural Information Processing Systems 21 (NIPS 2008), pp. 1249–1256 (2008)
11. Ramsey, J., Spirtes, P., Zhang, J.: Adjacency-faithfulness and conservative causal inference. In: Proceedings of Conference on Uncertainty in Artificial Intelligence (UAI 2006), pp. 401–408 (2006)
12. Reichenbach, H.: The Direction of Time. Dover Publications, Mineola (1956) (Republication of the work published by University of California Press, Berkely)
13. Richardson, T., Spirtes, P.: Ancestral graph markov models. Ann. Stat. **30**(4), 962–1030 (2002)
14. Spirtes, P., Glymour, C., Scheines, R.: Causation, Prediction and Search, 2nd edn. MIT Press, Cambridge (2000)
15. Xie, X., Geng, Z.: A recursive method for structural learning of directed acyclic graphs. J. Mach. Learn. Res. **9**, 459–483 (2008)
16. Yehezkel, R., Lerner, B.: Bayesian network structure learning by recursive autonomy identification. J. Mach. Learn. Res. **10**, 1527–1570 (2009)
17. Zhang, J.: On the completeness of orientation rules for causal discovery in the presence of latent confounders and selection bias. Artif. Intell. **172**, 1873–1896 (2008)
18. Zhang, J., Spirtes, P.: Detection of unfaithfulness and robust causal inference. Mind. Mach. **18**(2), 239–271 (2008)

Discriminative and Generative Models in Causal and Anticausal Settings

Patrick Blöbaum$^{(\boxtimes)}$, Shohei Shimizu, and Takashi Washio

The Institute of Scientific and Industrial Research, Osaka University, Osaka, Japan
bloebaum@ar.sanken.osaka-u.ac.jp

Abstract. Having knowledge about the real underlying causal structure of a data generation process has various implications for different machine learning problems. We address the idea of causal and anticausal learning with respect to a comparison of discriminative and generative models. In particular, we conjecture the hypothesis that generative models perform better in anticausal problems than in causal problems. We empirical evaluate our hypothesis with different real-world data sets.

Keywords: Causality · Causal and anticausal learning · Discriminative and generative models · Logistic regression · Naive bayes

1 Introduction

Choosing a suitable machine learning algorithm for a given problem can be a difficult task. In order to achieve the highest possible accuracy, the choice normally bases on the properties of the data including the number, variance or sparsity of the variables, as well as whether the prediction variable is a continuous number (regression) or a category (classification). Schölkopf et al. [1] proposed that, under certain assumptions, the underlying causal directions of variables have important implications for the prediction by differing between predicting in a causal and anticausal direction.

In order to explain our ideas, we will only regard two variable cases, but the ideas seem to be extendable to any number of variables. Therefore, there is only an input X, an output Y and we want to predict Y by X. In the underlying causal structure of a data generation process exists a cause C and an effect E. Between cause and effect is an unknown function or mechanism $E = \varphi(C, N_E)$ which transforms the cause into the effect given some noise N_E. The typical goal of a machine learning algorithm is to learn this mechanism in order to make new predictions. The term "mechanism" will be used for the function φ and for the conditional $P(E|C)$ in the following.

Schölkopf et al. proposed that differing between predicting the effect by the cause (causal direction) and predicting the cause by the effect (anticausal direction) leads to some interesting theoretical implications for different machine learning scenarios. One of the most important assumptions in this context is the independence between mechanism and input. Following this assumption,

© Springer International Publishing Switzerland 2015
J. Suzuki and M. Ueno (Eds.): AMBN 2015, LNAI 9505, pp. 209–221, 2015.
DOI: 10.1007/978-3-319-28379-1_15

the cause distribution $P(C)$ has no task relevant information about the underlying mechanism $P(E|C)$ and vice versa. Regarding this assumption, Schölkopf et al. particularly formulated the hypothesis that semi-supervised learning works better in anticausal or confounded problems than in causal problems, because additional information about the input does not help for predicting in causal direction. To support their claim, they provided some empirical results (see [1,2] for more details).

Inspired by their ideas, we investigate a further hypothesis regarding generative models. We hypothesize that generative models are more likely to perform better in anticausal or confounded problems than in causal problems. Generative models, in particular, try to model $P(X|Y)$ and using Bayes' rule for making predictions. Under the independence assumption, $P(X)$ has no task relevant information for modeling $P(Y|X)$ in causal problems, but it has relevant information in anticausal/confounded problems. Therefore, we conjecture that generative models are more likely to perform better in anticausal/confounded problems than in causal problems.

This is in particular interesting regarding the general belief that discriminative models outperform generative models in many application domains (e.g. in [3–5]). Due to some inaccurate modeling assumptions often made by generative models, the asymptotic error of generative models is greater than the asymptotic error of discriminative models [6], but since there is rarely enough data for a problem to generally say that the choice of a discriminative model is to be preferred, the underlying causal structure of a problem can give some good advice as to what kind of model may fit better. In our evaluations we compare the performances of generative and discriminative models in artificial and real-world datasets. In our future work we aim to provide a proof of the hypothesis supported by these results. In the following, we will call confounded problems also as anticausal problems, since the relevant properties of anticausal problems for our hypothesis are also valid for confounded problems. Regarding the notation, we will often differentiate between a cause C and an effect E. To denote if either the input X or the output Y is the cause or effect, we accordingly use C or E as index. For instance, if X is the cause, it is denoted as X_C and the effect Y as Y_E.

1.1 Causal Problems

In *causal problems*, we predict the effect by the cause. In general, there is an input X_C (cause C) and an output Y_E (effect E) and we aim to learn the mechanism φ (see Fig. 1). An example for a causal problem would be the "pole balancing" problem. In this problem, a pole is affixed to a cart via a joint and the cart is only allowed to move to the left or right. The classification task here is to predict which direction the cart should move in order to balance the pole. The classifier input is the current angle of the pole. In this case, we could think of some simple threshold rules as classification mechanism.

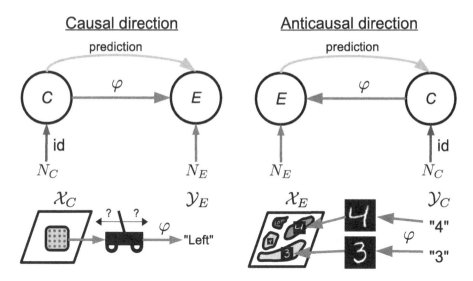

Fig. 1. An illustration on the difference between predicting in the causal and anticausal direction. Note the difference in the direction of φ. Causal direction: The input X_C causes the output Y_E. Therefore, we aim to predict the effect E by the cause C. Anticausal direction: The output Y_C causes the input X_E. Here, the cause C is predicted by the effect E.

1.2   Anticausal Problems

More interesting are *anticausal problems*, where we predict the cause by the effect (see Fig. 1). Thus, the output Y_C causes the input X_E, but we want to predict Y_C based on X_E. In fact, it seems that most classification problems have a similar structure and are therefore anticausal problems. The "MNIST" data set, for instance, is an anticausal problem. This data set consists of images of handwritten digits from 0 to 9 and the task is to classify these images. The underlying causal structure is anticausal, because the class label Y_C causes a motor pattern (mechanism φ) in our head to produce the image X_E.

In many cases, it can be easier to predict the effect by the cause. For example, given a regression problem with a causal structure like $X_E = Y_C + N_E$ with uniformly distributed error noise N_E. Here, we want to predict the cause Y_C by the effect X_E, but taking a look at Fig. 2 shows that predicting the effect X_E by the cause Y_C is a simpler task. Since this is a regression problem, it is possible to predict X_E by Y_C with a simple linear function, but predicting Y_C by X_E would require a non-linear function which is more difficult to fit [1]. Hence, in this respect it is easier to model $P(X_E|Y_C)$ first and then use Bayes' rule for modeling $P(Y_C|X_E)$ in order to predict Y_C.

These kinds of properties are also valid for confounded data [1].

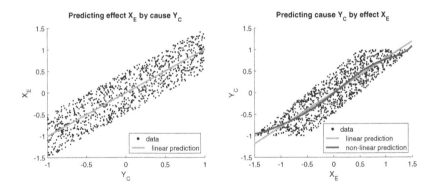

Fig. 2. Example for predicting the effect and predicting the cause: The causal structure of the underlying data generation process is $X_E = Y_C + N_E$, where Y_C and N_E are uniformly distributed. Note that X_E here is the effect and Y_C the cause. For a proper prediction of Y_C by X_E (right) a non-linear prediction is necessary. Since this is a regression problem, it is a much easier task to predict X_E by Y_C (left) with a simple linear function first and then using Bayes' rule for inferring Y_C. In this instance, knowledge about the underlying causal structure can be helpful for the prediction task. This figure is reproduced from [1].

1.3   Independence of Mechanism and Input

The independence of mechanism and input [7] is the most important assumption concerning our hypothesis and is similar to the autonomous data generation process [8]. Here, the definition of independence is different from the classical statistical definition (see [9] for more details). It is assumed that the cause distribution $P(C)$ is independent of the mechanism $P(E|C)$ and therefore has no information about it. In particular, changing $P(C)$ has no influence on the mechanism. This seems plausible taking a look at the "pole balancing" problem mentioned in Sect. 1.1. If the distribution of the angles has been changed to be gaussian instead of uniform, the threshold when the car should drive to the left or right would still be the same. On the other hand, $P(E)$ and $P(C|E)$ are dependent, because both have information about $P(E|C)$ and $P(C)$. As in the case of the "MNIST" data, an accurate estimation of $P(X)$ already reveals possible class boundaries and can help in the classification task. Therefore, there exists an asymmetry between cause and effect. To summarize the important properties of this assumption:

- $P(C)$ has no information about $P(E|C)$
- $P(E)$ has information about $P(E|C)$
- Changing $P(C)$ has no influence on $P(E|C)$ and vice versa

  $\Rightarrow P(X_C)$ has **no** information about $P(Y_E|X_C)$ in the **causal** case
  $\Rightarrow P(X_E)$ does **have** information about $P(Y_C|X_E)$ in the **anticausal** case
  $\Rightarrow$ There is an asymmetry between cause and effect.

1.4 Discriminative and Generative Models

Machine learning algorithms can generally be categorized as either a discriminative or generative model. There also exist hybrid models which try to combine the advantages of both, but we will not consider them further.

Discriminative models try to directly model $P(Y|X)$ by discriminating Y for any input X. Examples are Logistic Regression, Support Vector Machines or Perceptrons. On the other hand, generative models try to model the data generation $P(X,Y)$. This is done by estimating $P(X|Y)$, $P(Y)$ and particularly by taking $P(X)$ into account. The Bayes' rule is then applied for getting $P(Y|X)$. Typical examples are Naive Bayes, RBF Networks or Hidden Markov Models.

Every generative model addresses the problem of estimating $P(X,Y)$ differently. Several assumptions are necessary, such as the dependencies between variables and how to estimate the underlying distribution. In this work, we use the Naive Bayes (NB) classifier for our evaluations. In case for NB, the joint probability $P(X,Y)$ has the form

$$P(X,Y) = P(X|Y)P(Y).$$

In a multivariable case, the conditional $P(X|Y)$ is replaced by the product over all input variables conditioned by Y. Here, NB assumes that the variables are independent, which rarely holds true. Inferences for Y can then be made by conditioning $P(X,Y)$ on X, which leads to the Bayes' rule

$$P(Y|X) = \frac{P(X|Y)P(Y)}{P(X)}.$$

Besides modeling assumptions regarding the variable dependencies, further assumptions are necessary for estimating $P(X|Y)$. NB, for instance, assumes that the conditional(s) $P(X|Y)$ is gaussian distributed. As soon as one of these conditions are violated, the generative model becomes inaccurate. Note that NB actually does not utilize $P(X)$, only the conditional $P(X|Y)$ and the prior $P(Y)$ due to the maximum a posteriori estimation for inferring Y.

There exist many comparisons between discriminative and generative models such as in [6,10]. It is often pointed out that wrong modeling assumptions in generative models lead to higher asymptotic errors as compared to discriminative models even though generative models need less training data [6,10]. Therefore, the general belief in the machine learning community is that a discriminative model normally outperforms a generative model and is therefore preferred. We discuss this comparison under the aspect of causal and anticausal problems and especially point out that generative models can be a good choice in anticausal problems.

2 Hypothesis

Differing between causal and anticausal problems under the independence assumption leads to some interesting implications. Schölkopf et al. explicitly brought this to the context of semi-supervised learning [1,2]. They argued that additional input data sampled from $P(X_C)$ do not increase the performance in causal problems, because $P(X_C)$ has no information about the mechanism $P(Y_E|X_C)$ with respect to the independence assumption. On the contrary, additional samples can improve the performance in anticausal data seeing that $P(X_E)$ is the effect in this case and therefore has information about the mechanism.

As already mentioned, in most application tasks, the modeling assumptions by generative models are inaccurate with respect to the real underlying process. However, the assumptions are often still sufficient enough for predictions in several problems, but insufficient in others. We presume that these inaccurate modeling assumptions have a higher negative impact in causal data than in anticausal data. Therefore, we hypothesize that generative models in general perform better in anticausal problems than in causal problems.

Since generative models try to estimate $P(X)$, the encoded task relevant structures in anticausal problems can be utilized for making inferences based on $P(X_E|Y_C)$. We suppose that the modeling assumptions are more likely to be sufficient enough in anticausal data to exploit information based on $P(X_E)$. The distribution $P(X_E)$ depends on $P(Y_C)$, hence modeling the conditional $P(X_E|Y_C)$ reflect the actual causal relationship between cause and effect in a two variable case.

On the other side, under the independence assumption, X_C is independent of Y_E in causal problems. $P(X_C)$ provides no relevant information, hence no encoded information in $P(X_C)$ can be exploited for a proper estimation of $P(X_C|Y_E)$. The conditional $P(X_C|Y_E)$ here reflects a wrong causal relationship between cause and effect. Regarding this, it becomes even more difficult to find a proper model to explain (or estimate) $P(X_C|Y_E)$. Therefore, it is more likely that wrong modeling assumptions lead to an inaccurate estimation of $P(Y_E|X_C)$.

Note that we only conjecture that generative models tend to perform better in anticausal problems. This does not imply that they are better than discriminative models. However, it could be additionally argued that discriminative models are normally the better choice in causal problems, since the estimated $P(Y_E|X_C)$ is not negatively influenced by inaccurate estimations based on $P(X_C|Y_E)$ and $P(X_C)$ such as in generative models.

The idea of utilizing task relevant information based on $P(X)$ is also highly related to the following semi-supervised learning assumptions [11]:

Smoothness assumption: Two data points which are close to each other are more likely to share the same output.

Cluster assumption and low density separation: Points in the same cluster are more likely to be of the same class. This can be seen as a special case of the smoothness assumption. The cluster assumption can equivalently be formulated as low density separation which states that the decision boundary should lie in a region where the density $P(X)$ is low.

Manifold assumption: The (high-dimensional) data lie on a low-dimensional manifold. If we assume that the manifold is embedded in a high-dimensional data space, accurate density estimations of the input space can be useful for the classification or regression task.

These assumptions imply that there are task relevant information encoded in $P(X)$ and therefore an accurate estimation of $P(X)$ can be helpful for a proper estimation of $P(Y|X)$. In particular, we suppose that anticausal problems tend to fulfill these assumptions due to the property of $P(X)$ having information about $P(Y|X)$.

In order to clarify our idea, we once again bring in the context of the two examples for causal and anticausal problems given in Fig. 1. Having uniformly distributed input data in the "pole balancing" example, modeling the input distribution gives us no useful information about the mechanism (threshold) as to when the car should change direction. Therefore, $P(X_C)$ has no useful information. On the contrary, there are some clear clusters in the "MNIST" example. Even without seeing any class labels, we could already suppose that these clusters represent different classes. The input distribution has useful information here and taking $P(X_E)$ into account may be useful for the classification task.

Since $P(X_C)$ has absolutely no information about $P(Y_E|X_C)$ in the "pole balancing" example and "MNIST" fulfill all aforementioned assumptions, these two examples would be perfect cases which are rarely realistic. Nevertheless, we compared different input distributions of causal and anticausal real-world data sets (see Sect. 3 (Real-world data sets)) and noticed that the distributions seem tending to have these kind of properties.

3   Empirical Evaluations

For the investigation of our hypothesis we used Logistic Regression (LR) as discriminative model and NB as generative model in real-world and artificial data sets. Since NB is the generative counterpart of LR [10], these two classifiers offer the most fair comparison. If the intra-class variances are equal and the modeling assumption made by NB holds true, both models even converge to the same linear classifier [10]. In practice this rarely holds true, therefore the behavior is different. Even though NB does not take variable dependencies of the input into account, it still regards the data distribution and models the causal relationship between cause and effect. We also performed the comparison with more powerful classifiers and got the same tendency, therefore, we only present LR and NB.

We first utilized the same real-world data sets as mentioned in [1] and added a few additional. A proper categorization of a data set into causal or anticausal requires a precise documentation of how the data were acquired and nevertheless a clear categorization is often difficult. For example, it can be argued that the "MNIST" data set is anticausal because the drawer had the number in mind before drawing it or, on the contrary, it can be argued that the image was labeled based on the number that was recognized and is therefore causal. In this case, if we just focus on the process when the images were created, then the former seems more plausible. The categorization of data sets is another difficult topic, but a detailed discussion falls out of the scope of this work. Therefore, we followed the same categorization as that of Schölkopf et al. (see [1] for more details). Note that we are not interested in achieving a high classification performance. Instead we want to empirically evaluate our claim that generative models generally perform better in anticausal data than in causal data by providing a comparison of discriminative and generative models in causal and anticausal data.

Real-World Data Sets. The utilized real-world data sets are taken from the UCI repository and from the benchmark set in [11]. For the purpose of obtaining a fair comparison of discriminative and generative models in causal and anticausal data, we compared the average accuracy of LR and NB in 10 causal data sets and 18 anticausal data sets. We further provide an overview of the input spaces of each data set after applying the t-distributed stochastic neighbor embedding unsupervised dimensionality reduction algorithm [12]. This algorithm is especially suitable to get a proper two dimensional representation of high dimensional data. By this, we want to show that the properties of causal and anticausal data mentioned in Sect. 2 can be found in most real-world problems, though not in every one. Note that this dimensionality reduction algorithm is unsupervised so the resulting manifold tries to preserve as much relevant structure of the high dimensional space as possible without regarding discriminative information. An overview of all utilized real-world data sets with their according input space can be found in Fig. 3. Data sets "g241c" and "g241n" are artificial data sets, but we added them to the list since they were also used by Schölkopf et al. for their evaluations.

Each data set was evaluated with a 10-fold cross validation and averaged over 10 runs. The results are shown in Table 1. There is a clear performance gap between causal and anticausal data sets for NB. The average performance is around 14 % higher in anticausal data sets. Furthermore, NB even performed slightly better than LR in many cases. On the other hand, although NB performed better in some causal data sets, the performance differences between LR and NB is quite big for some data sets, such as in "chess", "TicTacToe", "UserModeling" or "wine". Besides NB, the performance of LR is in general also better in the anticausal data sets, which may indicate that anticausal problems can be easier than causal problems. However, the performance gap between causal and anticausal data set seems to be higher for NB than for LR. Note that a fair empirical comparison between discriminative and generative models is difficult due to the fundamental model differences. Furthermore, we only had 10 causal data sets available, hence the low accuracy of "UserModeling" has a big impact on the overall average accuracy of NB in causal data sets. Despite that, these results at least seem to support our hypothesis and do not contradict it.

Discussion About the Input Spaces. Aside from a performance evaluation, we are also interested in the actual input distribution of the different data sets so as to check how realistic the made assumptions are and to explain some

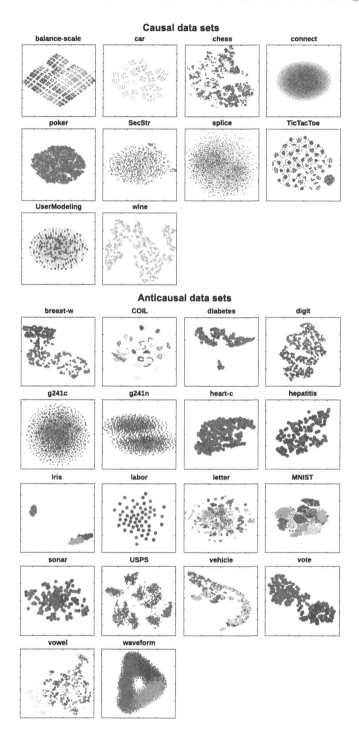

Fig. 3. An overview of the utilized real-world data sets with their according two dimensional inputs space after a unsupervised dimensionality reduction. Each color represents one class.

Table 1. A performance comparison of the average accuracy of Logistic Regression and Naive Bayes in causal and anticausal data.

Causal data sets	Logistic Regression	Naive Bayes
balance-scale	89.44 %	**91.44 %**
car	**93.28 %**	85.46 %
chess	**97.61 %**	87.71 %
connect	82.53 %	**84.75 %**
poker	**49.95 %**	46.19 %
SecStr	57.87 %	**65.84 %**
TicTacToe	**98.22 %**	69.91 %
UserModeling	**50.77 %**	15.57 %
wine	**53.83 %**	48.79 %
splice	90.56 %	**95.42 %**
	76.41 %	69.11 %
Anticausal data sets	Logistic Regression	Naive Bayes
breast-w	96.48 %	**97.57 %**
COIL	83.37 %	**86.49 %**
diabetes	**77.19 %**	75.22 %
digit	94.81 %	**95.61 %**
g241c	82.15 %	**87.06 %**
g241n	82.66 %	**86.76 %**
heart-c	83.31 %	**83.41 %**
hepatitis	82.55 %	**84.15 %**
iris	95.82 %	**96.59 %**
labor	**94.07 %**	93.57 %
letter	**77.27 %**	73.96 %
MNIST	90.74 %	**92.04 %**
sonar	72.51 %	**72.85 %**
USPS	**84.54 %**	82.22 %
vehicle	**79.72 %**	59.87 %
vote	**94.93 %**	89.69 %
vowel	68.02 %	**73.13 %**
waveform	**86.65 %**	80.01 %
	84.82 %	83.9 %

of the results. Therefore, we further provide an overview of the various input spaces after an unsupervised dimensionality reduction in Fig. 3. Even though the dimensionality reduction can probably not capture all relevant structures, it offers a good insight of the data distributions of real-world problems. The performance evaluations are based on the raw data, therefore, not all results can be explained by the low dimensional representation.

For the causal data sets, except for "splice", each data set requires a complex model for a proper estimation of $P(X_C|Y_E)$. In particular, "connect", "poker", "SecStr" and "UserModeling" seem to have no meaningful structures at all. Since NB is making inferences based on $P(X_C|Y_E)$ the individual class distributions are very important. Note that the classical NB is limited to model the class distribution only by a simple gaussian. Therefore, NB can be especially helpful in cases such as "splice" where the class distributions seem to be gaussian. In fact, the performance of NB in "splice" is significantly better than that of LR. On the other hand, modeling a class distribution by a simple gaussian in the other cases would not be particularly helpful. Nonetheless, e.g. for "TicTacToe", a more complex model such as a mixture of gaussians could reveal more helpful structures for the classification. The reason why NB performs significantly better than LR in "SecStr" is possibly due to a more meaningful class structure in the higher dimensional space.

In the anticausal data sets are more cases where a simple model for the estimation of a class distribution can already be sufficient, such as in "breast-w", "digit", "iris", "labor", "MNIST", "vote" or "waveform". In general, taking $P(X_E|Y_C)$ into account can be helpful in nearly every data set besides "diabetes" and "g241c". The better performance of NB in "g241c" can be again explained by the structures in the high dimensional space.

The comparison of the input spaces also seem to support the supposition that anticausal data tend to fulfill the smoothness, cluster and/or manifold assumption, since most of the anticausal data sets have these properties. This generally implies a more accurate estimation of $P(Y|X)$. For the causal data sets it looks like these assumptions are mostly violated. Further, it seems that more often a simple model for $P(X|Y)$ is sufficient in anticausal data than in causal data. Therefore, the drawbacks of generative models regarding wrong modeling assumptions are less crucial in anticausal data.

4   Conclusion

Having knowledge about the underlying causal structure of a problem can give helpful information for predicting a variable. We are interested in comparing discriminative and generative models in causal and anticausal settings and particularly formulate the hypothesis that generative models generally perform better in anticausal problems than in causal problems. In our empirical evaluations, we use Logistic Regression as a discriminative model and Naive Bayes as a generative model. The results seem to support our hypothesis. We further compare various input spaces of causal and anticausal real-world data sets in order to support our conjectures. The results are as follow:

- Discriminative and generative models perform better in anticausal problems.
- The performance gap between causal and anticausal problems is higher for generative models.
- Generative models even outperform discriminative models in some anticausal problems.

⇒ The performance gap between discriminative models and generative models is smaller in anticausal problems.

- Class distributions seem to be more complex in causal data.

⇒ Wrong modeling assumptions made by generative models have a higher negative impact in causal problems.

- It seems more likely that anticausal problems tend fulfill the smoothness, cluster and/or manifold assumption.

⇒ Anticausal problems tend to have simple and less overlapping class structures.
⇒ Simpler models for estimating $P(X_E|Y_C)$ are sufficient in anticausal problems.
⇒ Wrong modeling assumptions made by generative models have a smaller negative impact in anticausal problems.

Our results can give a new insight with respect to the general belief of preferring a discriminative over a generative model in most application domains.

In our future work, we are interested in additional evaluations, since we could only utilize a few real-world data sets in our experiments. Moreover, performing the same evaluations with more complex generative models could substantiate our hypothesis. If our claim holds true, we are also particularly interested in developing a mathematical proof by considering a deterministic causal process under the independence assumption.

Acknowledgement. This work was supported by JSPS KAKENHI #25240036, #26540116 and #24700275 and the Center of Innovation Program from the Japan Science and Technology Agency, JST.

References

1. Schölkopf, B., Janzing, D., Peters, J., Sgouritsa, E., Zhang, K., Mooij, J.M.: On causal and anticausal learning. In: Langford, J., Pineau, J. (eds.) Proceedings of the 29th International Conference on Machine Learning (ICML 2012), pp. 1255–1262. Omnipress, New York, July 2012
2. Schölkopf, B., Janzing, D., Peters, J., Sgouritsa, E., Zhang, K., Mooij, J.: Semi-supervised learning in causal and anticausal settings, Chap. 13, pp. 129–141. Festschrift in Honor of Vladimir Vapnik. Springer (2013)

3. Lasserre, J., Bishop, C.M.: Generative or discriminative? getting the best of both worlds. Bayesian Stat. **8**, 3–24 (2007)
4. Vapnik, V.N.: Statistical Learning Theory. Wiley-Interscience, New York (1998)
5. Nigam, K.: Using maximum entropy for text classification. In: IJCAI 1999 Workshop on Machine Learning for Information Filtering, pp. 61–67 (1999)
6. Liang, P., Jordan, M.I.: An asymptotic analysis of generative, discriminative, and pseudolikelihood estimators. In: Proceedings of the 25th International Conference on Machine Learning, ICML 2008, pp. 584–591. ACM, New York (2008)
7. Janzing, D., Schölkopf, B.: Causal inference using the algorithmic markov condition. IEEE Trans. Inf. Theory **56**(10), 5168–5194 (2010)
8. Pearl, J.: Causality: Models, Reasoning and Inference, 2nd edn. Cambridge University Press, New York (2009)
9. Daniusis, P., Janzing, D., Mooij, J., Zscheischler, J., Steudel, B., Zhang, K., Schölkopf, B.: Inferring deterministic causal relations. arXiv preprint arXiv:1203.3475 (2012)
10. Ng, A., Jordan, M.I.: On discriminative vs. generative classifiers: a comparison of logistic regression and naive bayes. Adv. Neural Inf. Proc. Syst. **14**, 841 (2002)
11. Chapelle, O., Schölkopf, B., Zien, A.: Semi-Supervised Learning, 1st edn. The MIT Press, Cambridge (2010)
12. van der Maaten, L., Hinton, G.: Visualizing data using t-SNE. J. Mach. Learn. Res. **9**, 2579–2605, 85 (2008)

A Non-Gaussian Approach for Causal Discovery in the Presence of Hidden Common Causes

Shohei Shimizu[✉]

The Institute of Scientific and Industrial Research, Osaka University,
8-1 Mihogaoka, Osaka, Ibaraki 5670047, Japan
sshimizu@ar.sanken.osaka-u.ac.jp
https://sites.google.com/site/sshimizu06/

Abstract. We discuss the problem of estimating the causal direction between two observed variables in the presence of hidden common causes. Managing hidden common causes is essential when studying causal relations based on observational data. We previously proposed a Bayesian estimation method for estimating the causal direction using the non-Gaussianity of data. This method does not require us to explicitly model hidden common causes. The experiments on artificial data presented in this paper imply that Bayes factors could be useful for selecting a better causal direction when using a non-Gaussian method.

Keywords: Causal discovery · Hidden common causes · Structural equation models · Non-Gaussianity

1 Introduction

We consider the problem of estimating causal relations based on observational data [2, 22, 33]. Assume that we are interested in the causal relations between two observed random variables x_1 and x_2. We are particularly interested in the causal direction between these two variables assuming one-way causation. We use the framework of structural causal models [22] to represent their causal relations. We want to estimate which of the following two models (Models 1 and 2) is better than the other by using a dataset of x_1 and x_2 randomly sampled from either of these two models:

$$\text{Model 1}: \begin{cases} x_1 = e_1 \\ x_2 = b_{21}x_1 + e_2 \end{cases}, \tag{1}$$

$$\text{Model 2}: \begin{cases} x_1 = b_{12}x_2 + e_1 \\ x_2 = e_2 \end{cases}, \tag{2}$$

where e_1 and e_2 are unobserved or hidden random variables, typically called error variables, exogenous variables, or external influences. b_{21} and b_{12} are constants that represent the magnitude of causation from x_1 to x_2 and from x_2 to x_1, respectively. For simplicity, here we assume that b_{21} and b_{12} are non-zero.

© Springer International Publishing Switzerland 2015
J. Suzuki and M. Ueno (Eds.): AMBN 2015, LNAI 9505, pp. 222–233, 2015.
DOI: 10.1007/978-3-319-28379-1_16

In Model 1, x_1 causes x_2 and the causal direction is $x_1 \rightarrow x_2$, whereas in Model 2, x_2 causes x_1 and the causal direction is $x_2 \rightarrow x_1$.

Note that these two models describe the data-generating process of x_1 and x_2 rather than simply defining the probability distribution of x_1 and x_2. For example, Model 1 states that the values of e_1 and e_2 are first generated and that the value of x_1 is observed as that of x_1 as it is; hence, the value of x_2 is generated as a linear combination of those of x_1 and e_2.

The major difficulty with estimating causal directions based on observational data is that the error variables e_1 and e_2 are dependent in general. Then, even if we know that the right causal direction is $x_1 \rightarrow x_2$, we cannot obtain the right estimate of the coefficient b_{21} by using the regression coefficient obtained when regressing x_2 on x_1. Such dependency between the error variables e_1 and e_2 is typically introduced by the unobserved variables that cause both x_1 and x_2. These unobserved variables are known as hidden common causes.

Assume that we have a single hidden common cause f_1 that makes e_1 and e_2 dependent. Then, we rewrite Models 1 and 2 as follows:

$$\text{Model 1'}: \begin{cases} x_1 = \underbrace{\lambda_{11}f_1 + e_1'}_{e_1} \\ x_2 = b_{21}x_1 + \underbrace{\lambda_{21}f_1 + e_2'}_{e_2} \end{cases}, \tag{3}$$

$$\text{Model 2'}: \begin{cases} x_1 = b_{12}x_2 + \underbrace{\lambda_{11}f_1 + e_1'}_{e_1} \\ x_2 = \underbrace{\lambda_{21}f_1 + e_2'}_{e_2} \end{cases}, \tag{4}$$

where λ_{11} and λ_{21} are constants that represent the magnitudes of causation. Now, the new error variables e_1' and e_2' are statistically independent. A well-known guideline [21, 24] is to observe the hidden common cause f_1, incorporate it into the model, and carry out three-variable analysis so that the error variables are independent. Although we should certainly follow this guideline, doing so can be hard since a large number of hidden common causes may exist and we often have no idea what they are.

In this paper, we discuss the problem of estimating the causal direction between two observed variables in the presence of hidden common causes. We use structural causal models [22] to represent causal relations and make causal inferences. We assume linear functional relations, acyclic causal relations, and the non-Gaussianity of the error variables. Further, we assume that the number of hidden common causes is unknown. Under these assumptions, we previously proposed a method for estimating the causal direction between two observed variables [26]. This method compares two models of two observed variables with opposite causal directions in a Bayesian model selection framework. The method does not require us to explicitly model hidden common causes and makes the number of hidden common causes remain unspecified. In the remainder of this paper, we first briefly review the method. Second, we consider using a set of prior distributions for cases with observed variables being standardized. Finally, we conduct experiments on artificial data. This paper thus supplements our previous work [26].

2   A non-Gaussian Causal Model with Hidden Common Cause Cases

We previously developed a linear structural causal model for causal discovery in the presence of hidden common causes [11]. This model is an extension of a linear non-Gaussian acyclic structural equation model known as LiNGAM [28,31]. Let us denote by $x_1, \cdots, x_p$ the observed variables, by $f_1, \cdots, f_Q$ the hidden common causes, and by $e_1, \cdots, e_p$ the error variables. All these are continuous variables. Then, we write the model as follows:

$$x_i = \mu_i + \sum_{j \neq i} b_{ij} x_j + \sum_{q=1}^{Q} \lambda_{iq} f_q + e_i, \tag{5}$$

where b_{ij} and λ_{iq} are constants that represent the magnitudes of causation and μ_i are intercepts. We assume that the causal relations are acyclic, i.e., there is no feedback relation. We further assume that the hidden common causes f_q ($q = 1, \cdots, Q$) and error variables e_i ($i = 1, \cdots, p$) are non-Gaussian and independent. Although the independence assumption on hidden common causes f_q looks strong, we can make this assumption without loss of generality under some common assumptions including linearity. See [11] for the details of the independence assumption on hidden common causes.

By using the model in Eq. (5), we compare the following two models with opposite directions of causation:

$$\text{Model 3}: \begin{cases} x_1 = \mu_1 + \sum_{q=1}^{Q} \lambda_{1q} f_q + e_1 \\ x_2 = \mu_2 + b_{21} x_1 + \sum_{q=1}^{Q} \lambda_{2q} f_q + e_2 \end{cases}, \tag{6}$$

$$\text{Model 4}: \begin{cases} x_1 = \mu_1 + b_{12} x_2 + \sum_{q=1}^{Q} \lambda_{1q} f_q + e_1 \\ x_2 = \mu_2 + \sum_{q=1}^{Q} \lambda_{2q} f_q + e_2, \end{cases} \tag{7}$$

Fig. 1 presents graphical representations of these two models. Note that we assume the number of hidden common causes Q to be unknown.

In [26], we related the model in Eq. (5) to a model having individual-specific intercepts instead of explicitly having hidden common causes. A major advantage of this approach is that we do not estimate the number of hidden common causes Q. To explain the idea, we first rewrite the model in Eq. (5) for observation l as follows:

$$x_i^{(l)} = \mu_i + \sum_{q=1}^{Q} \lambda_{iq} f_q^{(l)} + \sum_{j \neq i} b_{ij} x_j^{(l)} + e_i^{(l)} \tag{8}$$

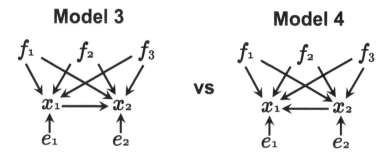

Fig. 1. Models 3 and 4: Two models with different causal directions in the presence of three hidden common causes

Now, let us denote the sums of the hidden common causes by $\tilde{\mu}_i^{(l)} = \sum_{q=1}^{Q} \lambda_{iq} f_q^{(l)}$. Then, we have the following model with individual-specific intercepts:

$$x_i^{(l)} = \mu_i + \underbrace{\tilde{\mu}_i^{(l)}}_{\sum_{q=1}^{Q} \lambda_{iq} f_q^{(l)}} + \sum_{j \neq i} b_{ij} x_j^{(l)} + e_i^{(l)}, \tag{9}$$

where μ_i are the intercepts common to all the observations and $\tilde{\mu}_i^{(l)}$ are the individual-specific intercepts. The distributions of $e_i^{(l)}$ $(l = 1, \cdots, n)$ are assumed to be identical for every i. In this model, the observations are generated from the model with no hidden common causes, possibly with different parameter values of the means $\mu_i + \tilde{\mu}_i^{(l)}$. This model has the intercepts μ_i and b_{ij} that are common to all the observations as well as the individual-specific intercepts $\tilde{\mu}_i^{(l)}$. This is similar to mixed models [5]. Thus, we call this a mixed-LiNGAM.

Now, the problem of comparing Models 3 and 4 in Eqs. (6) and (7) becomes that of comparing Models 3' and 4':

$$\text{Model 3'} : \begin{cases} x_1^{(l)} = \mu_1 + \tilde{\mu}_1^{(l)} + e_1^{(l)} \\ x_2^{(l)} = \mu_2 + \tilde{\mu}_2^{(l)} + b_{21} x_1^{(l)} + e_2^{(l)} \end{cases}, \tag{10}$$

$$\text{Model 4'} : \begin{cases} x_1^{(l)} = \mu_1 + \tilde{\mu}_1^{(l)} + b_{12} x_2^{(l)} + e_1^{(l)} \\ x_2^{(l)} = \mu_2 + \tilde{\mu}_2^{(l)} + e_2^{(l)} \end{cases}, \tag{11}$$

where $\tilde{\mu}_1^{(l)} = \sum_{q=1}^{Q} \lambda_{1q} f_q^{(l)}$ and $\tilde{\mu}_2^{(l)} = \sum_{q=1}^{Q} \lambda_{2q} f_q^{(l)}$ $(l = 1, \cdots, n)$.

We apply a Bayesian approach to compare Models 3' and 4' and estimate the possible causal direction between the two observed variables x_1 and x_2. We assume that the prior probabilities of the two candidate models are uniform. Then, we may simply compare the log-marginal likelihoods of the two models to assess their plausibility. The model with the larger log-marginal likelihood is considered as the closest to the true model [15].

3   Likelihood

Let $\mathcal{D}$ be the observed data set $[\boldsymbol{x}^{(1)^T}, \cdots, \boldsymbol{x}^{(n)^T}]^T$, where $\boldsymbol{x}^{(l)} = [x_1^{(l)}, x_2^{(l)}]^T$. Denote Models 3' and 4' by $M_{3'}$ and $M_{4'}$ and their log-marginal likelihoods by $\log p(\mathcal{D}|\boldsymbol{\theta}_r, M_r)$ ($r = 3', 4'$). Then, their log-marginal likelihoods are given by

$$\log p(\mathcal{D}|\boldsymbol{\theta}_r, M_r) = \log \prod_{l=1}^{n} p(\boldsymbol{x}^{(l)}|\boldsymbol{\theta}_r, M_r) \tag{12}$$

$$= \sum_{l=1}^{n} \log p(\boldsymbol{x}^{(l)}|\boldsymbol{\theta}_r, M_r) \tag{13}$$

$$= \begin{cases} \sum_{l=1}^{n} \log p_{e_1^{(l)}}(x_1^{(l)} - \mu_1 - \tilde{\mu}_1^{(l)}|\boldsymbol{\theta}_{3'}, M_{3'}) \\ + \sum_{l=1}^{n} \log p_{e_2^{(l)}}(x_2^{(l)} - \mu_2 - \tilde{\mu}_2^{(l)} - b_{21}x_1^{(l)}|\boldsymbol{\theta}_{3'}, M_{3'}) \quad \text{for } M_{3'} \\ \sum_{l=1}^{n} \log p_{e_1^{(l)}}(x_1^{(l)} - \mu_1 - \tilde{\mu}_1^{(l)} - b_{12}x_2^{(l)}|\boldsymbol{\theta}_{4'}, M_{4'}) \\ + \sum_{l=1}^{n} \log p_{e_2^{(l)}}(x_2^{(l)} - \mu_2 - \tilde{\mu}_2^{(l)}|\boldsymbol{\theta}_{4'}, M_{4'}) \quad \text{for } M_{4'} \end{cases} \tag{14}$$

The distributions of the error variables $e_1^{(l)}$ and $e_2^{(l)}$ are modeled by Laplace distributions with zero mean and variances of $\mathrm{var}(e_1^{(l)}) = h_1^2$ and $\mathrm{var}(e_2^{(l)}) = h_2^2$ ($h_1, h_2 > 0$) as follows:

$$p_{e_1^{(l)}} = Laplace(0, h_1/\sqrt{2}), \tag{15}$$

$$p_{e_2^{(l)}} = Laplace(0, h_2/\sqrt{2}). \tag{16}$$

Here, we simply use a super-Gaussian distribution, the Laplace distribution, to model $p_{e_1^{(l)}}$ and $p_{e_2^{(l)}}$. Super-Gaussian distributions have often been reported to work well in non-Gaussian estimation methods including independent component analysis [12] and linear non-Gaussian structural causal models if the actual error distributions are super-Gaussian [12,13].

4   Prior Distributions

The parameter vectors $\boldsymbol{\theta}_{3'}$ and $\boldsymbol{\theta}_{4'}$ in Eq. (14) are written as follows:

$$\boldsymbol{\theta}_{3'} = [\mu_i, b_{21}, h_i, \tilde{\mu}_i^{(l)}]^T \quad (i = 1, 2; l = 1, \cdots, n), \tag{17}$$

$$\boldsymbol{\theta}_{4'} = [\mu_i, b_{12}, h_i, \tilde{\mu}_i^{(l)}]^T \quad (i = 1, 2; l = 1, \cdots, n). \tag{18}$$

We first standardize the two observed variables x_1 and x_2 to have their means and variances zeros and ones before computing the log-marginal likelihoods. We prefer that the inference is not sensitive to the means and scales of the observed variables. Then, following [9], we model the prior distributions of the parameters common to all the observations as follows:

$$b_{12} \sim N(0,1)$$
$$b_{21} \sim N(0,1)$$
$$h_1 \sim \ln N(0,1)$$
$$h_2 \sim \ln N(0,1).$$

Further, we set the intercepts μ_i $(i = 1, 2)$ to be zeros following [9] since the observed variables have been standardized.

Next, we use an informative prior distribution for the individual-specific intercepts $\tilde{\mu}_i^{(l)}$ $(i = 1, 2; l = 1, \cdots, n)$. Those individual-specific intercepts $\tilde{\mu}_i^{(l)}$ are the sums of many non-Gaussian independent hidden common causes f_q and are dependent. The central limit theorem states that the sum of independent variables becomes increasingly close to the Gaussian [1]. Motivated by this observation, we approximate the non-Gaussian distributions of the individual-specific intercepts $\tilde{\mu}_i^{(l)}$ that are the sums of many non-Gaussian independent hidden common causes by using a bell-shaped curve distribution. We here model the prior distribution of the individual-specific intercepts by the multivariate t-distribution as follows:

$$\begin{bmatrix} \tilde{\mu}_1^{(l)} \\ \tilde{\mu}_{2,}^{(l)} \end{bmatrix} = \mathrm{diag}\left(\left[\sqrt{\tau_1^{indvdl}}, \sqrt{\tau_2^{indvdl}}\right]^T\right) \mathbf{C}^{-1/2}\mathbf{u} \tag{19}$$

where τ_1^{indvdl} and τ_2^{indvdl} are constants, $\mathbf{u} \sim t_\nu(\mathbf{0}, \boldsymbol{\Sigma})$, and $\boldsymbol{\Sigma} = [\sigma_{ab}]$ is a symmetric scale matrix whose diagonal elements are 1s. $\mathbf{C}$ is a diagonal matrix whose diagonal elements give the variance of elements of $\mathbf{u}$, i.e., $\mathbf{C} = \frac{\nu}{\nu-2}\mathrm{diag}(\boldsymbol{\Sigma})$ for $\nu > 2$. The degree of freedom ν is here taken to be eight.

The hyper-parameters are τ_1^{indvdl}, τ_2^{indvdl}, and σ_{21}. We take an empirical Bayesian approach to select the hyper-parameters. We test $\tau_i^{indvdl} = 0, 0.2^2, ..., 0.8^2, 1.0^2$ $(i = 1, 2)$ and $\sigma_{12} = 0, \pm0.3, \pm0.5, \pm0.7, \pm0.9$. We take the ordinary Monte Carlo sampling approach to compute the log-marginal likelihoods with 10,000 samples for the parameter vectors $\boldsymbol{\theta}_r$ $(r = 3', 4')$.

5   Experiments on Artificial Data

We generated data using the following non-Gaussian model with three hidden common causes:

$$x_1 = 5 + f_1 + f_2 + 1.5f_3 + e_1$$
$$x_2 = 10 + f_1 + 2f_2 + 0.5f_3 + 3x_2 + e_2.$$

We tested three distributions of the error variables e_1, e_2, and hidden common causes f_1, f_2, f_3: the Laplace distribution, the exponential distribution with the parameter value $1/\sqrt{2}$, and the uniform distribution. The Laplace distribution and exponential distribution have positive kurtoses and are super-Gaussian distributions, whereas the uniform distribution has negative kurtosis and is a sub-Gaussian distribution. The Laplace and uniform distributions are symmetric, whereas the exponential is asymmetric. The variances of e_1 and e_2 were set to 9 and those of f_q were ones. We permuted the variables according to a random ordering to hide the true orderings. We conducted 200 trials with sample sizes of 50, 100, 200, and 500.

We counted the numbers of successful discoveries of the causal directions and computed precisions. We also computed the Bayes factor. Let us denote by

K the Bayes factor of the two models compared, $M_{3'}$ and $M_{4'}$. For notational simplicity, we assume that we compute K so that the larger likelihood comes to the numerator and the smaller to the denominator. In [15], Kass and Raftery proposed that if $2 \log K$ is 0 to 2, the evidence is not worth more than a bare mention, if $2 \log K$ is 2 to 6, it is positive, if $2 \log K$ is 6 to 10, it is strong, and if $2 \log K$ is more than 10, it is very strong.

Tables 1, 2, 3 show the results when the actual error variables follow the Laplace, exponential, and uniform distributions, respectively. Table 4 shows the result when the actual distribution of each error variable was randomly selected from the three distributions for every trial. Overall, if the sample size increased or the Bayes factors rose, the numbers of successful discoveries and precisions improved. This finding implies that considering Bayes factors is useful when selecting a better model by using our method.

When the actual distribution was the Laplace or exponential, the performance seemed to be satisfactory (see Tables 1 and 2) because the Laplace and exponential distributions are super-Gaussian, as is the postulated distribution, the Laplace.

Table 1. Numbers of successful discoveries and precisions when the actual distributions are the Laplace. K is the Bayes factor.

	N. successes	N. findings	Precisions
$n = 50$			
$2\log K > 0$	143	200	0.71
$2\log K > 2$	11	13	0.85
$2\log K > 6$	0	0	N/A
$2\log K > 10$	0	0	N/A
$n = 100$			
$2\log K > 0$	142	200	0.71
$2\log K > 2$	52	60	0.87
$2\log K > 6$	0	0	N/A
$2\log K > 10$	0	0	N/A
$n = 200$			
$2\log K > 0$	161	200	0.81
$2\log K > 2$	114	127	0.90
$2\log K > 6$	20	21	0.95
$2\log K > 10$	1	1	1.00
$n = 500$			
$2\log K > 0$	167	200	0.83
$2\log K > 2$	144	164	0.88
$2\log K > 6$	108	114	0.95
$2\log K > 10$	56	57	0.98

When the actual distribution was the uniform, the performance (Table 3) was much worse than the cases with the actual distribution being the Laplace and exponential, because the uniform distribution is sub-Gaussian, unlike the Laplace.

When each of the actual error distributions was randomly selected, the performance again became worse than the Laplace and exponential distribution cases (but performance was not terrible). This finding occurs because two of the three distributions used in this experiment were super-Gaussian, as was the postulated error distribution.

6 Related Work

For the past 10 years, many semi-parametric methods for estimating causal directions under the assumption of no hidden common causes have been developed [4,6,10,13,14,23,27–30,35]. In contrast to non-parametric methods [22,33], semi-parametric methods make some assumptions on the function forms of

Table 2. Numbers of successful discoveries and precisions when the actual distributions are the exponential. K is the Bayes factor.

	N. successes	N. findings	Precisions
$n = 50$			
$2\log K > 0$	137	200	0.69
$2\log K > 2$	16	20	0.85
$2\log K > 6$	0	0	N/A
$2\log K > 10$	0	0	N/A
$n = 100$			
$2\log K > 0$	151	200	0.76
$2\log K > 2$	56	64	0.88
$2\log K > 6$	0	0	N/A
$2\log K > 10$	0	0	N/A
$n = 200$			
$2\log K > 0$	161	200	0.81
$2\log K > 2$	120	136	0.88
$2\log K > 6$	32	33	0.97
$2\log K > 10$	3	3	1.00
$n = 500$			
$2\log K > 0$	165	200	0.82
$2\log K > 2$	152	174	0.87
$2\log K > 6$	106	111	0.95
$2\log K > 10$	78	78	1.00

causal relations and/or the error distributions to make the models identifiable. Those semi-parametric methods have recently been applied to empirical research including economics [16,20], neuroscience [18,19], epidemiology [25], and chemistry [3]. See [31] for a review of semi-parametric methods and [32,34] for recent reviews of non-parametric methods. Links to most of the papers related to semi-parametric methods are available on the web.[1]

In practice, those methods assuming no hidden common causes seem to work well in those papers. However, what distinguishes observational studies from experimental studies is the existence of hidden common causes. Therefore, in some applications, empirical researchers hesitate to accept the estimation results of those methods that assume no hidden common causes.

We could take a non-Gaussian approach [11] that uses an extension of independent component analysis with more latent independent components than observed variables (overcomplete ICA [17]) to formally consider hidden common causes in semi-parametric methods. Unfortunately, however, current versions of

Table 3. Numbers of successful discoveries and precisions when the actual distributions are the uniform. K is the Bayes factor.

	N. successes	N. findings	Precisions
$n = 50$			
$2\log K > 0$	77	200	0.39
$2\log K > 2$	0	1	0.00
$2\log K > 6$	0	0	N/A
$2\log K > 10$	0	0	N/A
$n = 100$			
$2\log K > 0$	65	200	0.33
$2\log K > 2$	3	24	0.13
$2\log K > 6$	0	0	N/A
$2\log K > 10$	0	0	N/A
$n = 200$			
$2\log K > 0$	60	200	0.30
$2\log K > 2$	10	74	0.14
$2\log K > 6$	0	3	0.00
$2\log K > 10$	0	0	N/A
$n = 500$			
$2\log K > 0$	54	200	0.27
$2\log K > 2$	29	144	0.20
$2\log K > 6$	6	47	0.13
$2\log K > 10$	0	10	0.00

---

[1] https://sites.google.com/site/sshimizu06/home/lingampapers.

Table 4. Numbers of successful discoveries and precisions when the actual distributions are randomly selected from the Laplace, exponential, and uniform distributions. K is the Bayes factor.

	N. successes	N. findings	Precisions
$n = 50$			
$2\log K > 0$	119	200	0.60
$2\log K > 2$	12	14	0.86
$2\log K > 6$	0	0	N/A
$2\log K > 10$	0	0	N/A
$n = 100$			
$2\log K > 0$	118	200	0.59
$2\log K > 2$	48	63	0.76
$2\log K > 6$	4	4	1.00
$2\log K > 10$	0	0	N/A
$n = 200$			
$2\log K > 0$	122	200	0.61
$2\log K > 2$	71	109	0.65
$2\log K > 6$	18	22	0.82
$2\log K > 10$	1	1	1.00
$n = 500$			
$2\log K > 0$	136	200	0.68
$2\log K > 2$	123	168	0.73
$2\log K > 6$	80	102	0.78
$2\log K > 10$	41	45	0.91

the overcomplete ICA algorithms are computationally unreliable since they often suffer from local optima [7]. In [8], Henao and Winther proposed a Bayesian approach to estimate the model. Their method seems to work for larger numbers of variables than the overcomplete ICA-based method. However, both these methods need to explicitly model all the hidden common causes. This approach could sometimes be computationally tough since the number of hidden common causes can be large, while specifying the exact number of hidden common causes might also be challenging.

Thus, in [26], we proposed an alternative approach that does not require us to specify the number of hidden common causes or explicitly model them.

7   Conclusions

In this paper, we discussed a non-Gaussian approach for estimating causal directions in the presence of hidden common causes. The experiments on artificial data implied that looking at Bayes factors could be useful for selecting a better causal direction. We distribute the Python codes under the MIT license at https://sites.google.com/site/sshimizu06/mixedlingamcode.

Acknowledgments. This work was partially supported by JSPS KAKENHI Grant Numbers 24700275 and 24300106 and the Center of Innovation Program from the Japan Science and Technology Agency, JST.

References

1. Billingsley, P.: Probability and Measure. Wiley-Interscience, New York (1986)
2. Bollen, K.: Structural Equations with Latent Variables. Wiley, New York (1989)
3. Campomanes, P., Neri, M., Horta, B.A., Röhrig, U.F., Vanni, S., Tavernelli, I., Rothlisberger, U.: Origin of the spectral shifts among the early intermediates of the rhodopsin photocycle. J. Am. Chem. Soc. **136**(10), 3842–3851 (2014)
4. Chen, Z., Chan, L.: Causality in linear nonGaussian acyclic models in the presence of latent Gaussian confounders. Neural Comput. **25**(6), 1605–1641 (2013)
5. Demidenko, E.: Mixed Models: Theory and Applications. Wiley-Interscience, New York (2004)
6. Dodge, Y., Rousson, V.: Direction dependence in a regression line. Commun. Stat. Theor. Methods **29**(9–10), 1957–1972 (2000)
7. Entner, D., Hoyer, P.O.: Discovering unconfounded causal relationships using linear non-Gaussian models. In: Bekki, D. (ed.) JSAI-isAI 2010. LNCS, vol. 6797, pp. 181–195. Springer, Heidelberg (2011)
8. Henao, R., Winther, O.: Sparse linear identifiable multivariate modeling. J. Mach. Learn. Res. **12**, 863–905 (2011)
9. Hoyer, P.O., Hyttinen, A.: Bayesian discovery of linear acyclic causal models. In: Proceedings of 25th Conference on Uncertainty in Artificial Intelligence (UAI 2009), pp. 240–248 (2009)
10. Hoyer, P.O., Janzing, D., Mooij, J., Peters, J., Schölkopf, B.: Nonlinear causal discovery with additive noise models. Adv. Neural Inf. Process. Syst. **21**, 689–696 (2009)
11. Hoyer, P.O., Shimizu, S., Kerminen, A., Palviainen, M.: Estimation of causal effects using linear non-Gaussian causal models with hidden variables. Int. J. Approx. Reasoning **49**(2), 362–378 (2008)
12. Hyvärinen, A., Karhunen, J., Oja, E.: Independent Component Analysis. Wiley, New York (2001)
13. Hyvärinen, A., Smith, S.M.: Pairwise likelihood ratios for estimation of non-Gaussian structural equation models. J. Mach. Learn. Res. **14**, 111–152 (2013)
14. Hyvärinen, A., Zhang, K., Shimizu, S., Hoyer, P.O.: Estimation of a structural vector autoregressive model using non-Gaussianity. J. Mach. Learn. Res. **11**, 1709–1731 (2010)
15. Kass, R.E., Raftery, A.E.: Bayes factors. J. Am. Stat. Assoc. **90**(430), 773–795 (1995)
16. Lai, P.C., Bessler, D.A.: Price discovery between carbonated soft drink manufacturers and retailers: a disaggregate analysis with PC and LiNGAM algorithms. J. Appl. Econ. **18**(1), 173–197 (2015)
17. Lewicki, M., Sejnowski, T.J.: Learning overcomplete representations. Neural Comput. **12**(2), 337–365 (2000)

18. Liu, Y., Wu, X., Zhang, J., Guo, X., Long, Z., Yao, L.: Altered effective connectivity model in the default mode network between bipolar and unipolar depression based on resting-state fMRI. J. Affect. Disord. **182**, 8–17 (2015)
19. Mills-Finnerty, C., Hanson, C., Hanson, S.J.: Brain network response underlying decisions about abstract reinforcers. NeuroImage **103**, 48–54 (2014)
20. Moneta, A., Entner, D., Hoyer, P., Coad, A.: Causal inference by independent component analysis: theory and applications. Oxford Bull. Econ. Stat. **75**(5), 705–730 (2013)
21. Pearl, J.: Causal diagrams for empirical research. Biometrika **82**(4), 669–688 (1995)
22. Pearl, J.: Causality: Models, Reasoning, and Inference. Cambridge University Press, Cambridge (2000). (2nd ed. 2009)
23. Peters, J., Janzing, D., Schölkopf, B.: Causal inference on discrete data using additive noise models. IEEE Trans. Pattern Anal. Mach. Intell. **33**(12), 2436–2450 (2011)
24. Rosenbaum, P.R., Rubin, D.B.: The central role of the propensity score in observational studies for causal effects. Biometrika **70**(1), 41–55 (1983)
25. Rosenström, T., Jokela, M., Puttonen, S., Hintsanen, M., Pulkki-Råback, L., Viikari, J.S., Raitakari, O.T., Keltikangas-Järvinen, L.: Pairwise measures of causal direction in the epidemiology of sleep problems and depression. PloS ONE **7**(11), e50841 (2012)
26. Shimizu, S., Bollen, K.: Bayesian estimation of causal direction in acyclic structural equation models with individual-specific confounder variables and non-Gaussian distributions. J. Mach. Learn. Res. **15**, 2629–2652 (2014)
27. Shimizu, S., Hoyer, P.O., Hyvärinen, A.: Estimation of linear non-Gaussian acyclic models for latent factors. Neurocomputing **72**, 2024–2027 (2009)
28. Shimizu, S., Hoyer, P.O., Hyvärinen, A., Kerminen, A.: A linear non-Gaussian acyclic model for causal discovery. J. Mach. Learn. Res. **7**, 2003–2030 (2006)
29. Shimizu, S., Hyvärinen, A.: Discovery of linear non-Gaussian acyclic models in the presence of latent classes. In: Proceedings of 14th International Conference on Neural Information Processing (ICONIP 2007), pp. 752–761 (2008)
30. Shimizu, S., Inazumi, T., Sogawa, Y., Hyvärinen, A., Kawahara, Y., Washio, T., Hoyer, P.O., Bollen, K.: DirectLiNGAM: a direct method for learning a linear non-Gaussian structural equation model. J. Mach. Learn. Res. **12**, 1225–1248 (2011)
31. Shimizu, S.: LiNGAM: non-Gaussian methods for estimating causal structures. Behaviormetrika **41**(1), 65–98 (2014). Special Issue on Causal Discovery
32. Shpitser, I., Evans, R.J., Richardson, T.S., Robins, J.M.: Introduction to nested markov models. Behaviormetrika **41**(1), 3–39 (2014). Special Issue on Causal Discovery
33. Spirtes, P., Glymour, C., Scheines, R.: Causation, Prediction, and Search. Springer, New York (1993). (2nd ed. MIT Press 2000)
34. Tillman, R.E., Eberhardt, F.: Learning causal structure from multiple datasets with similar variable sets. Behaviormetrika **41**(1), 41–64 (2014). Special Issue on Causal Discovery
35. Zhang, K., Hyvärinen, A.: On the identifiability of the post-nonlinear causal model. In: Proceedings of 25th Conference on Uncertainty in Artificial Intelligence (UAI 2009), pp. 647–655 (2009)

Forest Learning Based on the Chow-Liu Algorithm and Its Application to Genome Differential Analysis: A Novel Mutual Information Estimation

Joe Suzuki$^{(\boxtimes)}$

Osaka University, Suita, Japan
suzuki@math.sci.osaka-u.ac.jp

Abstract. This paper proposes a new mutual information estimator for discrete and continuous variables, and constructs a forest based on the Chow-Liu algorithm. The state-of-art method assumes Gaussian and ANOVA for continuous and discrete/continuous cases, respectively. Given data, the proposed algorithm constructs several pairs of quantizers for X and Y such that each interval of the both axes contains the equal number of samples, and estimate the mutual information values based on the discrete data for the histograms. Among the mutual information values, we choose the maximum one, which is validated in terms of the minimum description length principle. Although strong consistency is not proved mathematically, the proposed method does not distinguish discrete and continuous values when dealing with data, and independence is detected correctly with probability one as the sample size grows. The obtained forest construction procedure is applied to genome differential analysis in which a discrete variable (wild and mutant phenotypes) affects gene expression values.

Keywords: Mutual information · Chow-Liu algorithm · Consistency · Forest graphical model · Gene differential analysis

1 Introduction

Let $X^{(1)}, \cdots, X^{(N)}$ be discrete random variables ($N \geq 1$). In 1968, Chow and Liu [3] considered to approximate a given distribution by some tree expressed as

$$P(X^{(j)}) \prod_{k \in V \setminus \{j\}} P(X^{(k)} | X^{(\pi(k))}) \tag{1}$$

when we choose $j \in V := \{1, \cdots, N\}$ as its root, where $\pi(k)$ is the parent of $k \in V \setminus \{j\}$. For example, suppose $N = 4$. The distributions

$$P(X^{(1)})P(X^{(2)}|X^{(1)})P(X^{(3)}|X^{(1)})P(X^{(4)}|X^{(1)})$$

$$P(X^{(1)})P(X^{(2)}|X^{(1)})P(X^{(3)}|X^{(1)})P(X^{(4)}|X^{(2)})$$

© Springer International Publishing Switzerland 2015
J. Suzuki and M. Ueno (Eds.): AMBN 2015, LNAI 9505, pp. 234–249, 2015.
DOI: 10.1007/978-3-319-28379-1_17

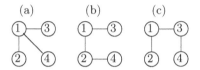

Fig. 1. Undirected graphs

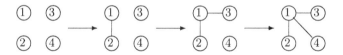

Fig. 2. The Chow-Liu algorithm: when $I(1,2) > I(1,3) > I(2,3) > I(1,4) > I(3,4) > I(2,4)$.

$$P(X^{(1)})P(X^{(2)}|X^{(1)})P(X^{(3)}|X^{(1)})P(X^{(4)}|X^{(3)})$$

are expressed by Fig. 1(a)–(c) using an undirected tree. We define the edge set E by the set of pairs $\{j,k\}$ if each of $\{j,k\} \in E$ is connected in the undirected tree. For examples, the edge sets of Fig. 1(a)–(c) are $E = \{\{1,2\},\{1,3\},\{1,4\}\}$, $E = \{\{1,2\},\{1,3\},\{2,4\}\}$, and $E = \{\{1,2\},\{1,3\},\{3,4\}\}$, respectively.

Let $I(j,k)$ be the mutual information between $X^{(j)}$ and $X^{(k)}$ ($j \neq k$). Chow and Liu [3] constructed a tree as follows: starting with $E = \{\}$ and $\mathcal{E} = \{\{i,j\}|i \neq j\}$, repeatedly choose a pair $\{j,k\}$ such that $I(j,k)$ is the largest among $I(j',k')$ with $\{j',k'\} \in \mathcal{E}$, delete it from $\mathcal{E}$, and add it to E only if adding it to E does not cause to generate a loop. The procedure is repeated until $\mathcal{E}$ is empty. For example, suppose $N = 4$ and $I(1,2)$ and $I(1,3)$ are the largest and the second largest mutual information values, respectively. Then, $\{2,3\}$ cannot be joined in the edge set E because loop $\{1,2,3\}$ will be generated. If $I(1,4)$ is the third largest, Fig. 2(a) is obtained through the procedure as in Fig. 2.

They proved that the obtained edge set E minimizes the Kullback-Leibler divergence from the original distribution $P(X^{(1)},\cdots,X^{(N)})$ among the distributions in the form of (1).

The same idea is applied to finding probabilistic relations among the N variables when the true distribution is not known and only a data set with n observations of them is available. In particular, they considered to obtain the maximum likelihood value of (1) given n examples

$$(X^{(1)} = x_{i,1},\cdots,X^{(N)} = x_{i,N})_{i=1}^{n},$$

by replacing the mutual information $I(j,k)$ by its likelihood, where each example consists of N variable values, and proved that the obtained edge set E maximizes the likelihood among the distributions in the form of (1).

In 1993, Suzuki [15] considered the problem in terms of the minimum description length (MDL) principle, and showed that the obtained graph should be a forest rather than a spanning tree when a subset of the variables is independent of another subset. Then, several authors revisited the same formula (6) such as P. Liang and N. Srebro (2004) [8], K. Panayidou (2010) [10], Edwards et al. (2010), [4].

In this paper, we propose a novel estimator of mutual information and apply it to the Chow-Liu algorithm. Thus far, the same problem has been considered by many authors, but we mainly deal with the case that discrete and continuous data are mixed.

This paper is constructed as follows. Section 2 explains the background and previous works on the problem, in Sect. 3, we propose a novel estimator of mutual information, and in Sect. 4, we show several data on experiments. In particular, in Sect. 4.2, we consider an application of the proposed method to genome differential analysis. In Sect. 5, we summarizes the result in this paper and state future tasks.

2 Background

2.1 Suzuki (1993)

Suppose we have two sequences x^n and y^n of length n that have been emitted by random variables X and Y (i.i.d.), respectively.

The maximum likelihood estimation of the mutual information of X and Y is obtained as follows: count the frequencies $c_X(x), c_Y(y), c_{XY}(x,y)$ of $X = x$, $Y = y$, and $(X, Y) = (x, y)$ in x^n and y^n, and plug-in them to the formula of mutual information:

$$I(X, Y) = \sum_x \sum_y P_{XY}(x, y) \log_2 \frac{P_{XY}(x, y)}{P_X(x) P_Y(y)} .$$

More precisely, the estimation can be expressed by

$$I_n = \sum_x \sum_y \frac{c_{XY}(x, y)}{n} \log_2 \frac{\dfrac{c_{XY}(x, y)}{n}}{\dfrac{c_X(x)}{n} \cdot \dfrac{c_Y(y)}{n}} . \tag{2}$$

However, we observe that the quantity is positive for any n even if X and Y are independent (note that mutual information between X and Y is zero if and only if X and Y are independent).

In 1993, Suzuki [15] considered another estimation based on the minimum description length (MDL) principle [12]. Given examples, the MDL chooses a rule that minimizes the total description length when the examples are described in terms of a rule and its exceptions. In this case, there are two candidate rules: whether X and Y are independent or not. When they are independent, for each of X and Y, we first describe the independent conditional probability values, and using them, the examples can be described. The total length will be

$$L^n(x^n) := -\sum_x c_X(x) \log_2 \frac{c_X(x)}{n} + \frac{\alpha - 1}{2} \log_2 n \tag{3}$$

plus

$$L^n(y^n) := -\sum_y c_Y(y) \log_2 \frac{c_Y(y)}{n} + \frac{\beta - 1}{2} \log_2 n \tag{4}$$

bits up to constants, where α and β are the cardinalities of X and Y, respectively. When they are not independent, we write the examples in

$$L^n(x^n, y^n) := -\sum_x \sum_y c_{XY}(x, y) \log_2 \frac{c_{XY}(x, y)}{n} + \frac{\alpha\beta - 1}{2} \log_2 n \tag{5}$$

bits up to constants. Hence, the difference $(3) + (4) - (5)$ divided by n is

$$J_n = I_n - \frac{(\alpha - 1)(\beta - 1)}{2n} \log_2 n. \tag{6}$$

It is known that $J_n \leq 0$ if and only if X and Y are independent for large n [16]. Figure 3 depicted a box plot of 1000 trials for the two estimations for $n = 100$ and $\alpha = \beta = 2$ for each of when X and Y are independent and when they are not independent. Suzuki [15] considered its modification to the AIC [1] (Akaike's Information Criterion) by replacing the second term in (6) by $(\alpha - 1)(\beta - 1)/n$.

When we apply (6) rather than (2) to the Chow-Liu algorithm, only $\{j, k\}$ with positive $J_n(j, k)$ are the candidates of the elements in E because otherwise the total $\sum_{\{j,k\} \in E} J_n(j, k)$ would not be maximized. Thus, we can obtain a forest (more than one root may exist) rather than a spanning tree: if two subtrees are independent, they will not be connected for (6) while (2) seeks only a spanning tree regardless of whether one part of the tree is independent from the other. We should also notice that the orders of $\{I_n(j, k)\}$ and $\{J_n(j, k)\}$ may be different,

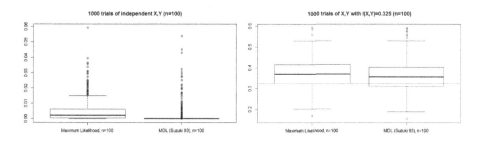

Fig. 3. The MDL computes the correct values while the maximum likelihood shows larger values than the correct ones.

Fig. 4. The spanning tree and forest obtained by the maximum likelihood (Left) and the MDL (Right) for the Asia data set in the bnlearn R package [13].

where $I_n(j, k)$ and $J_n(j, k)$ are the values of I_n and J_n when $X = X^{(j)}$ and $Y = X^{(k)}$, respectively, so that the edge set E obtained by $\{J_n(j, k)\}$ may not be a subset of the one obtained by $\{I_n(j, k)\}$.

Figure 4 depicts the obtained spanning tree and forest when using (2) and (6), respectively, for the Asia data set of the R bnlearn package [13].

More than ten years later after Suzuki [15], several authors revisited the same formula (6) such as P. Liang and N. Srebro (2004) [8], K. Panayidou (2010) [10], Edwards et al. (2010), [4].

2.2   Edwords et al's (2010)

Edwords et al's [4] considered the case such that some variables are discrete and others Gaussian.

Suppose the variables X and Y are Gaussian, and that examples $x^n = (x_1, \cdots, x_n)$ and $y^n = (y_1, \cdots, y_n)$ have been emitted from random variables X and Y, respectively. Then, the maximum likelihood estimation is given by

$$I_n = -\frac{n}{2} \log_2(1 - \hat{\rho}^2), \tag{7}$$

where $\hat{\rho}$ is the maximum likelihood estimation of the correlation coefficient. If we connect X and Y, one additional parameter will be required, so that Edwords et al. [4] proposed the quantity to estimate the mutual information

$$J_n = I_n - \frac{1}{2} \log_2 n. \tag{8}$$

Next, suppose the variables X and Y are Gaussian and discrete, respectively, and that examples x^n and y^n have been emitted from random variables X and Y, respectively. Then, the maximum log-likelihood of the conditional probability of X given $Y = y$ is

$$\frac{1}{2} n_y \log_2(2\pi e \hat{\sigma}_y^2),$$

where n_y is the number of occurrences of $Y = y$ in y^n, and

$$\bar{x}_y = \frac{1}{n_y} \sum_i x_i, \ \hat{\sigma}_y^2 = \frac{1}{n_y} \sum_i (x_i - \bar{x}_y)^2,$$

where the summand is over $i = 1, \cdots, n$ such that $y_i = y$, so that the resulting maximum likelihood estimator of mutual information is

$$\bar{x}_0 = \frac{1}{n} \sum_{i=1}^n x_i, \ \hat{\sigma}_0^2 = \frac{1}{n} \sum_{i=1}^n (x_i - \bar{x}_0)^2, \ I_n = \sum_y \frac{n_y}{n} \log_2 \frac{\hat{\sigma}_0^2}{\hat{\sigma}_y^2}$$

(ANOVA). If we connect X and Y, additional $2(\beta - 1)$ parameters will be required, so that Edwords et al's [4] proposed the quantity

$$J_n = I_n - (\beta - 1) \log_2 n,$$

Fig. 5. It is easy to obtain the likelihood for Gaussian and discrete variables X, Y (Left). But, how can we calculate it when a Gaussian variable X is between discrete variables Y, Z (Right)?

where β is the cardinality of the set in which Y takes values.

However, suppose that the Gaussian X is further connected to discrete Z as in Fig. 5, How can we obtain the maximum likelihood estimation as analyzed for the other cases? So, in order to avoid such an inconvenience of the proposed method, Edwords et al's [4] proposed to deal with the probability models such that no Gaussian variable is between any two discrete variables in the connected subtree of the forest. However, this assumption (they called such forests SD) avoids the challenging problem but makes the model selection much more restrictive.: for each of the connected subtrees, the discrete and Gaussian variables should be in upper and lower parts of the connected subtree, respectively.

2.3   Suzuki (2012)

Suzuki considered that the Eqs. (3), (4) and (5) have been obtained by the following procedure. For length (3), since the probability $\theta(x)$ of $X = x$ is unknown, we consider to weight the probability $\theta(x)$ using some weight $w(\theta)$ with $\theta = \theta(x)$ such as Direchlet distribution [6]. Then, given examples $x^n = (x_1, \cdots, x_n)$ emitted from X, we have a quantity defined by

$$Q^n(x^n) := \int_\theta \prod_{i=1}^n \theta(x_i)w(\theta)d\theta = \int_\theta \prod_x \theta(x)^{c_n(x)}w(\theta)d\theta,$$

where $c_i(x)$ is the number of occurrences of x in $(x_1, \cdots, x_i)$. It is known that (3) was obtained by expanding $-\log_2 Q^n(x^n)$ using Stirling's formula and neglecting its higher order terms.

We may assume that we have chosen $w(\cdot)$ such that

$$Q^n(x^n) = 2^{-L^n(x^n)} \quad,\quad Q^n(y^n) = 2^{-L^n(y^n)} \quad, \text{ and } Q^n(x^n, y^n) = 2^{-L^n(x^n, y^n)}$$

based on (3), (4), and (5), respectively, then we have

$$J_n = \frac{1}{n} \log_2 \frac{Q^n(x^n, y^n)}{Q^n(x^n)Q^n(y^n)}. \tag{9}$$

However, in this paper, for discrete data, we use the following exact formula with $w(\theta) \propto \prod_x \theta(x)^{-0.5}$ [6] because the sample sizes are small and approximations are not appropriate in many cases.

For variable X and its examples $(x_1, \cdots, x_n)$ of length n, Suzuki [16] generated a sequence of histograms, and obtained a quantized sequence for each histogram: $(x_1, \cdots, x_n) \mapsto (a_1^{(s)}, \cdots, a_n^{(s)})$, where $a_i^{(s)} \in \{1, \cdots, 2^s\}$ is the bin into which x_i falls for the s-th histogram. Because $a_s^n = (a_1^{(s)}, \cdots, a_n^{(s)})$ is a sequence each of which takes a finite number of values, we can define its Bayesian measure $Q_s^n(a_s^n)$ for each histogram s and its quantized sequence a_s^n. Suzuki [16] approximated the density at $x \in \mathbb{R}$ by dividing the probability of the bin including x by the width of the bin to obtain $g_s^n(x^n)$ that replaces $Q(x^n)$, and weighted them over weights $w_s > 0$ s.t. $\sum_s w_s = 1$. Suzuki [16] showed that the obtained the Bayesian measure $g^n(x^n) := \sum_s w_s g_s^n(x^n)$ satisfies consistency [16].

However, this method assumes an infinite number of histograms, and we are not certain how many histograms are required to obtain a level of correctness. Furthermore, compared with Silva, the variance of the sample size in each cluster is large. Sometimes, many clusters may have no sample, and it is hard to adjust the parameters.

3 Results

This paper solves the aforementioned problems by extending the MDL approach [15] to the continuous cases, and by improving the previous works by Silva et al. (2010) [14] and Suzuki (2012) [16].

3.1 Proposed Estimator of Mutual Information

The proposed estimation consists of the three steps:

1. prepare nested histograms based on Gessaman's construction,
2. compute the Bayesian estimator of mutual information for each histogram, and
3. choose the maximum value among the estimations w.r.t. the histograms.

Suppose we are given examples $x^n = (x_1, \cdots, x_n)$ and $y^n = (y_1, \cdots, y_n)$, and that they have been sorted as

$$\tilde{x}_1 \le \tilde{x}_2 \le \cdots \le \tilde{x}_n \text{ and } \tilde{y}_1 \le \tilde{y}_2 \le \cdots \le \tilde{y}_n. \tag{10}$$

First of all, we assume that no consecutive values are equal in each of the two sequences (10), which is true with probability one as $n \to \infty$ when the density function exists. Let $s := \lfloor 0.5 * \log_2 n \rfloor$, and for each $u = 1, \cdots, s$, we prepare histograms with 2^u bins for X, Y, and (X,Y). Let $t := n/2^u$. The sequences (10) are divided into clusters such as

$$(\tilde{x}_1, \cdots, \tilde{x}_{\lfloor t \rfloor}), \cdots (\tilde{x}_{\lfloor (j-1)t \rfloor +1}, \cdots, \tilde{x}_{\lfloor jt \rfloor}), \cdots, (\tilde{x}_{\lfloor (2^u-1)t \rfloor +1}, \cdots, \tilde{x}_n)$$

and

$$(\tilde{y}_1, \cdots, \tilde{y}_{\lfloor t \rfloor}), \cdots (\tilde{y}_{\lfloor (k-1)t \rfloor +1}, \cdots, \tilde{y}_{\lfloor kt \rfloor}), \cdots, (\tilde{x}_{\lfloor (2^u-1)t \rfloor +1}, \cdots, \tilde{y}_n).$$

Thus, we have quantized sequences $x^n \mapsto a_u^n = (a_1^{(u)}, \cdots, a_n^{(u)})$ and $y^n \mapsto b_u^n = (b_1^{(u)}, \cdots, b_n^{(u)})$ with $u = 1, \cdots, s$ using the clusters. For example, suppose we generate $n = 1000$ standard Gaussian random sequences x^n and y^n with correlation coefficient 0.8. The frequency distribution tables of a^n, b^n for $u = 3$ are both

1	2	3	4	5	6	7	8
125	125	125	125	125	125	125	125

and that of (a^n, b^n) for $u = 3$ looks like

	1	2	3	4	5	6	7	8
1	75	32	12	5	1	0	0	0
2	25	41	25	18	9	7	0	0
3	15	23	32	27	14	11	1	2
4	5	17	24	22	27	19	11	0
5	5	9	19	24	23	23	17	5
6	0	3	7	18	26	26	28	17
7	0	0	6	9	19	21	45	25
8	0	0	0	2	6	18	23	76

Thus, the distributions of a^n and b^n are uniform if n is a power of two. Compared with Suzuki (2012) [16], because enough samples are given to each cluster at least for one-dimensional X, Y if n is large, and the estimations are more robust.

Because the obtained sequences a_u^n and b_u^n are discrete, we can compute their Bayesian measures $Q_u^n(a_u^n)$, $Q_u(b_u^n)$, and $Q_u(a_u^n, b_u^n)$ similar to Suzuki (2012) [16].

$$J_n^{(u)} := \frac{1}{n} \log_2 \frac{Q_u^n(a_u^n, b_u^n)}{Q_u^n(a_u^n) Q_u^n(b_u^n)}. \tag{11}$$

Let (X_u, Y_u) and (X_v, Y_v) be the random variables for histograms u and v such that $u \le v$. Suppose that given examples a_v^n and b_v^n emitted from (X_v, Y_v), we wish to know whether (X_v, Y_v) are conditionally independent given (X_u, Y_u) based on the MDL principle. Then, we can answer the question affirmatively if we compare the values of description length and

$$- \log \frac{Q_v^n(a_v^n)}{Q_u^n(a_u^n)} - \log \frac{Q_v^n(b_v^n)}{Q_u^n(b_u^n)} \le - \log \frac{Q_v^n(a_v^n, b_v^n)}{Q_u^n(a_u^n, b_u^n)},$$

which is equivalent to $J_n^{(v)} \le J_n^{(u)}$. This means that according to the MDL principle, we can use the decision that (X_v, Y_v) are conditionally independent given (X_u, Y_u) if and only if $J_n^{(v)} \le J_n^{(u)}$. Hence, if u gives the maximum value of $J_n^{(u)}$, then we choose the histogram u. Thus, we propose the estimation given by $J_n := \max_{1 \le u \le s} J_n^{(u)}$.

Another interpretation is that if the sample size in each bin is too small, then the estimation is not robust. On the other hand, if the number of bins is too small, the approximation of the histogram is not appropriate. The two factors are balanced by the MDL principle.

For example, for $n = 1000$, the size will be $s = \lfloor 0.5 * \log_2(n) \rfloor = 4$. If we have the following four values:

```
u      J(u)
1   0.2664842
```

```
2    0.5077115
3    0.5731657
4    0.4601272
```

then the final estimation will be 0.5731657 ($u = 3$). Notice that there are other methods to find the maximum mutual information. For example, $s = 0.5 * \lfloor \log_a(n) \rfloor$ and a^u clusters for each of (X, Y) works if $a > 1$ (the smaller a, the larger s). For $a = 1.5$, we experimentally find (Fig. 6) that the value of $J(u)$ depicts a concave curve, i.e. the maximum value is obtained at the point $u = 5$ that the sample size of each bin (robustness of the estimation) and the number of bins (approximation of the histogram) are balanced.

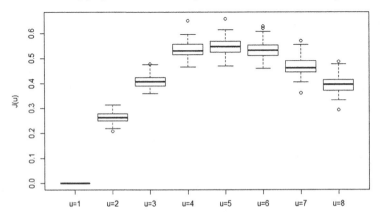

Fig. 6. The values of $J_n^{(u)}$ with $1 \le u \le s$: the maximum value is obtained at the point that the sample size of each bin and the number of bins are balanced.

Next, we consider the case that two values at consecutive locations are the same in one of the two sequences (10). Formally, we divide each cluster half at each stage $u = 1, 2, \cdots, s$, and if two values at consecutive locatios are equal and they need to be divided, then we choose another border: suppose k values are equal from the $(j + 1)$-th location:

$$\tilde{x}_j < \tilde{x}_{j+1} = \cdots = \tilde{x}_{j+k} < \tilde{x}_{j+k+1}.$$

and that we need to divide $(j+i)$-th and $(j+i+1)$-th positions ($1 \le i \le k-1$), we instead divide either between j-th and $(j + 1)$-th positions or between $(j + k)$-th and $(j+k+1)$-th positions, depending on whether $i < j+k/2$ or $i \ge j+k/2$. For example, if $n = 8$ and $x^8 = (2, 4, 1, 2, 3, 4, 3, 3)$, the cluster generating process for $(\tilde{x}_1, \cdots, \tilde{x}_8) = (1, 2, 2, 3, 3, 3, 4, 4)$ is as follows:

$$\{(1, 2, 2, 3, 3, 3, 4, 4)\} \rightarrow \{(1, 2, 2), (3, 3, 3, 4, 4)\} \rightarrow \{(1), (2, 2), (3, 3, 3), (4, 4)\}$$

In this way, even when the sequence x^n is discrete, we can obtain the quantization $x^n \mapsto a_u^n = (a_1^{(u)}, \cdots, a_n^{(u)})$. In particular, we have $a_u^n = x^n$ if u is not too small. The proposed scheme does not distinguish whether each of given two sequences is either discrete or continuous.

3.2   Theoretical Properties of the Proposed Estimator

In this subsection, we prove two fundamental claims:

1. For large n, the mutual information estimation of each histogram converges to the correct approximated value.
2. For large n, the estimation is either zero or negative if and only if X and Y are independent.

First of all, we have the following lemma from the law of large numbers:

Lemma 1. The $2^u - 1$ breaking points of histograms $u = 1, 2, \cdots$, converge to the correct values ($100 \times j/2^u$ percentile points, $j = 1, \cdots, 2^u - 1$) with probability one as the sample size n (hence its maximum depth s) grows. where we have assumed the value of a to be two for simplicity.

Let $I(X_u, Y_u)$ be the true mutual information w.r.t. the correct breaking points of the histogram $u = 1, \cdots, s$.

Lemma 2. The mutual information is monotonic w.r.t. the histograms: if $u \leq v$,

$$I(X_u, Y_u) \leq I(X_v, Y_v).$$

Proof. The difference from the right to the left is non-negative:

$$\sum_{X_v} \sum_{Y_v} P(X_u, Y_u) P(X_v, Y_v | X_u, Y_u) \log_2 \frac{P(X_v, Y_v | X_u, Y_u)}{P(X_v | X_u) P(Y_v | Y_u)} \geq 0.$$

This completes the proof.

Theorem 1. For large n, the mutual information estimation of each histogram converges to the correct approximated value.

Proof. Since each boundary converges to the true value in each histogram (Lemma 1), and the number of samples in each bin increases as n grows. Therefore, the estimation in histogram $u = 1, 2, \cdots$ converges to the correct mutual information value $I(X_u, Y_u)$.

Theorem 2. Suppose that $J_n^{(u)}$ is convex w.r.t. $u = 1, 2, \cdots$. With probability one as $n \to \infty$, $J_n \leq 0$ if and only if X and Y are independent.

Proof. Suppose X and Y are independent. this means $I(X, Y) = 0$ and $I(X_u, Y_u) = 0$ for all $u = 1, \cdots$ (Lemma 2). Then, for all $u = 1, 2, \cdots$, $J_n^{(u)} = 0$ with probability one as $n \to \infty$ (Theorem 1). On the other hand, suppose that X and Y are not independent. Since $I(X, Y) > 0$, we have $I(X_u, Y_u) > 0$ for at least one u. Thus, $J_n^{(u)}$ for some u is positive mutual information $I(X, Y)$ with probability one as $n \to \infty$ (Theorem 1), and $J_n^{(u)} > 0$. This completes the proof.

4   Experiments

4.1   Preliminary Experiments

Before constructing forests given examples, we estimated the value of mutual information.

If the random variables are known to be Gaussian a prior, it is considered to be easer to estimate the correlation coefficient and to compute the estimation based on (7) and (8) than the proposed method that works for every random variables, We compared the proposed algorithm with the Gaussian method.

1. X and Y obey the negative binomial distribution with parameters (P, w_x) and (P, w_y) such that X and Y are the numbers of occurrences before an event with probability P occurs w_x and w_y $(w_x \leq w_y)$ times, respectively. In particular, we set $P = 0.5$, $w_x = 3$, $w_y = 4$, and $n = 200, 500, 2,000$.
2. $X \sim \mathcal{N}(0, \sigma^2)$ with $\sigma^2 > 0$, $W \in \{-1, 1\}$ with probability 0.5, $Y = X + W$, and $n = 100$.
3. $X \in \{-1, 1\}$ with probability 0.5, $W \sim \mathcal{N}(0, \sigma^2)$ with $\sigma^2 > 0$, $Y = X + W$, and $n = 100$.

For the first experiment, the proposed method outperformed the Gaussian method, in particular for large n (Fig. 7). The negative binomial (NB) distribution extends the Poisson distribution in that the number of occurrences of the encountered event is not restricted to one. The NB is used as count data for modeling the RNA-sequence in the genome analysis [11,17]: how many times each gene is mapped from the sequence. The data is actually discrete but the number of values is not bounded, so no existing data processing can be applied. The proposed method does not distinguish discrete and continuous data and executes the same process.

For the second experiment, the proposed method outperformed the Gaussian method as well (Fig. 8(a)). This process could not realized in Edwards et al's result.

On the other hand, for the ANOVA case (Experiment 3), the values of mutual information obtained by the proposed mathod is closer to the true value than

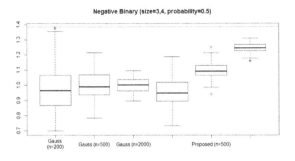

Fig. 7. Experiment 1

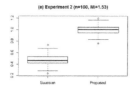
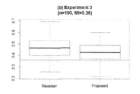

Fig. 8. Experiments 2 and 3

those obtained by the Gaussian (Fig. 8(b)). However, the difference is not so large compared with Experiment 2, which is due to the fact that the noise is Gaussian, and the Gaussian method is designed to deal with Gaussian noise.

4.2   Main Experiment

The first dataset comes from a gene expression study [9], whose purpose was to compare the gene expression profiles in tumours taken from two groups of breast cancer patient, those with and those without a mutation in the p53 tumour suppression gene. A dataset containing a subset of the study data is supplied along with the R library gRbase. The dataset has $n = 250$ observations and $N = 1001$ variables, comprising 1000 continuous variables and the class variable. There are 58 cases (with a p53 mutation) and 192 controls (without the mutation). They have been standardized to zero mean and unit variance.

Then, we installed the R limma package to execute the results standard difference analysis. Below are the top 10 genes that distinguish 250 samples with and those without a mutation in the p53 tumour suppression gene.

```
                  logFC     AveExpr           t       P.Value     adj.P.Val         B
A.202870_s_at 1.409476   0.00147564 10.226102  1.532066e-24  1.532066e-21  44.85602
A.209408_at   1.381635   0.00075120 10.024111  1.206754e-23  6.033771e-21  42.83468
A.212949_at   1.362423  -0.00143776  9.884723  4.897057e-23  1.362643e-20  41.46333
A.204822_at   1.360944   0.00180988  9.873987  5.450573e-23  1.362643e-20  41.35851
B.224428_s_at 1.348424   0.00102028  9.783154  1.342591e-22  2.685181e-20  40.47616
B.226936_at   1.343134   0.00235960  9.744770  1.960282e-22  3.267136e-20  40.10575
B.228069_at   1.337237   0.00107060  9.701992  2.983765e-22  4.262522e-20  39.69465
B.222958_s_at 1.319032   0.00087500  9.569904  1.079336e-21  1.349170e-19  38.43669
B.236641_at   1.305614   0.00194204  9.472556  2.753517e-21  3.059463e-19  37.52063
A.209642_at   1.301011   0.00238816  9.439162  3.788577e-21  3.788577e-19  37.20854
```

Figure 9 shows the distribution of the estimated mutual information values between the class variable and each of the gene expression data for the top 50 genes and for all of the 1000 genes, respectively. We see a significant difference of the estimated mutual information values between the top 50 and the others.

We executed the Chow-Liu algorithm for the two cases: for the top 50 genes (Fig. 10), and for all the 1000 genes (Fig. 11), and obtained two observations:

1. the top gene A.202870_s_at (108) was not connected to the class variable in the forest of the top 50 genes; and
2. only eight top 50 genes were selected in the neighbor of four from the class node in the forest of all the genes.

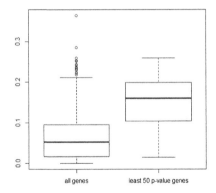

Fig. 9. The box plots of the estimated mutual information values between the class variable and each of the gene expression data for the least 50 p-value genes and for all of the 1000 genes.

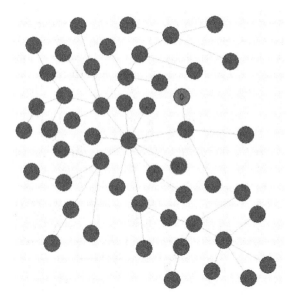

Fig. 10. The forest consisting of expression data of the top 50 genes and the class, marked by red. The class node is connected only to one gene A.202580_x_at (94) (Color figure online).

For the first observation, note that the mutual information of continuous variables (gene expression) is in many cases higher than that between discrete (class) and continuous (gene expression) variables as for the current data set. In particular, the mutual information is at most one bit for the latter case. As a result, the class variable has only one connection with the continuous variables.

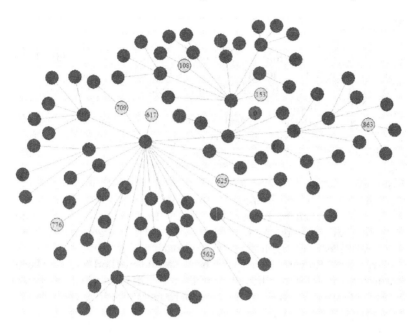

Fig. 11. The forest consisting of expression data of all the 1000 genes and the class (the subgraph consisting of genes within diameter four from the class node is shown, and (the class and top 50 genes are marked by "red" and "yellow", respectively)). The class node is connected only to A.215303_at (486) (Color figure online).

On the other hand, intuitively, the more mutual information, the less p-value as we have seen in Fig. 9. However, the two criteria are essentially different: in order to obtain the p-value, we assume a linear model such that $y = \beta_0 + x\beta_1$ with constants β_0 and β_1, where x and y are one of the gene expression variables and the class variable, respectively, and do hypothesis testing such that the null hypothesis is $H_0 : \beta_1 = 0$. The p-value is obtained assuming such a setting. On the other hand, the exact value of mutual information between two variables always exists, and the proposed and Edwards et al's [4] methods estimate the correct value.

In reality, the gene A.202580_x_at (# 94 in Fig. 10) has the largest mutual information 0.259614 and p-value 2.243162e-19 while the top gene A.202870_s_at (# 108 in Fig. 10) has the least p-value 1.532066e-24 and mutual information 0.229024. On the other hand, in Fig. 11, the gene A.202580_x_at (# 9 in Fig. 11) with a relatively large p-value 2.748447e-06 has the largest mutual information 0.259614 that is larger than that of the top gene A.202870_s_at (# 108 in Fig. 11).

In this sense, if one binary and other continuous variables exist, it seems that regression from the continuous to the binary may be more appropriate than the forest model although we followed Edwards et al's [4] and examined the validity of the approach.

5   Discussion

We proposed the Bayesian mutual information estimator to deal with both discrete and continous data. The estimator seeks the optimal balance between the number of samples in each bin and the approximation of the histogram in the sense of the MDL principle.

Then, we compared the proposed method with the Gaussian method that estimates the correlation coefficient between the two variables. In reality, if the difference between X and Y obeys the Gaussian, the existing method still shows better performance.

Finally, we tried to deal with genome differential analysis using the Chow-Liu algorithm based on the proposed mutual information estimator. The same problem was approached by Edwards et al. [4]. However, the mutual information estimators are different. As we tested in Sect. 4.2, the genome data are not Gaussian, and we condidered a different approach was required.

Compared with Edwards et al. [4] The proposed method has another merit that continuous variables may be between discrete variables when more than one discrete variable exis. We need to make experiments to verify the merit in the genome analysis.

The ultimate goal of this work is to propose a Markov network structure estimation based on maximizing the posterior probability (for example, Lee and Hastie [7], Cheng et al. [2]).

Acknowlegements. This work was partially supported by Advanced Research Networks A, Japan Society for the Promotion of Science (Takashi Suzuki, Osaka University).

References

1. Akaike, H.: Information theory and an extension of the maximum likelihood principle. In: 2nd International Symposium on Information Theory, Budapest, Hungary (1973)
2. Cheng, J., Levina, E., Zhu, J.: High-Dimensional Mixed Graphical Models (2013)
3. Chow, C.K., Liu, C.N.: Approximating discrete probability distributions with dependence trees. IEEE Trans. Inf. Theor. **IT–14**(3), 462–467 (1968)
4. Edwards, D., de Abreu, G.C.G., Labouriau, R.: Selecting high-dimensional mixed graphical models using minimal AIC or BIC forests. MBC Bioinform. **11**(18) (2010). doi:10.1186/1471-2105-11-18
5. Gessaman, M.P.: A consistent nonparametric multivariate density estimator based on statistically equivalent blocks. Ann. Math. Stat. **41**(4), 1344–1346 (1970)
6. Krichevsky, R.E., Trofimov, V.K.: The performance of universal encoding. IEEE Trans. Inf. Theor. **IT–27**(2), 199–207 (1981)
7. Lee, J.D., Hastie, T.J.: Learning the Structure of Mixed Graphical Models. J. Comput. Graph. Stat. **24**, 230–253 (2014)
8. Liang, P., Srebro, N.: Methods and experiments with bounded tree-width Markov networks. Technical report. MIT (2004)

9. Miller, L.D., Smeds, J., George, J., Vega, V.B., Vergara, L., Ploner, A., Pawitan, Y., Hall, P., Klaar, S., Liu, E.T., Bergh, J.: An expression signature for p53 status in human breast cancer predicts mutation status, transcriptional effects, and patient survival. Proc. Natl. Acad. Sci. USA **102**(38), 13550–13555 (2005)
10. Panayidou, K.: Estimation of tree structure for variable selection. Ph.D. thesis, University of Oxford (2010)
11. Garber, M., Grabherr, M.G., Guttman, M., Trapnell, C.: Computational methods for transcriptome annotation and quantification using RNA-seq. Nat. Methods **8**(6), 469–477 (2011)
12. Rissanen, J.: Modeling by shortest data description. Automatica **14**, 465–471 (1978)
13. Scutari, M.: Package ebnlearnf (2015). https://cran.r-project.org/web/packages/bnlearn/bnlearn.pdf
14. Silva, J., Narayanan, S.S.: Nonproduct data-dependent partitions for mutual information estimation: strong consistency and applications. IEEE Trans. Sig. Process. **58**(7), 3497–3511 (2010)
15. Suzuki, J.: A construction of Bayesian networks from databases on an MDL principle. In: The Ninth Conference on Uncertainty in Artificial Intelligence, Washington, D.C., pp. 266–273 (1993)
16. Suzuki, J.: The Bayesian chow-liu algorithm. In: The Proceedings of The Sixth European Workshop on Probabilistic Graphical Models, Granada, Spain (2012)
17. Wang, Z., Gerstein, M., Snyder, M.: RNA-Seq: a revolutionary tool for transcriptomics. Nat. Rev. Genet. **10**(1), 57–63 (2009)

Tips and Tricks for Building Bayesian Networks for Scoring Game-Based Assessments

Russell G. Almond[✉]

Educational Psychology and Learning Systems, College of Education,
Florida State University, Tallahassee, FL 32306, USA
ralmond@fsu.edu

Abstract. Game-based assessments produce multiple, dependent observations from student game play. Bayesian networks can model the dependence, but, typically, only a small amount of pilot data are available at
the time the network is constructed. This paper examines the process of
creating Bayesian network scoring models, focusing on several practical
techniques that have been used in the construction of models for *Physics
Playground*. In particular, the following techniques are helpful: (1) The
use of evidence-centered assessment design to define latent competency
variables and observable indicator variables. (2) The use of correlation
matrixes to uncover and validate the conditional independence structure
of the Bayes net. (3) The use of discrete IRT models to create large
portion of the Bayesian networks from a single spreadsheet. (4) Adjusting the Bayes net parameters using both hand tuning and a generalized
EM algorithm, creating networks which are a mixture of expert opinion
and data. (5) Using expected classification accuracy matrixes to judge
assessment validity and reliability. (6) Using evidence balance sheets to
identify unusual subjects and observable indicators.

Keywords: Bayesian networks · Evidence-centered assessment design ·
Prior information · Weight of evidence · Classification consistency

1 Introduction

One measurement model frequently used in educational assessments is item
response theory (IRT) [12]. In its usual form, IRT is a naive Bayes model with
the observable outcome variables (the responses to the test items) taken as conditionally independent given a single latent proficiency variable. In high-stakes
assessments, test construction process ensures that this conditional independence
assumption is approximately true.

As computers, tablets and smart phones become increasingly common in the
classroom, assessments are being built around simulations and games [11,26].
These technology platforms allow for a richer environment to gather evidence
about complex 21st century skills [16,23,25]. In these assessments, tasks measure multiple latent proficiency variables using multiple dependent observations.
Therefore, a new class of measurement models is needed to draw inferences

© Springer International Publishing Switzerland 2015
J. Suzuki and M. Ueno (Eds.): AMBN 2015, LNAI 9505, pp. 250–263, 2015.
DOI: 10.1007/978-3-319-28379-1_18

about the latent proficiencies. Bayesian networks provide a coherent framework for constructing these measurement models [10].

Building Bayesian networks to score complex assessments is challenging. Typically, the size of the pilot samples is small (several hundred students at best) and does not approach the 1500 sample size required to estimate a binomial proportion with a margin of error of 3 percentage points. Thus, a successful Bayesian network will be a mixture of expert opinion and limited data.

This paper explores some techniques that facilitate that process, emphasizing work that I did with Val Shute, Yoon Jeon Kim, Mathew Ventura and others in designing a scoring model for the game *Physics Playground* [8,15,16,25]. *Physics Playground* is a two-dimensional physics game, inspired by the commercial game *Crayon Physics Deluxe*. In each level of the game, players try to move a ball to a target (a balloon) using simple machines (i.e., ramps, levers, pendulums and springboards) that they draw on the screen. All objects in the two dimensional world obey the laws of physics. The goal of the project was to use the game to assess three different aspects of proficiency: qualitative physics [22], persistence, and creativity.

2   Evidence-Centered Assessment Design

Almond and Mislevy [9] noticed that a Bayesian network model for an assessment can be divided into two parts. One part was a complete Bayesian network which described knowledge about the status (on several proficiency variables) of the student, which they called the *student model*. The second part was a collection of Bayesian network fragments which related observable outcome variables associated with a particular assessment task (or item) to the proficiency variables, which they called *evidence models*. The evidence models were fragments because they contained pointers to the proficiency variables rather than the variables themselves. Inference in this collection of networks consisted of a series of steps:

1. The evidence model for the task the student just completed was retrieved from a database and adjoined to the student model.
2. The observable outcome variables were instantiated to certain values based on the student's performance on the task and the evidence propagated to the student model variables.
3. The variables unique to the evidence model were removed after their evidence had been absorbed.

Evidence-centered assessment design (ECD) [10,19] expands on this idea giving these models more context. In this system, the conceptual assessment framework for an assessment consists of a collection of design object (called models). The main ECD design objects are:

Student Proficiency Model. A statistical model (e.g., a Bayesian network) describing the proficiency variables, their relationships, and how to calculate scores given the proficiency variables.

Task Models. A description of a class of tasks that could be presented to an examinee to assess knowledge about some aspect of proficiency. This includes a description of the *work products* an examinee produces by interacting with the task.

Evidence Models. A description of how to update the proficiency model given the work product from a particular task model. This consists of two pieces: *rules of evidence*, which describe how to set the values for observable variables given the work product, and a *statistical model* (e.g., a Bayesian network fragment) which describes how the proficiency model should be updated given the observable values.

Assembly Model. A description of how many tasks or how much evidence is required to make well grounded (valid) inferences about an examinee. Part of this model is ensuring that there is a sufficient mixture of tasks that the latent construct is well defined.

Because the proficiency variables are latent, they must be carefully defined by the design team. One tool that has proved useful for this task is a *construct map* [14, 30]. In a construct map the variable representing the construct of interest is drawn on a vertical axis and the designer describes characteristics of an examinee or a performance at high, moderate and low levels of the construct.

Physics	Task Characteristics	Persistence	Response Characteristics
Complete	Can solve problems involving both linear and angular momentum.	Strong	Will spend a long time attempting difficult problems.
Partial	Can only solve problems involving linear momentum.	Moderate	Will spend a moderate amount of time on difficult problem.
Minimal	Has difficulty solving all problems.	Weak	Gives up quickly when faced with a difficult problem.

Fig. 1. Construct maps for Physics and Persistence

Figure 1 present construct maps for *Physics Playground* (for a more complete description see [25]). Note that levels of the construct can be defined by characteristics of the task (e.g., whether or not the task has a certain feature) or the performance (e.g., whether or not a certain method was used to solve the problem). It is important to consider both possibilities as often the characteristics of the tasks need to be manipulated so that the resulting instrument will have the desired psychometric properties [2].

Note that moving from any level on the construct map to a higher level involves an additional *claim* that can be made about the student. The central feature of evidence-center design is that at this stage is critical to think about what would constitute evidence that those claims do or do not hold. In particular, the design team needs to define *observable outcome variables* which will provide evidence for or against the claim when observed in certain contexts (i.e., within a given task). Providing operation definitions for the variables is a critically important step in Bayesian network construction when the network will be embedded in another application.

Consider the claim "Understands angular momentum" which is associated with high levels of the Physics construct. One possible piece of evidence is that the player used a pendulum in the solution to a problem. This induces a subproblem: how to identify an object drawn on the screen as a pendulum. The design team then needed to work out logic to identify features of objects so that *Physics Playground* could log when student used a pendulum, lever, springboard or ramp.

3   Inverse Correlation Matrix

The construct map is useful for defining a single proficiency variable, but not the relationship between variables. In education, a typical source of knowledge about the relationship among proficiency variables is factor analytic studies [4]. While factor analyses produce correlation matrixes for the latent variables, Bayesian networks focus on modeling the inverse correlation matrix [29]. In particular, zeros in the inverse correlation matrix correspond to conditional independencies. The R package CPTtools [5] contains code to assist in building network structures from correlation matrixes.

If the design team has access to data which measures the desired proficiency variables (either a pilot sample, or a prior study using similar variables), then a factor analysis can be used to generate a covariance matrix. Similarly, if only the loadings from the factor analysis are available, then factor scores can be computed for each latent variable. The design team can then directly assess conditional independence or calculate the correlation matrix and proceed as before.

The design team needs to be cautious about the inclusion of latent variables for which it is impossible to compute a factor score because there are no observable variables connected to them. These situations can lead to multimodal likelihood surfaces which can makee identification of the latent variable from data difficult [3].

Figure 2 shows the initial Bayesian network for the Physics portion of the *Physics Playground*, with four observable outcome variables at the bottom. In this configuration, information is available the four lowest level proficiency variables, corresponding to knowledge of the four simple machines (ramps, levers, pendulums and springboards). Note however, that there are no directly observable nodes connected to the three higher level nodes representing knowledge of

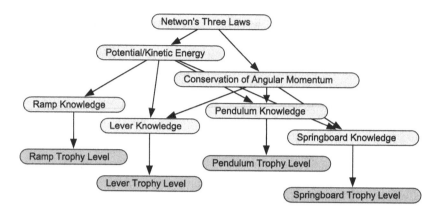

Fig. 2. Physics proficiency model and generic observables

Newton's three laws, potential/kinetic energy and angular momentum. In the case of the highest level node, an external physics test (used to gather validity evidence) provides information about the state of knowledge of students in the sample. This can be used in model checking and parameter estimation.

The two mezzanine level nodes (energy and angular momentum) present a problem. There is no direct evidence available about them within the system, nor is there an external measure available to establish their value for model validation or parameter estimation. As reporting the value of these nodes was not a firm requirement, the nodes were dropped. Figure 3 shows the result. Note that without the external post-test data, there still might be difficulty identifying the highest level node from data alone.

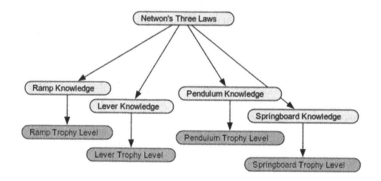

Fig. 3. Simplified physics proficiency model and generic observables

4   Discrete Item Response Theory

Assessment designs task models describe a family of related tasks and not just a single task. In other words, a task model describes a class of tasks, and individual tasks are instances of that task model. Correspondingly, the evidence model also describes a class of task-specific *link models* [10]. The link models usually inherit the graphical structure from the evidence model, but have task-specific values for the conditional probability tables (CPTs).

In *Physics Playground* all the tasks corresponded to game levels. The work product in each case was identical, a transcript of the objects that the player had creating in the process of coming to a solution as well as information about the final status (e.g., whether the level was solved, how much time was spent, which simple machines were used). Thus, each task supported the same list of observables and the same general evidence model could be used for each task. However, the tasks varied in both which agents could be used in a solution as well as other characteristics that made the harder or easier [2].

As the graphical structure for each level was identical, and there were over 70 levels, the easiest way to manage the work was not to use a graphical representation, but rather a Q-matrix [4,28]. Rows is a Q-matrix represent observable variables (or sometimes tasks) and columns represent proficiency variables. An element q_{jk} is one if Proficiency k is thought to influence performance on Observable j, in other works, if there is an edge between the two variables in a Bayesian network. Looking at the pattern of ones and zeros in the Q-matrix yields information about how much evidence is available to determine the value of each of the latent proficiency variables.

The Q-matrix together with the proficiency model gives the graphical structure of the Bayesian network. To complete the evidence models, a CPT needs to be constructed for each observable variable. As the pilot tests typically have small sample sizes, a mixture of data and expert opinion is needed. Furthermore, in most cases the CPT should be monotonic: increasing the proficiency level should increase the probability of better responses.

IRT [12] offers a class of models that are monotonic in the latent variable. Mapping each level of the proficiency variables onto a point on a continuous scale allows IRT models to be pressed into service as parametric models for CPTs [7]. As the scale of the latent variable in IRT is usually taken to be a unit normal, assigning values to the levels of the proficiency variables based on quantiles of the unit normal [10] puts the parameters of the model on the same scale used in IRT.

The software packages CPTtools [5] and Peanut [1,6] lay out a general framework for parameterized CPTs where both child and parent variables describe ordered categories. Let Y be the child variable, and let its states be $y_0, \ldots, y_S$. Assume that there are K parent variables, $X_1, \ldots, X_K$, and let i' be an index over the possible configurations of the parent variables (i.e., the rows of the CPT).

Mapping. For each configuration, i', let $\tilde{\theta}_k(i')$ be the *effective theta*, that is a real number, corresponding to the state that Variable X_i takes on in Configuration i'. This defines a mapping between the configurations and real vectors of effective thetas, $\boldsymbol{\theta}(i') = (\tilde{\theta}_1(i'), \ldots, \tilde{\theta}_K(i'))$.

Combination Rule. A combination function is applied to each row of the CPT to produce a table of effective thetas, $z_{i's} = Z_s(\boldsymbol{\theta}(i'))$. Note that there is one combination function for each column (although $Z_0(\cdot)$ is usually a constant).

Link. A link function, $g_s(z_{i'0}, \ldots, z_{i'S})$, is evaluated at each row to produce the conditional probabilities for Row i' of the CPT for Variable Y.

The combination functions, $Z_s(\cdot)$, are usually chosen to reflect a design pattern representing cognitive experts thoughts on how the referenced proficiencies impact performance [1,7,10]. The two most commonly used design patterns are the *compensatory* and *conjunctive* patterns. In the compensatory pattern, more of one skill compensates for lack of the other. The compensatory rule resembles a linear regression:

$$Z_s(\tilde{\boldsymbol{\theta}}(i')) = \frac{1}{\sqrt{K}} \sum_k \alpha_{sk}\tilde{\theta}_k(i') - \beta_s.$$

In the IRT language, the slope parameters, α_{sk} are known as *discriminations* and the intercept is known as a *difficulty* (note the negative sign). In the conjunctive design pattern, all skills are thought to be necessary for the solution, so the examinee's performance quality is driven by the weakest skill. The conjunctive combination rule is thus a minimum, and it makes more sense to have a separate difficulty (effective demand) for each skill and a common discrimination (slope):

$$Z_s(\tilde{\boldsymbol{\theta}}(i')) = \alpha_s \min_k(\tilde{\theta}_k(i') - \beta_{sk}).$$

If Y only has two states, then a natural link function is the inverse logistic function

$$\Pr(Y = y_1 | \mathrm{pa}(Y)) = \mathrm{logit}^{-1}\left(1.7Z_1(\tilde{\boldsymbol{\theta}}(i'))\right) = \frac{exp\left(1.7Z_1(\tilde{\boldsymbol{\theta}}(i'))\right)}{1 + exp\left(1.7Z_1(\tilde{\boldsymbol{\theta}}(i'))\right)}, \quad (1)$$

where the factor of 1.7 is used to make the logistic curve approximate the normal ogive.

There are several possible generalizations of the basic IRT model to polytomous responses. The generalized partial credit link function [20] is the most flexible [1]. Assume that the task assigned to the examinee is one that requires several steps, and that Y represents how many of the necessary steps the examinee completed. If Y has S steps, then the possible scores are $0, \ldots, S$. For $s > 0$, let $P_{s|s-1}(\tilde{\boldsymbol{\theta}}(i')) = \Pr(Y \geq y_s | Y \geq y_{s-1}, \tilde{\boldsymbol{\theta}}(i')) = \mathrm{logit}^{-1}(1.7Z_s(\tilde{\boldsymbol{\theta}}(i')))$; that is, let $P_{s|s-1}(\tilde{\boldsymbol{\theta}}(i'))$ be the probability that the examinee completes Step s, given that the examinee has completed steps $0, \ldots, s-1$. Note that $\Pr(Y \geq y_0) = 1$, and so let $P_{0|-1} = 1$, and $Z_0(\tilde{\boldsymbol{\theta}}(i')) = 0$. The probability that an examinee whose configuration of parent variables is i' will achieve Score s on the item is then:

$$\Pr(Y = s|\tilde{\boldsymbol{\theta}}(i')) = \frac{\prod_{r=0}^{s} P_{r|r-1}(\tilde{\boldsymbol{\theta}}(i'))}{C},$$

where C is a normalization constant, which collapses to:

$$\Pr(X_j = s|\tilde{\boldsymbol{\theta}}(i')) = \frac{\exp\left(1.7 \sum_{r=0}^{s} Z_{jr}(\tilde{\boldsymbol{\theta}}(i')[\mathbf{q}_j r])\right)}{\sum_{R=0}^{S_j} \exp\left(1.7 \sum_{r=0}^{R} Z_{jr}(\tilde{\boldsymbol{\theta}}(i')[\mathbf{q}_j r])\right)}. \tag{2}$$

Because the generalized partial credit link places no restrictions on the combination functions $Z_s(\cdot)$, it is extremely flexible. In particular, it allows different functional forms and different combinations of parent variables to be selected for each state transition.

The other advantage of this discrete IRT modeling approach is that it can be described a relatively small number of parameters: a combination rule, a link function, plus discrimination and difficulty parameters as appropriate for the number of parent variables and the number of states of the child variables [1,4]. These can be placed in spreadsheet, which is basically the Q-matrix described earlier with discrimination parameters replacing the ones, and extra columns added to describe which combination rules and which link function should be used as well as the difficulties.

This was how the model construction was done for *Physics Playground*. A spreadsheet was used to capture the relevant proficiencies, difficulties and discriminations for each game level. Code in the R language [24] using the *CPTtools* package was used to generate conditional probabilities which were put into a network for testing. The networks using the "expert" numbers were first used for scores on the pilot testing. Later, the expert numbers were used to generate priors for learning the CPTs from data.

5   Hybrid Learning Algorithms

A commonly used parameterization for Bayesian networks is the *hyper-Dirichlet* law [27]. In this law, each row of each CPT is given an independent Dirichlet distribution. Each variable, Y, has as its parameter a matrix A_Y, whose rows correspond to the possible configurations of pa(Y) and whose columns correspond to the possible states of Y. Again, let i' index the configurations of pa(Y) and s index the states of Y. Then $a_{i's}$ is the parameter for Configuration i' and State s.

Assume that both Y and pa(Y) are fully observed in the pilot data. Let X_Y be a table of counts where $x_{i's}$ indicates the number of times that $Y = y_s$ when pa(Y) = i'. In this circumstance, the posterior distribution for the CPT will also be a hyper-Dirichlet distribution with parameters $\tilde{a}_{i's} = a_{i's} + x_{i's}$. Multiplying the "expert" CPTs produced in the previous section by a set of weights, $w_{i'}$, one for each row of the CPT, produces prior distributions for the CPTs. In *Physics Playground* weights of around 10 seemed to work well.

In educational testing, the parent variables are often correlated in the population of interest. For example, if there are two parent variables representing two skills which are moderately correlated in the population, then individuals who are high on Skill 1 and low on Skill 2 will be rare. That means that there will be less data available for the corresponding row of the CPT.

The solution is to turn once again to the parametric models of the previous section. These should smooth the distribution, using information from rows with more observations to inform the distribution of rows with fewer observations. Note that the matrix of counts, X_Y, is a sufficient statistic for the CPT. Therefore, the parametric form can be fit to a table of counts using a simple gradient decent algorithm. This is implemented in `CPTtools` [5]. A quasi-Bayesian estimate can be formed by using the posterior pseudo-counts, $\tilde{A}_Y$ in place of the actual counts.

The preceeding ignores the fact that the proficiency variables are latent. If all of the CPTs are parameterized using the hyper-Dirichlet law, then there is a fairly straightforward EM algorithm which will estimate the parameters [27]. It alternates between calculating the expected values for the latent variables and pulling them into tables of pseudo-counts which are used to perform the conjugate updating for the Dirichlet distributions. This algorithm is implemented in many commercial Bayesian network tools.

Note that the hyper-Dirichlet EM algorithm produces a table of pseudo counts, A_Y^* for each variable, Y [1]. This table of pseudo-counts can be input into the gradient decent algorithm to get new parameters for the parametric model. Alternating these two steps creates a higher level EM algorithm. Note that in this circumstance, the inner hyper-Dirichlet EM algorithm only needs to be run for a few steps. The *Peanut* package implements this algorithm [6].

6   Evidence Balance Sheets

Game-based assessments are complex systems, and consequently when they are first deployed they are likely to produce unexpected behaviors. These unexpected behaviors could be a result of problems with the scoring models, or the designs of the game tasks. One challenge in the development of such systems is isolating the cause of anomalous behavior, often called *debugging* the system.

One tool that is useful for debugging is the *evidence balance sheet* [8,18]. The evidence balance sheet is a graphical display that tracks the evidence provided by each source for some hypothesis of interest. The hypothesis can be any binary proposition of interest; for example, "The student's proficiency is above a threshold." The weight of evidence is described as the change in the log odds for the hypothesis after learning the evidence. If H is the hypothesis of interest and E_t is the observed evidence from Task t, then the weight of evidence is:

$$\text{WOE}(H : E_t) = \log \frac{P(H|E_t)}{P(\neg H|E_t)} - \log \frac{P(H)}{P(\neg H)}. \tag{3}$$

Let $E_1, \ldots, E_T$ be the evidence from a sequence of tasks. The conditional weight of evidence from the last task in the sequence is defined as:

$$\mathrm{WOE}(H : E_T | E_1, \ldots, E_{T-1}) = \log \frac{P(H | E_1, \ldots, E_\mathrm{T})}{P(\neg H | E_1, \ldots, E_\mathrm{T})} - \log \frac{P(H | E_1, \ldots, E_{T-1})}{P(\neg H | E_1, \ldots, E_{T-1})}.$$

These add together in the obvious way, that is:

$$W(H : E_1, \ldots, E_T) = \mathrm{WOE}(H : E_T | E_1, \ldots, E_{t-1}) + \mathrm{WOE}(H : E_{T-1} | E_1, \ldots, E_{T-2})$$
$$+ \cdots + \mathrm{WOE}(H : E_2 | E_1) + \mathrm{WOE}(H : E_1)$$

To form the evidence balance sheet plot the conditional weight of evidence for each task in sequence. The evidence should be stronger in the beginning and then tail off. Sudden jumps in evidence mean that something unexpected happened and they are worth pursuing.

After the *Physics Playground* pilot test, the design team constructed weight of evidence balance sheets for each student and used them to screen for game levels with unexpected behaviors [8]. Figure 4 shows the balance sheet for one student, the two problematic game levels are "jar of coins" and "Jurassic park". Examining the replay videos revealed that the player, rather than using knowledge of physics to solve the problem, simply created objects to move the ball around the screen into position. As a consequence, the design team added object limits to discourage this kind of solution.

7   Reliability and Validity

Reliability and validity are two of the big ideas from the field of psychological measurement [10,13]. To the extent that a Bayesian network is built to measure a latent construct, the designers need to think about the reliability and validity of the system they have built.

Reliability comes from classical test theory, and can be explained through the simple classical test theory model:

$$X = T + E \tag{4}$$

Here X is the observed score, T is the true score and E is measurement error. *Reliability* is the fraction of the variance of the observed score that is explained by variation in the true score. In other words, high reliability means the instrument has low measurement error.

Something similar to reliability can be calculated through a simple simulation experiment [10]. First, use the Bayesian network to generate random data for both proficiency and observable variables. Second, mask the proficiency variables and estimate the proficiency variables from the observables. The true proficiency variables together with the Bayesian network output produce a confusion matrix for each proficiency variable. (Alternatively, if the Bayes net output consists of marginal distributions over the proficiency variables, this produces an expected confusion matrix).

WOE for student S259 , PhysicsUnderstanding > Low

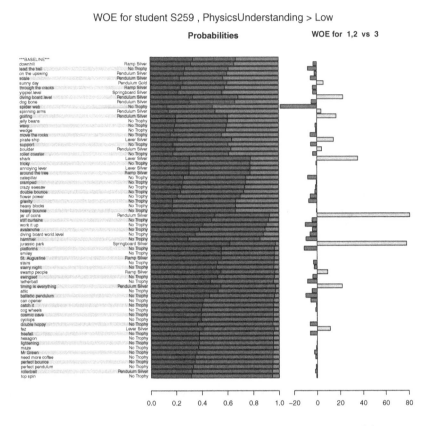

Fig. 4. Weight of evidence balance sheet for student S259 [8]

Although there are several different ways of summarizing a confusion matrix [10], I prefer to use Goodman and Kruskal's lambda:

$$\lambda = \frac{\sum_n a_{n,n} - \max_n a_{n,+}}{1 - \max_n a_{n,+}}, \tag{5}$$

where $a_{n,n}$ an element of the confusion matrix (in probability form) and $\max_n a_{n,+}$ is the probability of the modal category. Thus, lambda compares the observed agreement with the agreement that would be expected if all individuals were simply classified at the modal category, a likely alternative procedure if the assessment was not used.

What can be done if a system has low reliability? According to classical test theory, two factors drive reliability: test length and evidence quality. In the Bayesian network context, increasing the test length means increasing the number of observable variables. In the game context, this would mean adding game levels or letting the players play longer. Improving the evidence quality means controlling factors that otherwise would allow the observable variables to vary. For example, making it harder for the player to use solutions that don't require the targeted proficiencies. ([2] shows other ways to increase the evidentiary strength in the context of game based assessments).

While reliability is a measure of the internal self-consistency of the assessment system, validity is a measure of how well the assessment measures the constructs it was designed to measure. There are many ways to approach validity [13]. The most commonly used methods can be divided into two groups: methods that document the evidentiary argument of the assessment, and methods that look at the relationship between a new measure and other existing measures of the same construct.

For the documentation approach, evidence-centered assessment design has a clear advantage. In particular, it forces the design team to explicate the evidentiary argument as they are building the assessment. The task of building a validity argument then consists of editing already gathered information rather than going out and gathering new information. Furthermore, making sure that all tasks have an evidentiary rationale early on in the design process makes it less likely that the tasks will need to be removed later because they do not provide adequate evidence.

External validity measures are harder because they require that some group of examinees be given both the new assessment and the comparison measure. As the Bayesian network is often a classification system, this means that the validity study must come up with a *gold standard* classification. This is often quite expensive. The set of labeled cases can then be used to build a confusion matrix or expected confusion matrix [10]. As the goal now is to compare two rating systems (the gold standard and the assessment system), Cohen's kappa is the preferred statistic.

In *Physics Playground*, the external measure was a test of intuitive Physics given as both pretest and posttest and created by the design team. The correlation between the expected Physics proficiency level from the game and the combined pretest and posttest score was disappointingly low (around .4) even when the network was calibrated using the pretest and posttest items to measure the latent variables. The problem was that the pretest and posttest were too hard for the target population (middle school students). A new study was planned with a new posttest.

8   Concluding Remarks

It has been over 25 years since the landmark works of Pearl [21] and Lauritzen and Spiegelhalter [17] established Bayesian networks as a field in the intersection of Statistics and Computer Science. Although much progress has been made since then on both representational and algorithmic issues, there is still no well defined approach to model construction. This papers surveys some of the methods which were useful in one application area, designing scoring models for game-based assessments, in hopes that the techniques would translate easily to other areas of application.

Unfortunately, there has not been enough space to provide complete descriptions of all of the techniques. Hopefully the references will be sufficient to allow interested readers to follow up on the details. In particular, the new book [10] summarizes many of the details, and the software packages *CPTtools* and *Peanut* implement many of the algorithms. The software is freely available at http://pluto.coe.fsu.edu/RNetica/.

While there are many reasons people build Bayesian networks, they are often variations of the theme of trying to make inferences about something unobserved from something that is observed. Here, I think the field of psychological measurement has a lot to say to Bayesian network designers. In particular, the concepts of reliability and validity should be explored any time one is building a measurement system. Hopefully this work will encourage other Bayesian network designers to consider those issues more carefully in their work.

Acknowledgments. Many aspects of the *Physics Playground* examples are based on work of the *Physics Playground* team, Val Shute, P.I. The team includes Yoon Jeon Kim (who did much of the knowledge elicitation work for the Bayesian network), Matthew Ventura, Matthew Small, Don Franceschetti, Lubin Wang, and Weinan Zhao. Work on *Physic Playground* was supported by the Bill & Melinda Gates Foundation U.S. Programs Grant Number #0PP1035331, *Games as Learning/Assessment: Stealth Assessment*. Any opinions expressed are solely those of the author.

References

1. Almond, R.: An irt-based parameterization for conditional probability tables. In: Agosta, J.M., Carvalho, R.N. (eds.) Bayesian Modelling Application Workshop at the Uncertainty in Artificial Intelligence (UAI) Conference, Amsterdam, The Netherlands, July 2015. Additional material available at http://pluto.coe.fsu.edu/RNetica/

2. Almond, R.G., Kim, Y.J., Velasquez, G., Shute, V.J.: How task features impact evidence from assessments embedded in simulations and games. Meas. Interdisc. Res. Perspect. **12**(1–2), 1–33 (2014). with Discussion

3. Almond, R.G., Yan, D., Hemat, L.A.: Parameter recovery studies with a diagnostic Bayesian network model. Behaviormetrika **35**(2), 159–185 (2008)

4. Almond, R.G.: I can name that Bayesian network in two matrixes. Int. J. Approximate Reasoning **51**, 167–178 (2010)

5. Almond, R.G.: CPTtools: R code for Constructing Bayesian Networks. Florida State University, College of Education, 0–3.2 edition, June 2015. Open source software package

6. Almond, R.G.: Peanut: an object-oriented framekwork for parameterized Bayesian Networks. Florida State University, College of Education, 0–1.3 edition, July 2015. Open source software package

7. Almond, R.G., DiBello, L., Jenkins, F., Mislevy, R.J., Senturk, D., Steinberg, L.S., Yan, D.: Models for conditional probability tables in educational assessment. In: Jaakkola, T., Richardson, T. (eds.) Artificial Intelligence and Statistics 2001, pp. 137–143. Morgan Kaufmann (2001)

8. Almond, R.G., Kim, Y.J., Shute, V.J., Ventura, M.: Debugging the evidence chain. In: Almond, R.G., Mengshoel, O. (eds.) Proceedings of the 2013 UAI Application Workshops: Big Data meet Complex Models and Models for Spatial, Temporal and Network Data (UAI2013AW). CEUR Workshop Proceedings, Aachen, vol. 1024, pp. 1–10 (2013)

9. Almond, R.G., Mislevy, R.J.: Graphical models and computerized adaptive testing. Appl. Psychol. Meas. **23**, 223–238 (1999)
10. Almond, R.G., Mislevy, R.J., Steinberg, L.S., Yan, D., Williamson, D.M.: Bayesian Networks in Educational Assessment. Springer, New York (2015)
11. DiCerbo, K.E., Behrens, J.T.: Implications of the digital ocean on current and future assessment. In: Lissitz, R.L., Jiao, H. (eds.) Computers and their Impact on State Assessment, pp. 273–306. Lawrence Erlbaum Associates (2012)
12. Hambleton, R.K., Swaminathan, H., Rogers, H.J.: Fundamentals of Item Response Theory. Sage, Newbury Park (1991)
13. Kane, M.T.: Validation. In: Brennan, R.L. (ed.) Educational Measurement, 4th edn. pp. 17–64. American Council on Education/Praeger (2006)
14. Kennedy, C.A., Wilson, M.: Using progress variables to map intellectual development. Paper Presented at MARCES Conference, October 2006
15. Kim, Y.J., Almond, R.G., Shute, V.J.: Applying evidence-centered design for development of game-based assessments in Physics Playground. Int. J. Test. (in press). Special issue on cogntive diagnostic modeling
16. Kim, Y.J., Shute, V.J.: Opportunities and challenges in assessing and supporting creativity in video games. In: Green, G., Kaufmann, J. (eds.) Video Games and Creativity. Elsevier, (to appear, in press)
17. Lauritzen, S.L., Spiegelhalter, D.J.: Local computation with probabilities on graphical structures and their application to expert systems (with discussion). J. Roy. Stat. Soc. Ser. B **50**, 205–247 (1988). Reprinted in Shafer and Pearl (1990)
18. Madigan, D., Mosurski, K., Almond, R.G.: Graphical explanation in belief networks. J. Comput. Graph. Stat. **6**(2), 160–181 (1997)
19. Mislevy, R.J., Steinberg, L.S., Almond, R.G.: On the structure of educational assessment (with discussion). Meas. Interdisc. Res. Perspect. **1**(1), 3–62 (2003)
20. Muraki, E.: A generalized partial credit model: application of an em algorithm. Appl. Psychol. Meas. **16**, 159–176 (1992)
21. Pearl, J.: Probabilistic Reasoning in Intelligent Systems: Networks of Plausible Inference. Morgan Kaufmann, San Mateo (1988)
22. Ploetzner, R., VanLehn, K.: The acquisition of informal physics knowledge during formal physics training. Cogn. Instruction **15**(2), 169–205 (1997)
23. Quellmalz, E.S., Davenport, J., Timms, M.J., Buckley, B.C.: Quality science simulations for formative and summative assessments. Technical report, WestEd (2009)
24. R Core Team. R: A Language and Environment for Statistical Computing. R Foundation for Statistical Computing, Vienna, Austria (2014)
25. Shute, V.J., Ventura, M.: Stealth Assessment in Digital Games. MIT, Cambridge (2013)
26. Shute, V.J., Ventura, M., Bauer, M.I., Zapata-Rivera, D.: Melding the power of serious games and embedded assessment to monitor and foster learning: Flow and grow. In: Ritterfeld, U., Cody, M.J., Vorderer, P. (eds.) Serious Games: Mechanisms and Effects, pp. 295–321. Routledge, Taylor and Francis, Mahwah (2009)
27. Spiegelhalter, D.J., Lauritzen, S.L.: Sequential updating of conditional probabilities on directed graphical structures. Networks **20**, 579–605 (1990)
28. Tatsuoka, K.K.: Rule space: an approach for dealing with misconceptions based on item response theory. J. Educ. Meas. **20**, 345–354 (1983)
29. Whittaker, J.: Graphical Models in Applied Multivariate Statistics. Wiley, New York (1990)
30. Wilson, M.: Constructing Measures: An Item Response Modeling Approach. Psychology Press, New York (2005)

Author Index

Printed in the United States
by Baker & Taylor Publisher Services